620

BTEC Level 3 National in

Engineering

Orders: please contact Bookpoint Ltd, 130 Milton Park, Abingdon, Oxon OX14 4SB. Telephone: +44 (0)1235 827720. Fax: +44 (0)1235 400454. Lines are open from 9.00 a.m. to 5.00 p.m., Monday to Saturday, with a 24-hour message-answering service. You can also order through our website www.hoddereducation.co.uk

If you have any comments to make about this, or any of our other titles, please send them to educationenquiries@hodder.co.uk

British Library Cataloguing-in-Publication Data
A catalogue record for this title is available from the British Library.

ISBN: 978 1 444 11053 1

First Edition Published   2010
Impression number   10 9 8 7 6 5 4 3 2 1
Year   2015, 2014, 2013, 2012, 2011, 2010

Hachette UK's policy is to use papers that are natural, renewable and recyclable products and made from wood grown in sustainable forests. The logging and manufacturing processes are expected to conform to the environmental regulations of the country of origin.

Cover photo © John Gollings/Arcaid/Corbis
Artwork by Barking Dog Art
Typeset by Fakenham Photosetting Ltd, Fakenham, Norfolk
Printed in Italy for Hodder Education, a Hachette UK Company, 338 Euston Road, London NW1 3BH.

endorsed by
edexcel

BTEC Level 3 National in
Engineering

Simon Clarke
David Wyatt
Brian Crossland
Edited by Mike Deacon

HODDER
EDUCATION

Acknowledgements

The authors and publishers would like to thank the following for use of copyrighted material in this volume:

Figure 1.1 © iStockphoto.com; Figure 1.2 © Lewis W. Hine/George Eastman House/Getty Images; Figure 1.4 © Marcin Balcerzak/iStockphoto.com; Figure 1.7 © Paul Gibbings – Fotolia.com; Figure 1.9 © iStockphoto.com; Figure 1.10 © Tammy Bryngelson/iStockphoto.com; Figure 1.14 © Lisa F. Young/iStockphoto.com; Figure 1.15 taken from the Health and Safety Executive website at http://www.hse.gov.uk/risk/template.pdf and reproduced under the terms of the Click-Use licence; Figure 1.20 © Mikael Damkier/iStockphoto.com; Figure 1.21 © iStockphoto.com; Figure 1.24 © iStockphoto.com; Figure 1.25 taken from the Health and Safety Executive website at http://www.hse.gov.uk/costs/incost_calc/includes/INDG355967B2F967DForm.pdf and reproduced under the terms of the Click-Use licence; Figure 1.27 Courtesy of Stewart Superior; Figure 2.1 © Cindy Singleton/iStockphoto.com; Figure 2.15 Integrated Dispensing Systems www.fluidmetering.co.uk; Figure 2.18 © 2010 Bob Formisano (http://homerepair.about.com/). Used with permission of About, Inc. which can be found online at www.about.com. All rights reserved; Figure 2.38 reproduced with kind permission by Scott Ambler; Figure 2.46 reproduced by kind permission of Roy Beardmore; Figure 2.49 reproduced by kind permission of Roy Beardmore; Figure 2.60 © Dorling Kindersley/Getty Images; Figure 2.72 © zhang bo/iStockphoto.com; Figure 2.73 © iStockphoto.com; Figure 2.74 © Jacobs Stock Photography/Getty Images; Figure 2.75 © Viorika Prikhodko/iStockphoto.com; Figure 2.76 © iStockphoto.com; Figure 2.77 Courtesy of Professional Engineering Publishing; Figure 2.79 © Arkadiusz Fajer/iStockphoto.com; Figure 2.80 Courtesy of Unaflex, Inc.; Figure 2.81 © Guy Erwood – Fotolia.com; Figure 2.82 © 1997 C Squared Studios/Photodisc/Getty Images; Figure 2.83 © Johnny Lye – Fotolia.com; Figure 2.85 © Stephen Coburn – Fotolia.com; Figure 2.87 © Jérôme SALORT – Fotolia.com; Figure 2.90 © Alexey – Fotolia.com; Figure 2.91 © Akhilesh Sharma – Fotolia.com; Figure 2.92 © Tom Merton/OJO Images/Getty Images; Figure 2.93 © GUSTOIMAGES/SCIENCE PHOTO LIBRARY; Figure 2.95 © Alix Minde/PhotoAlto/Corbis; Figure 3.1 © Octus – Fotolia.com; Figure 3.2 © Trombax – Fotolia.com; Figure 3.4 © Joss – Fotolia.com; Figure 3.9 © David H. Lewis/iStockphoto.com; Figure 3.10 © Hans F. Meier/iStockphoto.com; Figure 3.11 © iStockphoto.com; Figure 3.12 © iStockphoto.com; Figure 3.13 © iStockphoto.com; Figure 3.14 © ROSLAN RAHMAN/AFP/Getty Images; Figure 3.15 © Michael Mathis/iStockphoto.com; Figure 3.18 © Lester Lefkowitz/Getty Images; Figure 3.20 © Andres Rodriguez – Fotolia.com; Figure 3.22 Courtesy of The MathWorks, Inc; Figure 3.23 © iStockphoto.com; Figure 3.24 © iStockphoto.com; Figure 4.1 © Stockdisc/Corbis; Figure 4.2 © RIA NOVOSTI/SCIENCE PHOTO LIBRARY; Figure 4.3 © Andrew Ammendolia / Alamy; Figure 4.4 © MELBA PHOTO AGENCY / Alamy; Figure 4.6 © Lance Bellers/iStockphoto.com; Figure 4.7 © Natalia Siverina/iStockphoto.com Figure 4.14 © Photodisc/Getty Images; Figure 4.15 © PhotoAlto; Figure 4.28 © Ingram Publishing Limited; Figure 4.29 © Ingram Publishing Limited; Figure 4.30 © Lai Leng Yiap – Fotolia.com; Figure 5.61 © SSPL via Getty Images; Figure 6.1 © Lorelyn Medina/iStockphoto.com; Figure 6.2 © iStockphoto.com; Figure 6.13 © Sean McBride/iStockphoto.com; Figure 6.58 © Angelo Marcantonio/iStockphoto.com; Figure 6.79 © GIPHOTOSTOCK/SCIENCE PHOTO LIBRARY; Figure 6.111 © Marek Tihelka – Fotolia.com; Figure 6.120 © Marek Tihelka – Fotolia.com.

Every effort has been made to trace and acknowledge copyright. The publishers will be glad to make suitable arrangements with any copyright holders whom it has not been possible to contact.

Contents

By the end of this unit you should be able to:

- understand the key features of health and safety legislation and regulations
- know how to identify and control hazards in the workplace
- be able to carry out a risk assessment and identify control measures
- understand the methods used when reporting and recording accidents and incidents.

In order to pass this unit, the evidence you present for assessment needs to demonstrate that you can meet all of the above learning outcomes for this unit. The criteria below show the levels of achievement required to pass this unit.

To achieve a pass grade you need to:	To achieve a merit grade you also need to:	To achieve a distinction grade you also need to:
P1 explain the key features of relevant regulations on health and safety as applied to a working environment in two selected or given engineering organisations	**M1** explain the consequences of management not abiding by legislation and regulations and carrying out their roles and responsibilities in a given health and safety situation	**D1** justify the methods used to deal with hazards in accordance with workplace policies and legal requirements
P2 describe the roles and responsibilities, under current health and safety legislation and regulations, of those involved	**M2** explain the importance of carrying out all parts of a risk assessment in a suitable manner	**D2** determine the cost of an accident in the workplace from given data.
P3 describe the methods used to identify hazards in a working environment	**M3** explain how control measures are used to prevent accidents.	
P4 describe how hazards which become risks can be controlled		
P5 carry out a risk assessment on a typical item/area of the working environment		
P6 suggest suitable control measures after a risk assessment has been carried out and state the reasons why they are suitable		
P7 explain the principles that underpin reporting and recording accidents and incidents		
P8 describe the procedures used to record and report accidents, dangerous occurrences or near misses.		

Introduction

Health and safety is of prime importance to all people working in manufacturing or engineering workplaces. Engineering can be, potentially, one of the most dangerous occupations workers can undertake because of the nature of the machinery, equipment, tools and techniques used. All workers should be protected from danger and risks to their health from whatever cause. In addition, visitors to the organisation, customers and the public should also be protected from risk.

Health and safety is regulated by law and it is the responsibility of all organisations to be aware of current legislation, the particular regulations that apply to the industry they are involved in, the materials they use and the environment in which those employees will be working. This unit will introduce you to the importance of health and safety regulations, how to identify hazards in the workplace, how/why we undertake risk assessments and how to report accidents and incidents that do take place.

> **Key Points**
>
> The European Commission reports that, although improvements are being made, every year nearly 5,200 workers lose their lives as a result of a work-related accident. In addition, there are approximately 4.8 million accidents per year and two-thirds of these accidents lead to an absence from work of more than three days.

This section of the book is organised as follows; each of the subsections can be readily linked to the learning outcome (LO), pass (P), merit (M) and distinction (D) criteria.

Section/content	LO	P	M	D
1.1 Understanding the key features of health and safety legislation and regulations	1	1, 2	1	
1.2 Knowing how to identify and control hazards in the workplace	2	3, 4		
1.3 Carrying out a risk assessment and identifying control measures	3	5, 6	2, 3	1
1.4 Understanding the methods used when reporting and recording accidents and incidents	4	7, 8		2

This section will cover the following grading criteria:

P1   P2   M1

To ensure that the safety of all people in the workplace is taken seriously, there are laws which are specifically designed to minimise the risks to employees and members of the public. In Britain the Health and Safety at Work Act 1974 outlines an employer's duties in ensuring health and safety 'so far as is reasonably practicable'. The law requires employers to appreciate that they have a duty of care, to ensure that all potential workplace risks are minimised.

Figure 1.1 Scale of justice

Key features of legislation and regulations

By the end of this section you should have developed a knowledge and understanding of the main aspects of relevant health and safety regulations in order to apply them in a working environment. You will be required to apply this understanding to two engineering organisations.

Laws are used to ensure that all organisations and individuals are regulated for everybody's benefit. The penalties for breaking the law vary and are usually pursued through court action. In the United Kingdom laws are introduced through Acts of Parliament.

If we consider UK laws, we talk about Acts of Parliament; consequently, there are key pieces of legislation for us to consider, such as the Factories Act 1961 and the Health and Safety at Work Act 1974 (HASWA). The most important piece of legislation for employers is the HASWA, but this cannot list every detail or requirement. Regulations are issued by government that effectively tell industry how to implement or use the HASWA for the safety of their employees and the public in general. Regulations do vary as other countries can have different laws and residents of Europe, including those in the UK, are also subject to European law.

Key Terms

Hazard – Something that has the potential to cause harm or be harmful.

Accident – An unplanned or uncontrolled event resulting in an injury, near miss, death or damage to property.

Risk – The chance of something being harmful.

Reasonably practicable – A balance between the feasibility or cost and the potential risk.

Make the Grade   P1

Grading criterion P1 requires you to explain the key features of relevant regulations on health and safety as applied to a working environment. You will either select or be given two engineering organisations in which to put this knowledge into context.

In the following sections we are going to explore the key features of both legislation and regulations that apply in the engineering workplace.

Legislation

Legislation concerning health and safety in the UK can be traced back to the early nineteenth

3

century, when acts were introduced to protect children working in the cotton mills. During the remainder of the nineteenth century, further acts were gradually introduced to improve conditions for particular groups such as women and children, the most important one being the Factory and Workshop Act 1878. This consolidated earlier legislation and was steadily built upon and improved. The final amendments to this law were made as late as 1961.

Key Term

Legislation – A law or laws which have been made by a legislative body such as Parliament.

Figure 1.2 Early nineteenth-century child labour in a textile mill

Factories Act 1961

The Factories Act of 1961 is the culmination of a series of Acts of Parliament dating back to 1802. Prior to this date, and throughout the nineteenth century, the Industrial Revolution changed the working landscape of Britain. Large textile mills were appearing and some mill owners were employing women, children and orphans to work in dark and dangerous conditions for long hours and low wages. Campaigns including the abolition of slavery and other welfare reforms led to the Health and Morals of Apprentices Act. This was to become the first Factory Act and was specifically aimed at child workers in cotton and woollen mills. It was the culmination of a movement originating in the 1700s, where reformers had tried to push several acts through Parliament to improve the health of the workers and apprentices. The Act had the following provisions:

- all rooms must be ventilated and, twice a year, should be disinfected with lime
- children should have two sets of clothing
- under 9 years old, children are not allowed to work; instead, they should be attending primary schools that the mill owners must set up
- children can only work between 6 a.m. and 9 p.m. and not exceeding 12 hours a day
- between the ages of 9 and 13, children are only allowed to work 8 hours at a time
- between 14 and 18 years old, boys and girls are permitted to work a maximum of 12 hours
- during the first four years of working, children should study reading, writing and arithmetic (the 3Rs)
- boys and girls must be allocated separate dormitories
- there should be a maximum of two children per bed
- Sunday school must be established, with an hour of Christian religious study on Sunday mornings
- factory owners are responsible for treating any infectious diseases
- owners of factories must obey the law.

Activity 1.1
Looking at the list of responsibilities outlined in the 1802 Factory Act, can you think of any situations, anywhere in the world, where the Act is not being adhered to today?
As an extension activity, can you think why this happens?

Although the original Factory Act was largely ignored by employers, fines of between £2 and £4 were enforceable. Perhaps the most significant impact of this law is that it established the first significant steps towards recognising the rights of workers and the responsibilities of factory owners.

Following on from the 1802 Act and throughout the nineteenth and twentieth centuries, Parliament passed a whole series of factory and workshop acts which gradually improved the safety and working conditions for factory workers. This culminated in the Factories Act 1961. This piece of legislation consolidated the majority of previous acts into one comprehensive piece of legislation. The Factory Act is still an important law and prosecutions under the Factory Act still take place. However, because many people do not work in factories, the more general HASWA has largely superseded this piece of legislation.

Health and Safety at Work Act 1974

In the UK the most comprehensive piece of legislation relating to health and safety is the HASWA. It sets out the requirements for any organisation to ensure high standards of health and safety in the workplace. The main focus of this Act is on the employer; however, it also considers landlords, tenants, suppliers, employees and members of the public.

For an employer, the purpose of the Act is to:

- make health and safety everybody's responsibility, encourage the awareness of safety issues, and promote teamwork to ensure that health and safety issues are promoted and that people work together in a safe manner
- ensure the health, safety and welfare of all employees by promoting safe systems of work, safe workplaces and safe facilities with sufficient amenities
- provide safe access to the workplace and safe exit from the workplace
- control the storage and use of explosive, highly flammable or dangerous substances
- control the emissions of noxious fumes
- provide information, guidance, training and supervision to employees to provide for their health and safety in the workplace
- produce and keep updated a health and safety policy in all organisations with five or more employees
- provide safe machinery, plant, equipment, and so on, and provide safe methods for the handling, storing and transporting of materials
- consider the welfare and safety of visitors/members of the public in the same way as for employees.

For an employee, the purpose of the Act is to:

- take reasonable care of themselves
- display a duty of care to others who might be affected by their actions
- cooperate with the employer on health and safety issues
- ensure that no damage or danger is caused by tampering with any provisions made to aid health and safety – for example, removing guards, removing signs, and so on.

Employment Act 2002

The Employment Act 2002 was introduced to recognise the rights of different employees. The main points of the Act are:

- employees on fixed-term contracts must be treated in the same way as those on permanent contracts
- employers should operate a disciplinary and grievance procedure, regardless of the number of employees
- maternity leave to be increased to six months' paid with a further six months' unpaid if required – this also applies to adoptive parents
- two weeks' paternity leave for working fathers
- flexible working requests by the parents of young or disabled children must be considered by the employer
- the government must reimburse employers for maternity, paternity or adoptive care costs, which can be up to 100 per cent of the cost for small employers
- union learning representatives should be established, with a right to paid time off
- at employment tribunals, for equal pay cases, a questionnaire procedure should be adopted to allow comparison with other employees' pay rates.

Fire Precautions Act 1971

The purpose of the Fire Precautions Act is to require certain premises to be checked for safety by the fire brigade and be given a fire certificate. The Act came into force following a fire at a hotel in Saffron Walden (Essex) that killed 11 people. Although 17 people were rescued, the scale of the disaster forced the government to reconsider the elements of the Factories Act and the Offices, Shops and Railway Premises Act that covered fire safety. These elements were brought together in the new Act to include hotels and guest houses.

Key Term

Fire certificate – An annual certificate issued to a public and/or industrial building that indicates that fire protection systems and fire safety measures are in good working order and properly maintained.

Typically, the fire brigade will ensure that the following are in place before issuing a fire certificate:

- people working in a premises requiring a fire certificate should be trained in the use of firefighting equipment and understand the fire evacuation procedure

- firefighting equipment such as fire extinguishers should be readily available and maintained in good working order
- fire escape routes should be provided and kept available, without routes being obstructed or doors locked
- all people employed to work on the premises should receive instructions and have training in what action to take in the event of a fire.

Figure 1.3 Fire warning sign

The local fire authority issues fire certificates to premises where sleeping accommodation is available above the first floor, for more than six staff or guests. This has been extended to premises where more than 20 people are employed, or more than ten if anywhere but at ground level. In addition, factories and warehouses where flammable or explosive material is stored also come under the Act.

Although not all premises require a fire certificate, there are fire safety requirements and the fire brigade has the power to shut down premises if they feel that there is a serious risk of danger due to fire hazard. This is known as a prohibition notice.

Key Term

Prohibition – An order or law that forbids something.

Activity 1.2
Research the HASWA and list five key points from the Act.

Key Points

Many public venues, such as cinemas, nightclubs and live venues, are exempt from the Act and are the responsibility of the local authority. The local authority has the enforcement powers necessary to close down a venue, and the fire brigade often carries out inspections on behalf of the local authority.

Educational establishments such as schools, colleges and universities are not covered by the Fire Precautions Act.

Regulations

In UK law, although legislation is made through Acts of Parliament, this is a cumbersome and time-consuming process. The Acts can be very general and cover a wide range of areas, such as the HASWA. Regulations have to be made on the same subject as the act, but tend to be more specific. Regulations are published and displayed in the House of Commons and are available for scrutiny for a period of time; however, if insufficient members of Parliament object and demand a vote, then regulations are passed.

Regulations are legally binding and should be treated as law, so from a health and safety perspective most regulations are made under the HASWA, following recommendations from the Health and Safety Executive (HSE) or following directives from the European Commission (EC).

Regulations can apply to specific industries such as mining or offshore industries. However, many regulations, such as manual handling, apply to all industries and sectors.

Activity 1.3
This is an activity which requires you to produce a comparison between health and safety laws and health and safety legislation. Carry out appropriate research from a variety of sources and use a comparison table to list your findings.

Key Term

European Commission (EC) – The European Commission is that part of the European Union that is responsible for introducing legislation and ensuring that it is implemented.

Employment Equality (Age) Regulations 2006

These regulations make it illegal to discriminate against trainees, job seekers or employees because of their age.

Management of Health and Safety at Work Regulations 1999

These regulations require employers to carry out risk assessments and review them from time to time. Employers must make arrangements to put in place risk prevention measures and, where necessary, to change working practices to minimise risk. In addition, employers must ensure that training is given to avoid hazardous situations occurring.

Provision and Use of Work Equipment Regulations (PUWER) 1998

These require employers to ensure that all machinery and equipment are safe to use.

In addition, records have to be kept to demonstrate that regular checks have been made to ensure the safety of equipment.

Control of Substances Hazardous to Health (COSHH) Regulations 2002

These require employers to assess the risks from substances that are hazardous to health and to take precautions to ensure that they are handled safely.

Lifting Operations and Lifting Equipment Regulations 1998

These regulations are designed to reduce the risk to health and safety from the use of any kind of lifting equipment used in the workplace. Lifting equipment includes cranes, lifts, ropes, hoists and associated equipment.

Manual Handling Operations Regulations 1992

These regulations relate to moving objects by hand or by using bodily force.

Figure 1.4 Manual handling

Personal Protective Equipment (PPE) at Work Regulations 1992

These regulations require employers to provide appropriate protective equipment and protective clothing for employees and visitors/customers.

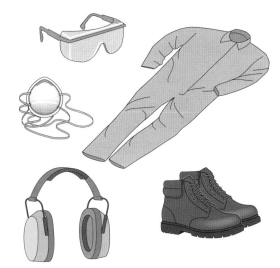

Figure 1.5 Personal protection

Confined Spaces Regulations 1997

These regulations refer to any space of an enclosed nature where there is risk of death or serious injury due to the dangerous conditions or the presence of substances that are hazardous. Examples of enclosed spaces include:

- open-topped chambers
- silos
- vats
- sewers
- unventilated or poorly ventilated rooms
- freezers

7

- ductwork
- enclosed drains.

The typical dangers to guard against could include:

- lack of oxygen
- poisonous gases
- liquids or gases suddenly filling the space
- fires and/or explosions
- high levels of dust or other contaminants
- extreme temperatures, which could be hot or cold.

Electricity at Work Regulations 1989

These regulations require all people in control of electrical systems to ensure that they are maintained in a safe condition and are safe to use.

Health and Safety (Display Screen Equipment) Regulations 1992

These regulations outline the procedures to be undertaken when working with visual display units such as computer monitors.

Control of Noise at Work Regulations 2005

These regulations require employers to take action to protect employees and others in the workplace from excessive noise.

Reporting of Injuries, Diseases and Dangerous Occurrences Regulations (RIDDOR) 1995

These regulations require employers, the self-employed and people in control of premises to report work-related deaths, major injuries, work-related diseases and dangerous occurrences (known as near misses). These incidents should be reported to the appropriate authorities via the Incident Contact Centre (ICC).

Workplace (Health, Safety and Welfare) Regulations 1992

These regulations cover a variety of issues related to the working environment, such as ventilation, lighting, heating, seating and workstations.

Working Time Regulations 1998

This is a good example of the law being amended due to European legislation. The regulations implement the European Working Time Directive and set the following standards:

- a maximum average working week of 48 hours per week over a 17-week period
- maximum average night-work of 8 hours in any 24-hour period, again averaged over 17 weeks
- a daily rest allowance of 11 consecutive hours
- weekly rest of 24 hours
- a minimum daily rest break of 20 minutes provided that 6 hours have been worked.

There are minor variations to this for young people.

> **Key Point**
>
> When the Working Time Directive was first introduced into EC law, the British government secured an 'opt-out' clause. This allowed employees to work longer hours if they signed an agreement to do so.

Health and Safety (First Aid) Regulations 1981

These regulations require employers to make adequate and appropriate first aid provision for employees at all times. Included within the regulations are requirements for training, equipment and facilities, as well as for informing all employees of first aid facilities and procedures.

Figure 1.6 Ear protection sign

Figure 1.7 First aid box

Figure 1.8 CE mark

The Health and Safety Information for Employees Regulations 1989

These regulations require employers to display a poster telling employees everything they need to know with respect to health and safety. This should indicate the protective clothing, procedures and safety measures for their work area.

> **Key Point**
>
> Not every possible risk has to have a safety sign; this would result in there being so many signs that people would not take them seriously. Safety signs are required when there is a significant risk that cannot be avoided.

Supply of Machinery (Safety) (Amendment) Regulations 2005 (SI 2005/831)

This regulation updates the original regulation of 1992 and requires that machinery meets minimum standards of safety. The machinery should have certification to demonstrate that it conforms to the standard, and a CE mark that should be fixed to the machine.

Activity 1.4

Which regulation would cover each of the following situations?

- A machine which has been installed without sufficient guarding
- A workshop with no trained first-aiders
- An aeroplane hanger with no ear defenders provided for maintenance crew
- A welding and fabrication site with no safety shoes or protective goggles provided for visitors
- A workshop with no noticeboards or posters visible

As an extension to this activity you should consider what might happen if management do not put in place measures to protect workers against the hazards listed above.

Roles and responsibilities of those involved

By the end of this section you should have developed a knowledge and understanding of the roles and responsibilities of employers, employees, the Health and Safety Executive (HSE) and other parties.

It is often thought that the employer is responsible for health and safety. However, this is not entirely true. Clearly, employers have a role to play in ensuring that they provide safe working conditions for their employees; however, there are also responsibilities for employees, the HSE and others.

Key Term

Health and Safety Executive (HSE) – The HSE is responsible for protecting people against risks to health and safety arising from work activities. They provide advice, guidance and information; however, they are also a government agency and are responsible for regulating and enforcing legislation and regulations regarding health and safety.

Make the Grade P2

Grading criterion P2 requires you to describe the roles and responsibilities of those involved under current health and safety legislation and regulations. You will be required to explain who is responsible for what, and this might require you to do some research on bodies such as the HSE.

Key Point

It is often said that it is *everybody's* responsibility to maintain effective healthy and safety in the workplace.

In the next section, we shall look at each of the important groups of people that have a responsibility for ensuring health and safety in the workplace.

Employers

The HASWA outlines the duties of employers in its different sections. The key responsibilities are to:

- maintain the health and safety of employees
- maintain the health and safety of non-employees
- control atmospheric emissions
- provide necessary PPE
- maintain safe premises
- take responsibility for the design, manufacture, import and supply of safe substances.

Employers are required to do whatever is reasonably practicable to conform to the following requirements as part of their duty of care:

- maintain a safe workplace

- prevent health risks
- maintain plant and machinery in a safe manner and ensure that it is operated in a safe way
- ensure that all materials are stored and handled safely
- ensure that first aid facilities are maintained
- ensure that training and supervision are adequately maintained and that all personnel are informed of hazards and dangerous materials
- ensure that adequate facilities for ventilation, lighting, washing, and so on are maintained
- ensure that escape routes and other emergency evacuation procedures are enabled
- ensure that equipment meets the appropriate safety standards and is guarded and maintained
- control any potential exposure to substances that may damage health
- protect against flammable, explosive, electrical, noise and radiation risks
- ensure that precautions are in place to prevent injury due to manual handling
- provide protective clothing/equipment and training in its use
- maintain warning signs and ensure that they are correct, visible and up to date
- ensure that all accidents and incidents are reported in the correct manner.

Key Point

You will notice that the key responsibilities of employers are, effectively, the same as the regulations in the previous section.

Employees

Health and safety is the responsibility of all and this includes employees. For the purposes of the HASWA, employees are classed as any person who is permanently or temporarily on the payroll or anyone who is regularly on the premises; this could include students on work experience or part-time staff or trainees.

Employees have the following responsibilities, to:

- take reasonable care of their own and other employees' health and safety
- cooperate with their employers
- carry out activities in accordance with training and instructions
- inform the employer of any serious risks.

Larger organisations have a nominated health and safety officer, whose primary responsibility is to promote and enforce good practice. In addition, an organisation that has trade union representation will normally have a union official whose responsibility is to ensure the health and safety of the union's members.

> **Activity 1.5**
>
> A work experience student is tasked to use electrical test equipment. List three safety issues which the employer might face, and explain how he or she should ensure the student's safety.

Health and Safety Executive

The HSE is responsible for preventing death, injury and ill health to all workers and those affected by work activities. It is a government body which provides advice and guidance on matters of health and safety and is responsible for advising on legislation and regulations. However, it is also responsible for enforcing health and safety laws and it has considerable powers. The enforcement powers are separated from the HSE and are usually split between three different bodies:

- HSE – primarily responsible for manufacturing facilities, factories, railways, mines and quarries, offshore installations, agricultural industries and nuclear facilities
- local authority – responsible for shops, offices, warehouses and premises where no machinery is used
- fire authority – primarily focusing on fire prevention, issuing of fire certificates and ensuring means of escape; where a process might have a high risk of fire, such as a petrochemical plant, the fire authority would become involved there too.

> **Activity 1.6**
>
> Which government body would you contact if you had a concern about:
>
> - a blocked fire exit?
> - a broken guard on a drilling machine?
> - food not being properly cooked?

Guidance notes and booklets

The HSE publishes notes and guidance. It also has a helpline and a website, with freely available resources to encourage health and safety in the workplace and to support business, workers and the wider community. There are many documents, forms and procedures which organisations must follow in order to conform to the various regulations. This information is available to download from the HSE website or, in some cases, can be completed online.

Span of authority

The HSE has a range of powers enshrined in law. These include powers to:

- enter and search premises
- direct that premises be left undisturbed for the purpose of investigation
- take measurements, photographs and recordings
- take samples of articles or substances found in premises or nearby
- dismantle and test any article appearing to have caused or be likely to cause danger
- detain items for testing or for use as evidence
- interview any person
- require the production and inspection of any documents
- require the provision of facilities and assistance for the purpose of the investigation.

Figure 1.9 Inspection

Right of inspection

Although the HSE is a supportive organisation whose primary function is to promote health and safety in the workplace, it also acts as an enforcing agency. The HSE employs inspectors who have the powers to enter premises, which can include the homes of self-employed people. Inspectors can enter premises at any time, although usually they will be visiting to investigate an incident or accident; they also carry out routine health and safety inspections.

When inspectors investigate an accident, they:

- investigate the cause of the accident

- advise on whether action is needed to prevent it happening again
- look for where health and safety laws may have been broken.

If the law has been broken, the consequences for businesses can be severe. Inspectors can:

- prosecute the individuals or the company
- issue a prohibition notice or improvement notice
- issue a written or verbal informal warning.

If appropriate, the HSE will contact the police and liaise with them in pursuing criminal proceedings.

Others

It is not just employees that organisations have a duty of care towards. People not employed by an organisation can still be affected by operations carried out by the engineering industry. Non-employees can be visitors, people who call on businesses, customers or contractors.

Examples of non-employees affected by health and safety are as follows:

- local residents could be affected if poisonous gases were ventilated into the atmosphere
- contractors could be carrying out work in a factory without wearing the appropriate PPE
- customers could be visiting a company and not know what to do if the fire alarm sounds
- students could be on a company visit or work placement and not be aware of the risk of noise and the correct use of ear defenders.

Make the Grade

Grading criterion M1 requires you to explain the consequences of management not abiding by legislation and regulations and carrying out their roles and responsibilities in a given health and safety situation. In the case of an employer not providing appropriate PPE, what consequences might occur? Try to find out what the HSE might do, what the employer's responsibilities are and what an employee should do.

1.2 Knowing how to identify and control hazards in the workplace

This section will cover the following grading criteria:

The Workplace (Health, Safety and Welfare) Regulations 1992 require organisations to maintain a safe workplace. There can be health hazards such as lighting, ventilation or confined space; similarly, there are safety issues such as trip or slip hazards or risk of fire. Organisations must ensure that they know how to deal with these potential hazards to ensure safe working practices.

Within the workplace

By the end of this section you should have developed a knowledge and understanding of how to identify hazards in the workplace.

The workplace is normally defined as the premises where an employee carries out his/her work. This can be factories, schools, offices, hospitals, hotels, theatres, clubs, and so on. However, it can also include common parts of shared buildings, including paths on industrial estates or business parks, and temporary workplaces such as construction sites.

Key Points

As you become familiar with a working environment, you can become complacent and not realise how dangerous something is becoming until an accident occurs. For this reason, it is important to systematically identify any hazards on a regular basis.

Make the Grade   P3

Grading criterion P3 requires you to describe the methods used to identify hazards in a working environment. This criterion is all about *how* you identify hazards and not the hazards themselves.

The next section looks at what methods can be used to ensure that hazards are identified in the correct manner.

Methods to identify hazards

Key Terms

Hazard – Anything that could cause harm.

Risk – How likely it is that a hazard will be harmful and the amount of harm the hazard will cause.

It is important to identify hazards before they cause accidents. In the workplace, you may be familiar with a wire that trails between two machines, so always step over it, or you might know that a doorway is low, so always duck as you go underneath it. However, if you are showing somebody around the workplace and you forgot to mention these hazards, an accident could easily occur. The question is, how to identify and eliminate or minimise these hazards. The trailing lead could be covered with a specially designed cover or could even be re-routed. A low doorway might not be so easy, and here it might be appropriate to display highly visible safety signage.

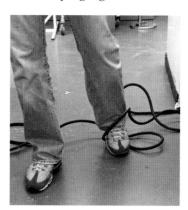

Figure 1.10 Tripping hazard

Typical ways that employers can identify hazards include:

- walk around the building – a site survey or inspection can be used to inspect the whole

13

work area; however, this can be less detailed than other techniques

- talk to the workforce – often the employees of an organisation are aware of potential risks, but might not have thought to discuss them with employers or managers
- safety audit – this can be very time-consuming and can involve a team of staff; the purpose is to check each and every appliance, fixture and fitting for any risk or danger associated with it
- examine accident and health records – by looking at the records of incidents/near misses or the health records of staff working in particular areas, it is often possible to determine where a hazard is causing problems, and actions can be taken to minimise or remove the risk
- consider working methods – often the way in which a job is undertaken can make it hazardous; for example, the use of trailing leads on power tools can be eliminated if cordless appliances are used
- checks on the environment – usually determined by measurement, this can be regularly measuring air quality, water quality, temperature, and so on, to ensure that no workers or members of the public are being exposed to hazards due to their working environment.

In addition, checks should be based on welfare provision such as toilets, canteen facilities and first aid. Fire precautions, appropriate PPE, machine guarding, electrical safety and access to the workplace are all factors to be considered within the workplace.

Keeping a record of all safety inspections and audits is important, as these can be referred to when undertaking new inspections or if an incident has occurred.

Working environment

By the end of this section you should have developed a knowledge and understanding of the inherent dangers presented in the working environment.

The working environment refers to the place where you work and the conditions that exist there. The happier people are in their environment, the more likely they are to be effective workers, and providing a healthy and safe environment is an important step in ensuring that you have a happy and productive workforce.

> **Key Points**
>
> Just because you cannot see an immediate hazard does not mean that a workplace is risk-free. Many environmental hazards are invisible, such as heat, noise, lack of ventilation or fumes.

In this section you are going to consider some of the environmental issues that arise in different situations when at work.

Consideration of the workplace and its potential for harm

Confined spaces

Working in a confined space does not necessarily mean that you cannot move around; what is intended is to provide ease of access and to allow people to move around and to get to their desk/machine/computer, and so on. A minimum of 11 cubic metres is the requirement for one person and this can be multiplied by the number of people normally working in a given space. Three metres is normally considered the maximum height when calculations are performed, and you have to consider furniture, machinery and other room contents.

For example, an inspection room measures 3 m by 2 m and has a roof height of 4 m. Inside the room there are two cupboards 2 m high, measuring 0.5 m by 0.5 m.

The following calculation would be used:

$3\,m \times 2\,m \times 3\,m = 18\,m^3$
(maximum allowable for height)

Cupboard volume: $2 \times 0.5\,m \times 0.5\,m \times 2\,m = 1\,m^3$

The overall room volume should be considered to be $17\,m^3$; consequently, only one person can work in the room.

> **Activity 1.7**
>
> In a group of three, take measurements of a typical room – for example, a classroom or workshop. Calculate the maximum number of people who could occupy that room.

Working over water or at heights

The danger of working at height or over water is considerable and requires a high degree of protection. Typical precautions when working over water should include:

- a barrier/fence on any scaffolding or platform, to prevent anyone falling into the water
- a skilled operator should be provided for electrical or mechanical hoists and the equipment checked and maintained before use
- danger signs should be clearly posted
- lighting of the whole area should be provided, particularly over the surface of the water
- personnel should wear life jackets or similar aids to buoyancy when working over water
- safety equipment, such as lifebelts/lines and a boat, if necessary, must be provided and checked for suitable use
- no electrical equipment or leads should be able to touch the water, even if dropped
- all persons working on the site should have emergency training in rescue and raising the alarm
- working alone is not permitted.

Similar precautions are taken for working at height and particularly with the use of ladders:

- ladders should be checked for damage before use
- ladders should extend 1.1 m above the platform they lead to
- a maximum of 6 m is allowed between platforms or levels
- where ladders pass through floors, the hole should be as small as possible

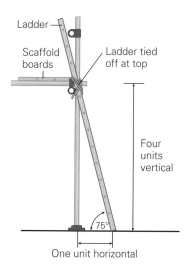

Figure 1.11 Safe use of ladders

- safety hoops should be fitted to ladders if they are more than 2.5 m high
- ladders are for access only and should not be used for working from.

In addition, there are requirements for crawling boards when employees are crawling on roofs that are not designed to support a person's weight; these roofs should be clearly marked as fragile, with appropriate safety signs.

Electrical hazards

Electricity has its own particular dangers and these are generally covered by the Electricity at Work Regulations. Electricity can cause explosions or fires, as well as the danger of electrocution. Consequently, employers are required to check all electrical appliances, including portable appliances, on a regular basis. The checking of appliances, known as Portable Appliance Testing (PAT), has to be logged and carried out regularly. Appliances are normally labelled with a sticky label, which shows the date of conformity and the due date for the next inspection.

The regulations are specific and require a competent person to carry out all activities associated with electrical appliances, which can include:

- changing fuses and light bulbs
- moving portable electrical equipment
- using extension leads
- switching on/off at the main isolation point
- fitting plugs.

Chemicals

Many substances used in engineering can be dangerous or harmful to health.

The Control of Substances Hazardous to Health (COSHH) Regulations give clear details of how hazardous chemicals should be used, handled and stored. The regulations outline how to deal with chemical spills or leaks and the degree of toxicity of different materials. In addition, with particular chemicals, the maximum exposure limit is indicated; some chemicals can be a minor irritant and should be handled with care, while others have a maximum exposure time of only a few minutes, with very specific protective clothing being required.

Where an organisation uses hazardous chemicals, they should carry out a COSHH assessment. This is intended to:

- identify and label all substances

15

- indicate the nature of the hazard – for example, poisonous, carcinogenic, and so on
- indicate the level of risk if persons come into contact with the substance
- provide for medical procedures and appropriate first aid if persons are exposed.

Additional precautions, such as secure storage and adequate signs, are required if chemicals are present at the work site.

Key Term

Carcinogenic – A substance that can cause cancer if persons are exposed to it, particularly over long periods of time.

Noise

Noise is measured in decibels (dB) and excessive noise can cause serious damage to hearing. Perhaps of more importance than the noise level is the amount of time you are exposed to loud noises. The Noise at Work Regulations 1989 guide employers as to what actions to take, including safety signage where loud noises are likely. In addition, the guidelines indicate where the provision of hearing protection to reduce noise levels is necessary.

Typical examples of noise levels are as follows:

150 dB   jet engine at 30 m/loud rock music
140 dB   gunfire at 1 m
130 dB   aeroplane taking off (threshold of pain)
120 dB   loud car stereo
110 dB   chainsaw
100 dB   pneumatic drill
90 dB   power tools
80 dB   busy street
70 dB   vacuum cleaner

60 dB   normal conversation
50 dB   heavy rainfall
40 dB   quiet office
30 dB   whispering
20 dB   quiet library
10 dB   quiet breathing
0 dB   hearing threshold

Figure 1.12 Ear protection

If noise reaches 85 dB, employers need to have the risk assessed, warn workers about noise and provide hearing protection. This is known as first action.

If the noise reaches 90 dB, employers have to do all that they can to reduce the noise, provide hearing protection and warning signage. This is known as second action.

In organisations where noise is an issue, hand signals are often used – for example, crane operators have a recognised code so that specific hand signals are interpreted in the same way all the time.

Figure 1.13 Confusing hand signals

16

Hazards which become risks

By the end of this section you should have developed a knowledge and understanding of how hazards can become risks and of the difference between hazards and risks.

People often confuse hazards and risks. A hazard is something that has the potential to cause harm and can include substances, people, working methods or machines. A risk is a measure of how likely it is that a specific hazard will cause harm and whom it will affect.

> **Key Points**
>
> The level of risk depends upon severity, likelihood and scale.
>
> - Severity – can range from a minor injury to death
> - Likelihood – can range from extremely unlikely to highly likely
> - Scale – can range from affecting just one person to affecting everybody in the workplace or surrounding area

When a can of petrol is safely stored in a locked cupboard, although there is a hazard as the fuel is flammable, there is very little risk. If the can is removed and the petrol is poured into a container to degrease engine parts, the risk is greatly increased as there is a danger of the fuel being spilled or ignited. This is a good example of when a hazard can become a risk.

Make the Grade   P4

Grading criterion P4 requires you to describe how hazards that become risks can be controlled. You will need to describe what constitutes a hazard, explain how it becomes a risk and then explain what you would do to control that risk.

Identification of trivial or significant risks

Identifying the severity of a risk is something that is of vital importance when carrying out risk assessments. This is something you will explore in the next section; however, it is useful to use a grid approach to determine whether a risk is trivial or significant.

Probability	Slightly harmful	Harmful	Very harmful
Unlikely	Trivial risk	Acceptable risk	Moderate risk
Likely	Acceptable risk	Moderate risk	Substantial risk
Very likely	Moderate risk	Substantial risk	Intolerable risk

Table 1.1 Risk matrix

From Table 1.1, you can see that a risk that is unlikely and slightly harmful is classified as trivial, whereas a very likely and very harmful risk is intolerable.

Choosing appropriate control methods

It is important to ensure that the correct action is taken, depending on the degree of risk.

The following steps are usually advised:

- trivial risk – no further steps
- acceptable risk – no additional risk control or measures to reduce the risk are required
- moderate risk – in this case, further risk controls are required; alternatively, measures should be taken to reduce the level of risk; the hazard should also be closely managed
- substantial risk – in this case, significant risk controls should be implemented; the risk should be reduced, but if the risk cannot be reduced, strict management of the hazard should include reducing the personnel exposed to the hazard and the time that they are exposed to it
- intolerable risk – all work involving the hazard should cease immediately and be prohibited in future.

Electrical safety

Electrical injuries can vary depending on circumstances: whether the current is a.c. or d.c., the size of the voltage, what protective measures are in place, and so on. The most likely effects are:

- electric shock
- electrical burns

17

- muscle control problems
- burns.

Electric shock

The human body uses electrical signals from the brain to control all the muscles and, consequently, movement. An accidental voltage that is applied between two parts of a person's body can disrupt these signals and can stop the heart, prevent breathing or cause spasms in the muscles. Even static electricity can be dangerous, as the smallest spark can ignite gas, paint fumes or other flammable vapours.

Electrical burns

Electrical burns are normally due to high voltages rather than domestic supply. The electric current flowing through the body heats the surrounding body tissue, causing often severe burns.

Muscle control problems

An electric shock can cause spasms in muscles that can break bones. This often means that a person who accidently grips a live connection cannot let go, or may be thrown from a height or into machinery.

Burns

Unlike electrical burns, a person who gets too close to electrical equipment can be burned very easily. Think of an electric cooker and you get the idea of how hot metal can become when electricity flows through it.

Residual current device (RCD)

An RCD is a device that will disconnect or 'trip' when the current flowing in two parts of a circuit is not the same or balanced; so if somebody accidentally touches a live part of a circuit, causing

Figure 1.14 Electrician working on circuit-breakers

the electricity to flow through them and to the ground, the return part of the circuit will be out of balance and the RCD will cut the power supply.

Overloading a circuit, by having too many appliances running at one time or by having loose wires, can cause problems with overheating or units cutting out. Circuits are protected by fuses or circuit-breakers. However, machines often have additional fail-safe features, so if the current is lost to a particular component, it is designed to stop working in a safe manner. An example of this might be a door that will always fail open, to prevent somebody being trapped inside.

Mechanical safety

Mechanical safety is often of concern with rotating machinery, sharp edges and heavy-duty equipment. Fail-safe mechanisms are employed in a similar way to electrical devices – for example, a machine will not operate if the guard is not in place. Similarly, sensors are often used to prevent parts of the body or other foreign objects from becoming caught in a machine. These sensors are often light beams; if something breaks the beam, a signal triggers a switch and the machine will stop.

Key Term

Guards – Devices that prevent personnel from getting close to dangerous machinery.

Safety devices

Safety devices are provided to prevent injury if something goes wrong. An example is an emergency stop, which can be activated if clothing gets tangled in a machine. An RCD is an example of a safety device in an electrical circuit. The handbrake on a car is often referred to as the safety brake. When designing safety into a system, it is important to ensure that devices are fail-safe.

Key Term

Fail-safe – A device that will prevent danger if a failure in a system occurs. An example is a fuse that stops current flowing if an electrical fault occurs. Similarly, if a problem occurs in a cooling system, a valve should fail open to allow water to keep flowing, rather than allowing overheating.

Activity 1.8

Explain how each of the following hazards can become risks:

• A chemical liquid, stored in a closed bottle, in a locked cupboard

• An electrical cable used as an extension lead and wrapped on a drum

• A mechanical press fitted with a safety guard

As an extension activity, you should consider how each of these activities can be controlled.

1.3 Carrying out a risk assessment and identifying control measures

This section will cover the following grading criteria:

P5 **P6** M2 M3 D1

Any hazard in the workplace can be harmful. According to the HASWA, it is the responsibility of employers to assess how likely it is that a hazard will be harmful and the amount of harm the hazard will cause. This is why organisations undertake risk assessments and introduce measures to control or eradicate that risk.

Risk assessments

By the end of this section you should have developed a knowledge and understanding of why organisations undertake risk assessments and how to carry one out.

There are five steps involved in undertaking a risk assessment; these focus on the process, from identifying hazards in the first instance to reviewing and revising your risk assessments.

Key Points

Risk assessments should be carried out by trained personnel. In a large organisation this should be the health and safety representative; however, in a smaller organisation this might be an employer or other responsible person.

There are a variety of standard templates that can be used to allow risk assessments to take place. You do not need to design your own. Organisations such as the HSE have standard forms, and many employers have templates that focus on the key hazards to look out for. It is not necessary for the process to be complex or difficult. In many workplaces the risks are well known and understood; it is often only necessary to ensure that effective precautions have been taken to minimise the risk.

Make the Grade — P5

Grading criterion P5 requires you to carry out a risk assessment on a typical item/area of the working environment. An item could be a particular machine or piece of equipment, while an area might typically be a workshop or production area.

Risk Assessment

Company name: _____   Date of risk assessment: _/_/_

Step 1: What are the hazards?
_____
_____
_____

Step 2: Who might be harmed and how?
_____
_____
_____

Step 3: What are you already doing? What further action is necessary?
_____
_____
_____

Step 4: How will you put the asessment into action?
_____
_____
_____

Step 5: Review date: _/_/_

Figure 1.15 Risk assessment template

Items/area to be assessed

Risk assessments require you to examine what, in your workplace, could cause harm to people. This allows you to consider the precautions in place and whether further measures should be taken; remember, all employees and other interested parties have the right to be protected from harm.

An area to be assessed will usually be the work area in question. If you are considering a large work area, such as a large production facility or warehouse, it might be sensible to split this up into smaller areas, for simplicity. For example, in a large manufacturing plant, you would expect risk assessment to be carried out for each production cell or zone.

However, a risk assessment can be equally appropriate for a large piece of machinery or for an item used in engineering. An example of this could be a hydraulic press, a robot or a chemical treatment tank.

Five steps

Once you have decided the item or area you are going to assess for risk, you can decide on how you are going to undertake your risk assessment. The usual process is to break the assessment down into five steps.

Five steps of risk assessment:

1. Look for the hazards.
2. Decide who is likely to be harmed and how.
3. Evaluate the risks and decide whether sufficient precautions have been taken or whether further action is required.
4. Record your findings.
5. Review your assessment and make revisions if necessary.

Look for hazards

Working in the same environment every day can lead to hazards being overlooked, so it is important to be systematic about the process.

- Walk around – what hazards can you spot?
- Ask around – talk to colleagues, co-workers and other interested parties. What do they know that is potentially harmful that you might have missed?
- Consider the less obvious hazards such as long-term health risks.
- Refer to records of ill health for the area.
- Consult the HSE – look at the guidance documents, website or talk to them directly.

- Consult interested parties – these could be trade associations, trade unions, health and safety officers, and so on.
- Check information sheets, materials specifications, manufacturers' instructions, etc – this information can reveal unknown hazards such as harmful chemicals.
- Refer to accident records for the item or area.

Figure 1.16 Timber offcuts present a fire risk

Decide who is likely to be harmed and how

It is often useful to group workers to identify the risk of hazard that they may be subject to. For example, computer operators may be at risk of repetitive strain injury (RSI) or eye strain.

It is important to remember:

- young workers, trainees, pregnant mothers, people with disabilities and others who may be at particular risk
- visitors, contractors, cleaners, facilities staff and others who may not be in the workplace all the time
- members of the public who may be exposed to risk – this would be important to consider with site work, for example.

Staff, colleagues and co-workers should be consulted to ensure that nobody has been overlooked.

Evaluate the risks and decide whether sufficient precautions have been taken or whether further action is required

Once you have noted the hazards, the next step is to decide on a course of action. Remember that the law requires the employer to do what is 'reasonably practicable' to protect people from harm. It is likely that you have already taken precautions and these may be adequate.

The minimum requirement is to ensure that all legal requirements have been met. For example, if noise levels are excessive, are workers using the correct ear protection, is all dangerous machinery correctly guarded?

Industry standards and good practice are often more stringent than the minimum legal requirements. You should be doing everything you can to minimise risk. Consider the following questions:

1. Can I remove the hazard altogether?
2. If not, how can I control the risks?

Often, this will require an action list and the following principles are recommended:

- Try a less risky option – could larger computer monitors reduce the risk of eye strain, for example?
- Prevent access to the hazard – improve guarding, for example.
- Organise work to prevent or reduce exposure to the hazard – design walkways to divert people around dangerous machinery.
- Supply PPE – this could be footwear, clothing, goggles, etc.
- Provide adequate welfare requirements – washing and toilet facilities, first aid require-ments, etc.

Often, simple steps can improve health and safety without significant cost – for example, fitting a handrail to a short flight of steps.

Record your findings

It is important to write down the results of your risk assessment and share them with staff or colleagues. This is partly to ensure that you have assessed all risks correctly and not missed anything. The other benefit is that an employer should share the responsibility with employees and vice versa.

Many elements of a risk assessment may be a confirmation of suitable measures or precautions – for example, danger from rotating machinery might be highlighted, with interlocked guards provided as protection.

HSE inspectors will expect evidence that:

- a suitable assessment was carried out
- anyone who might be affected was consulted
- significant hazards were dealt with, considering the implications
- sensible precautions were taken and any remaining risk is minimal
- all staff and their representatives were consulted.

The written record is a useful document to show HSE inspectors, but it is also helpful to refer to the next time someone needs to do a risk assessment in the same or similar work areas. The risk assessment can highlight many things. If this is the case, it is useful to produce an action plan of how you intend to tackle things. Obviously, you should prioritise the most important risks and tackle these first, but use the plan as a checklist and tick off risks as they are dealt with.

Review your assessment and make revisions if necessary

It would be unusual for a work area to remain the same for an extended period of time; new machines are introduced, staff change, different substances are used and the hazards associated with all of these can be different to when the original risk assessment took place. Consequently, it is a good idea to have a fixed date when risk assessments will be updated. This might be once a year, but obviously in some workplaces it might need to be more frequent, and a risk assessment should certainly be carried out if major changes occur, such as new machinery or changes to premises.

Key Points

Although it is not a legal requirement, if a company has fewer than five employees, it is still considered good practice to keep a written record of risk assessment and share this with employees.

Make the Grade — M2

Grading criterion M2 requires you to explain the importance of carrying out all parts of a risk assessment in a suitable manner. This should follow on from the risk assessment you will carry out to provide evidence for P5. You will need to explain the purpose of each step of the risk assessment and the method that should be used.

Activity 1.9
Carry out a risk assessment for a piece of equipment such as a photocopier, VDU or electric drill.

Use of control measures

By the end of this section you should have developed a knowledge and understanding of what control measures can be taken as a result of a risk assessment and why.

Completing a risk assessment is only the first step in ensuring a safe working environment. The action you take following the completion of a risk assessment determines how safe people will be in the workplace.

Key Point
You will need to identify a control measure against each substantial risk you identify from a risk assessment.

This section introduces you to the methods that can be used to control specific areas of risk, from removing the risk entirely, to what training, protective clothing, and so on, can be used to minimise that risk.

Make the Grade

Grading criterion P6 requires you to suggest suitable control measures after a risk assessment has been carried out, and to state the reasons why they are suitable. This criterion links with P5. You should look at the assessment you have carried out and recommend actions you would take to control the risks and why.

There are a variety of ways of controlling hazards in the workplace, depending on the nature and level of risk.

Remove need (design out)

Sometimes a risk can be removed entirely by redesigning a work area or removing a dangerous substance or component. You might find that a hazardous chemical used for cleaning and degreasing can be replaced with a non-toxic solution or a process not involving chemicals. Alternatively, a risk associated with noise or heat from a nearby building/installation can be removed by insulating the building.

Use of recognised procedures

It may be that you find a risk is present because the correct procedures are not being followed. For example, a dangerous chemical may not be being stored or used properly. Alternatively, an adhesive might be being used in a confined space and giving off fumes. Clearly, the solution might be as simple as following the manufacturer's procedures, such as using in a well-ventilated area or using appropriate breathing equipment.

Substance control

COSHH regulations give guidance on handling and using dangerous substances. Although manufacturers often give guidelines for the use of various substances, it is important to refer to these regulations, as they highlight:

- what is a substance hazardous to health
- what you need to do
- identifying hazards and assessing risk
- exposure limits
- safety data sheets
- control measures
- personal protective equipment
- monitoring
- health surveillance
- training
- emergencies.

Guarding

Where a machine or process is hazardous, guards are often the only way to protect people from danger. Guards are used to ensure that access to rotating components, moving parts or other dangerous equipment is prevented.

Features of guards are:

- fixed or removable

- interlocked – so that the machine can only operate when a guard is in place
- fit for purpose
- not easy to override
- do not allow access to dangerous parts
- do not create other problems, such as the potential for trapped limbs or hair.

Figure 1.17 Machine guards

Lifting assessments and manual handling assessments

Manual handling is defined as any supporting or transporting of a load by hand or using bodily force. This can include lifting, pushing, pulling, carrying, moving or putting down. Poor manual handling technique often leads to back pain, as well as severe pain in the wrists, arm, neck, etc. Employers should strive to:

- avoid manual handling or lifting where there is a risk of injury
- assess the injury risk from manual handling that is unavoidable
- reduce the risk of injury from manual handling hazards.

To reduce the risk, the first thought should be: can the load be moved to suit the person, rather than having the person move to the load? A good example of this is in the maintenance of cars. Instead of bending or lying on the floor, technicians can raise the vehicle to a comfortable height using a car lift.

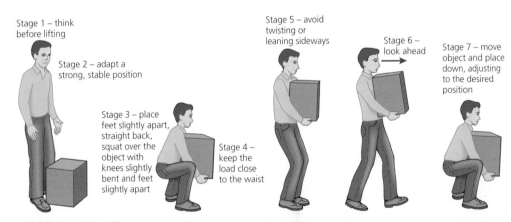

Figure 1.18 Manual handling

Figure 1.19 Lifting devices

24

Figure 1.20 Using a car lift

Other aids to lifting or handling can be provided, such as hoists, sack trucks or conveyors. All workers should have access to suitable lifting equipment and be trained in its use.

Manual Handling Operations Regulations 1992 offer comprehensive advice on manual handling. However, sometimes all that is required is training on manual handling to ensure that all employees know how to lift and carry safely and sensibly.

Regular inspection

Although risk assessments are, or should be, carried out on a regular basis, there are still circumstances where regular inspections are essential. In the nuclear industry, workers are regularly checked for exposure to radiation; similarly, in a mechanical workshop, guards should be checked for damage or missing parts. It is often employees who notice this, and team meetings/briefings should include an opportunity for health and safety concerns to be raised.

Figure 1.21 Water leak

Use of PPE

The use of PPE is often regarded as the last option, if no other control measures are feasible. This might seem unusual, but there are a variety of reasons for this:

- PPE normally only protects the individual wearing it
- PPE can be difficult to maintain and it is easy to damage or degrade its performance owing to wear and tear
- PPE can make it difficult to do the job effectively, as it can restrict mobility, vision, hearing or awareness
- PPE is often not fitted correctly or is compromised by operators not using it properly.

PPE regulations require that people are trained in the correct use of PPE and that employers must ensure that sufficient PPE is available and fitted correctly. A variety of considerations will arise from completing a risk assessment, and may include who is exposed and for how long. This should guide you in deciding the required PPE.

Figure 1.22 Eye protection

> **Activity 1.10**
> Suggest PPE for the following hazards highlighted during a risk assessment:
> - handling hot metal ingots
> - using a grinding wheel
> - working with hydraulic oil
> - working in a fume-filled room
> - using a welding kit.
> As an extension activity, can you explain why you have picked each particular item of PPE.

Training of personnel

One of the main results of carrying out a risk assessment is often the need for staff training. This is an important way of ensuring that safe

25

working practices are embedded in the workplace. Although most organisations ensure that first aid, fire, safe evacuation and emergency training take place, there are many other elements that can be missed.

Do personnel know how to:

- use PPE correctly?
- use standard procedure when handling dangerous substances?
- operate the machines, tools and equipment in their work area correctly and safely?
- use safe techniques for lifting and manual handling?
- use new techniques or materials correctly?
- operate safely when moved to an unfamiliar work area?

Even with well-trained staff, it is important to have refresher training activities to ensure that everyone is up to date. It is not just new members of staff who have health and safety training needs.

Other personal procedures for health, safety and welfare

Sometimes, what appear to be the most obvious health and safety procedures can be overlooked. It is important that:

- care is taken to ensure that rubbish is disposed of correctly
- work areas are kept clean and tidy
- staff regularly wash their hands
- food and drink are kept away from work areas
- clean, well-lit toilets and washing facilities are provided
- if appropriate, specialist hand-cleaning facilities are provided
- adequate provision of drinking water is made available
- accommodation for clothing is provided if workers need to change into uniforms or PPE
- laundry facilities are available for work clothing and/or PPE
- facilities are available for eating/drinking and breaks.

Make the Grade — M3

Grading criterion M3 requires you to explain how control measures are used to prevent accidents. If you have carried out a risk assessment (P5) and suggested suitable control measures (P6), then you are well on the way to achieving this criterion. You will need to add an explanation to each of your control measures, saying how it will prevent an accident.

Figure 1.23 Protective equipment signs

Make the Grade — D1

Grading criterion D1 requires you to justify the methods used to deal with hazards in accordance with workplace policies and legal requirements. To justify the methods, you are going to have to show why each method is the best solution compared with other options, and show how your methods meet regulations.

Activity 1.11

Suggest control measures for the following issues raised during a risk assessment:

- risk of falling from a platform
- risk of electric shock from exposed wiring
- risk of fire from sparking electric drills in a gas storage facility
- risk of RSI from constant pneumatic drill use.

As an extension activity, explain why each control measure will prevent accidents.

1.4 Understanding the methods used when reporting and recording accidents and incidents

This section will cover the following grading criteria:

Despite all efforts by employers and employees, accidents do happen. Accident prevention is not only good health and safety practice, it is also financially beneficial, as there is often a significant cost to the employer in lost production, skills and knowledge if an employee is lost to the business through injury or due to an accident.

It is important to learn from accidents to prevent them from happening again; consequently, any accident or near miss should be considered an opportunity to improve health and safety.

> **Key Point**
>
> It is a legal requirement to report specific accidents, injuries and dangerous occurrences to the HSE.

Principles

By the end of this section you should have developed a knowledge and understanding of the principal reasons for reporting accidents and injuries.

Reporting accidents or near misses is an important part of an organisation's health and safety policy. If you do not have a record of when, how and why incidents occur, it is difficult to take the correct, targeted improvements necessary to protect the health and safety of all persons who could potentially be affected.

> ## Key Term
>
> **Near miss** – In health and safety terms, a near miss can be defined as an incident that could have caused injury or damage.

> **Make the Grade**   P7
>
> Grading criterion P7 requires you to explain the principles that underpin reporting and recording accidents and incidents. You will need to understand how recording accident data can improve health and safety, which links to grading criterion P8, where you will describe accident reporting procedure.

In this section we shall investigate the reasons for reporting accidents and the issues that surround accidents and incidents in the workplace.

Why employers keep records of serious accidents, incidents and emergencies

Employers keep records of accidents for a variety of reasons. There are the legal requirements, which we will discuss later, but the general welfare of employees, together with the various costs to the business, are the reasons why employers should learn from accidents and try to improve health and safety accordingly.

> **Key Points**
>
> Measuring noise levels is a good example of why employers keep records. By recording increasing noise levels, it is possible that you can predict when they will become excessive and do something about it before it becomes a health hazard.

Serious accidents

Serious accidents are specifically defined in RIDDOR regulations and include fractures, loss of sight or hearing, and other accidents that result in employees having time off or being incapacitated.

Incidents

Incidents may or may not be dangerous to employees. They can include gas leaks, machine damage and near misses.

Figure 1.24 Protection from noisy machinery

Emergencies

Emergencies are likely to be due to major alerts, such as fire alarms or dangerous chemical spills. Organisations need to plan for these to ensure that they have:

- the right equipment to deal with the emergency (for example, a spill), including protective equipment and decontamination products
- the right procedures to deal with a casualty
- the right people trained to take action
- the right arrangements to deal with the waste created.

Responsibilities of competent persons

All engineers would consider themselves competent people – after all, who would describe themselves as incompetent? However, for the purposes of health and safety, a competent person is somebody who has had suitable training. For example, if an employer is concerned about employees being subjected to excessive noise, they might require measurements of noise levels being made on an hourly or daily basis. To ensure that this is done by a competent person, they would have undertaken required training in the correct use of the sound-measuring equipment.

Cost of accidents

Every accident, however minor, costs an organisation money. Even a minor cut requires an employee to be away from the workplace, if only

for a short period of time. In addition, a first-aider could be away from the workplace, tending to the cut, and other employees could be affected by one person's absence.

A more serious injury or accident could have much greater financial consequences and could include:

- wages for the injured person
- loss of production
- extra wages for replacement staff
- loss of time for managers/administrators dealing with the injury and organising replacement staff
- production delays
- damage to plant, machinery, tools, buildings, and so on.

If the injury is caused by negligence, the HSE could prosecute and this would entail extra cost, including:

- legal expenses
- fines
- court costs.

Employers take out insurance to protect themselves against these costs. However, these costs are normally the immediate injury cost, plus any damage and ill-health issues – clearly not the total cost to the business. In addition, the cost of insurance premiums could increase and the reputation of the business could be affected.

Incident cost calculator

The HSE has an incident cost calculator that allows employers to determine the cost of an accident; this is available as an online calculator from the HSE website or can be a paper-based technique. It breaks down the individual costs as follows:

Dealing with incident

- First aid
- Cost of transporting injured person to hospital
- Making the area safe
- Putting out fires
- Staff downtime

Investigation costs

- Staff time to report and investigate incident
- Meeting to discuss incident
- Time spent with HSE/local authority inspector
- Any consultant fees

Getting back to business

- Rescheduling work
- Recovering production
- Site clean-up
- Product rework
- Equipment or tool hire

Business costs

- Salary costs of injured person
- Salary costs of replacement workers
- Lost work time
- Overtime costs
- Recruitment costs
- Contract penalties
- Cancelled or lost orders

Action to safeguard future business

- Reassuring customers
- Alternative source of supply for customers

Sanctions and penalties

- Compensation claims
- Legal fees
- Staff time spent on legal issues
- Fines and costs
- Increase in insurance premiums

It is clear that there are many more costs when an accident occurs than the immediate cost of the accident.

The Incident Cost Calculator

Date and time of incident ....
Description of incident ....
Name of person involved ....

Dealing with incident (immediate action)

Examples	Time spent	Cost (£)
First-aid treatment		
Taking injured person to hospital/home		
Making the area safe		
Putting out fires		
Immediate staff downtime (eg work activity stopped)		
Other		

Investigation of incident

Examples	Time spent	Cost (£)
Staff time to report and investigate incident		
Meetings to discuss incident etc		
Time spent with HSE/local authority inspector		
Consultant's fees to assist company in investigation		
Other		

Getting back to business

Examples	Time spent	Cost (£)
Assessing/rescheduling work activities		
Recovering work/production (including staff costs)		
Cleaning up site and disposal of waste, equipment, products etc		
Bringing work up to standard (eg product reworking time/costs)		
Repairing any damage/faults		
Hiring or purchasing tools, equipment, plant, services etc		
Other		

Business costs

Examples	Time spent	Cost (£)
Salary costs of injured person while off work		
Salary costs of replacement workers		
Lost work time (people waiting to resume work, delays, reduced productivity, effects on other people's productivity etc)		
Overtime costs		
Recruitment costs for new staff		
Contract penalties		
Cancelled and/or lost orders		
Other		

Action to safeguard future business

Examples	Time spent	Cost (£)
Reassuring customers		
Providing alternative sources of supply for customers		
Other		

Sanctions and penalties

Examples	Time spent	Cost (£)
Compensation claim payments		
Solicitor's fees and legal expenses		
Staff time dealing with legal cases		
Fines and costs imposed due to criminal proceedings		
Increase in insurance premiums		
Other		

Other

Examples	Time spent	Cost (£)

Total		

Figure 1.25 Incident cost calculator

Activity 1.12

During a production operation, an employee trapped his or her hand in a piece of machinery. Fortunately, a colleague pressed the emergency stop and called for help. The following table outlines the health and safety procedure:

Action	Number of people	Time spent (hours)
First aid	2	0.5
Transport to hospital	1	4
Making the area safe	4	1
Work activity stopped	6	1
Time spent investigating	2	2
Time spent with HSE inspector	1	2
Repairing faults	1	1
Salary of person off work	1	32
Salary cost of replacement staff	1	32

Using the cost calculator, determine the total cost, assuming £30/hr as an average rate.

Make the Grade

Grading criterion D2 requires you to determine the cost of an accident in the workplace from given data. You will need to consider carefully all the information you are given and try to think of all the different things that have to be considered and what the cost of each action is. By totalling all these elements, you will arrive at the total cost.

Trends

Recording information can show when trends start to occur before an event happens. A good example of this is the use of seismographs to record earth tremors; this can help authorities to predict when an earthquake might happen. Alternatively, records of pollen levels are used to predict when sufferers will need to take their hay fever medication.

Major causes

The first trend to consider might be the major causes of accidents, injuries or health and safety issues. For example, collecting data over a long period shows us that exposure to the sun causes skin cancer, or that falls are the major cause of accidents for people working with ladders or scaffolding.

Fatal and serious injuries

Trends in fatal and serious injuries from all engineering sectors can be studied to inform businesses of the risks involved. Fatalities increase in certain conditions, and by observing these trends, an employer will try to prevent these conditions from arising. An example might be deaths due to adverse weather on an offshore installation. By halting work when the weather becomes extreme, an employer might save lives.

Methods of classification

Incidents are classified as:

- deaths
- major injuries
- reportable over-three-day injuries
- injuries to members of the public
- certain prescribed diseases
- gas incidents
- dangerous occurrences (near misses).

Statistics

The principle of using data to predict trends is a very important one in health and safety terms. Organisations should record near misses as if they were accidents. It is inevitable if that if a particular work area has an ever-increasing number of near misses, then an accident is sure to follow.

Collating statistics over a long period of time can show up some other interesting trends, such as an increase in accidents at a particular time of day, when people's concentration is lower, or at a certain time of year, due to cold or hot conditions.

Recording and reporting procedures

By the end of this section you should have developed a knowledge and understanding of how to report injuries, near misses and accidents.

All incidents, whether accidents causing injury to workers or damage to equipment/facilities, should be recorded. This is in order to meet legal requirements. However, it is also essential to record near misses, as this helps an organisation to focus on where health and safety issues are occurring without injury or damage necessarily taking place. In addition, it is also a legal requirement to record near misses.

Key Points

There are three important reasons why you should report accidents:

1. It is a legal requirement.
2. It helps an organisation see where it needs to concentrate on health and safety.
3. It is required for insurance purposes.

Make the Grade

Grading criterion P8 requires you to describe the procedures used to record and report accidents, dangerous occurrences or near misses. This means that you will be explaining where you would write down details of these incidents and who you would tell.

There are important steps that should be taken to report incidents, and regulations that have to be followed.

Regulations on accident recording and reporting

The most important piece of legislation covering accident reporting is RIDDOR 1995. These regulations require an employer to report incidents to the HSE. This can be via the phone, the internet or post/fax. There is even an ICC set up to receive this information.

RIDDOR requires the reporting of deaths, major injuries and injuries that involve over three days off work or incapacity. However, other categories, including injuries to members of the public, certain prescribed diseases and gas incidents, are also reportable.

Deaths

If an employee, self-employed person or member of the public is killed, the enforcing authority has to be notified immediately. This should be by phone and will need to be followed by the completion of a report of injury (F2508) form.

Major injuries

The procedure for major injuries is exactly the same as for deaths: a phone call should be followed by the F2508 form. The following is a list of what constitutes a major injury:

- fracture, other than to fingers, thumbs and toes
- amputation
- dislocation of the shoulder, hip, knee or spine
- loss of sight (permanent or temporary)
- chemical or hot metal burn to the eye or any penetrating injury to the eye
- injury resulting from an electric shock or electrical burn leading to unconsciousness, or requiring resuscitation or admittance to hospital for more than 24 hours
- any other injury, leading to hypothermia, heat-induced illness or unconsciousness; or requiring resuscitation or admittance to hospital for more than 24 hours
- unconsciousness caused by asphyxia or exposure to a harmful substance or biological agent
- acute illness requiring medical treatment, or loss of consciousness arising from absorption of any substance by inhalation, ingestion or through the skin
- acute illness requiring medical treatment where there is reason to believe that this resulted from exposure to a biological agent or its toxins or infected material.

31

Reportable over-three-day injuries

If an accident at work does not fall into the major injury category, but requires an employee or self-employed person to be off work for more than three days, this must be reported within ten days. Although the incident is not considered a major injury, the F2508 form must be completed or the ICC contacted by phone. If the injured person remains at work, but cannot complete their normal job for over three days, this is also notifiable.

Reportable diseases

The full list of reportable diseases is outlined in the regulations. If a doctor notifies you that you have an employee who suffers from a work-related disease that is reportable, you should contact the enforcing authority, this time using a F2508A form. Reportable diseases include:

- certain poisonings
- some skin diseases, such as work-related dermatitis, skin cancer, chronic ulcer
- lung diseases, such as work-related asthma, farmer's lung, asbestosis, mesothelioma, pneumoconiosis
- infections such as leptospirosis, hepatitis, tuberculosis, anthrax, legionellosis, tetanus
- other conditions, such as work-related cancers, certain musculoskeletal disorders, decompression illness and hand-arm vibration syndrome.

Reportable gas incidents

Organisations that process, distribute, supply or import flammable gas have a responsibility to report death or serious injury connected with the gas, in the same way as for other deaths or serious injuries.

There is also a responsibility which falls on installers of gas equipment or appliances, because the design, installation, servicing and construction/modification could cause:

- an accidental gas leak
- incomplete combustion of the gas
- inadequate ventilation of the gas fumes.

This responsibility is to provide details of any potentially dangerous equipment and is one of the reasons why only gas installers registered with the Council for Registered Gas Installers (CORGI) should be used by employers.

Activity 1.13

State the health and safety reporting procedure you would follow if:

- an employee is killed at work
- a contractor breaks two fingers after falling on company property
- an employee breaks a leg in a car accident on the way home from work
- an employee brings a doctor's note stating that their asthma is work-related.

Procedures to deal with near misses or dangerous occurrences

Incidents that do not result in death or serious injury could still be considered a dangerous occurrence. Accidents such as a building collapsing, an exploding gas cylinder, a part coming off a machine, a falling object on a building site, or a wheel coming off a car could easily be deadly accidents, and it might be luck that prevents death or serious injury. It is the responsibility of an employer to report these kinds of near misses to the appropriate authorities, using the F2508 form, as with other serious incidents.

Figure 1.26 Near miss

Reportable dangerous occurrences can be classified as:

- collapse, overturning or failure of load-bearing parts of lifts and lifting equipment

- explosion, collapse or bursting of any closed vessel or associated pipework
- failure of any freight container in any of its load-bearing parts
- plant or equipment coming into contact with overhead power lines
- electrical short circuit or overload causing fire or explosion
- any unintentional explosion, misfire, failure of demolition to cause the intended collapse, projection of material beyond a site boundary, injury caused by an explosion, or accidental release of a biological agent likely to cause severe human illness
- failure of industrial radiography or irradiation equipment to de-energise or return to its safe position after the intended exposure period
- malfunction of breathing apparatus while in use or during testing immediately before use
- failure or endangering of diving equipment, the trapping of a diver, an explosion near a diver, or an uncontrolled ascent
- collapse or partial collapse of a scaffold over 5 m high, or erected near water where there could be a risk of drowning after a fall
- unintended collision of a train with any vehicle
- dangerous occurrence at a well (other than a water well)
- dangerous occurrence at a pipeline
- failure of any load-bearing fairground equipment, or derailment or unintended collision of cars or trains
- a road tanker carrying a dangerous substance overturns, suffers serious damage, catches fire or the substance is released
- a dangerous substance being conveyed by road is involved in a fire or released
- the following dangerous occurrences are reportable except in relation to offshore workplaces: unintended collapse of any building or structure under construction, alteration or demolition where over 5 tonnes of material falls, a wall or floor in a place of work, any false work
- explosion or fire causing suspension of normal work for over 24 hours
- sudden, uncontrolled release in a building of: 100 kg or more of flammable liquid, 10 kg of flammable liquid above its boiling point, 10 kg or more of flammable gas, or 500 kg of these substances if the release is in the open air
- accidental release of any substance that may damage health.

Company procedures

It is of paramount importance that organisations comply with the regulations regarding the reporting of incidents and deaths/injuries. However, it is not always as clear when a near miss occurs; indeed, the employer may not be aware of it happening. For example, an employee might accidently open a valve, which causes flammable gas to be released. The gas soon escapes into the atmosphere and the employee might fear being disciplined, so decides to ignore the incident. This could become a habit, and one day a spark might cause an explosion.

It is important that employers reinforce to all employees that the reporting of incidents is seen as a positive way of improving health and safety and not as a disciplinary issue.

Employers have to provide employees with information including:

- emergency procedures
- health and safety risks and procedures to eliminate or minimise these risks
- data sheets from manufacturers, designers, suppliers, etc, including substance-handling guidance.

Safety representatives are entitled to see all this information, as well as information on accidents. This information is kept in an accident book.

A typical company process could be:

1. The employee reports an incident in the accident book.
2. The supervisor notes the accident and checks that completion of the accident book is correct.
3. The safety manager reports the incident in line with the RIDDOR regulations and investigates the incident, recording findings in the accident book.
4. Senior management takes action that is 'reasonably practicable' to prevent the incident from reoccurring.

Accident book

To comply with RIDDOR regulations, organisations can record incidents in an accident book. This is a document supplied by the HSE that organisations use to record their accident information as part of their management of health and safety. Each page has a section allowing the person involved to complete details of the accident or to check and sign that the details of an incident are correct if they cannot complete it themselves. The accident book is the key document that allows managers an overview of the safety in their organisation. It can pinpoint where the most accidents occur, what causes accidents and the nature of each incident.

> ### Activity 1.14
> Following a pipe blockage, a pressurised container ruptures, causing gas to escape to the atmosphere.
>
> Investigate the nature of the incident and what reporting/recording of the incident should take place. Draw a flow chart to show how it should be dealt with.

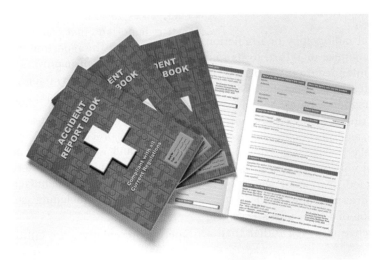

Figure 1.27 Accident book

To achieve a pass grade you will have:	To achieve a merit grade you will also have:	To achieve a distinction grade you will also have:

P1 explained the key features of relevant regulations on health and safety as applied to a working environment in two selected or given engineering organisations

M1 explained the consequences of management not abiding by legislation and regulations and carrying out their roles and responsibilities in a given health and safety situation

D1 justified the methods used to deal with hazards in accordance with workplace policies and legal requirements

D2 determined the cost of an accident in the workplace from given data.

P2 described the roles and responsibilities, under current health and safety legislation and regulations, of those involved

M2 explained the importance of carrying out all parts of a risk assessment in a suitable manner

P3 described the methods used to identify hazards in a working environment

M3 explained how control measures are used to prevent accidents.

P4 described how hazards which become risks can be controlled

P5 carried out a risk assessment on a typical item/area of the working environment

P6 suggested suitable control measures after a risk assessment had been carried out and stated the reasons why they were suitable

P7 explained the principles that underpin reporting and recording accidents and incidents

P8 described the procedures used to record and report accidents, dangerous occurrences or near misses.

By the end of this unit you should be able to:

- interpret and use engineering sketches/circuit/network diagrams to communicate technical information
- use verbal and written communication skills in engineering settings
- obtain and use engineering information
- use information and communication technology (ICT) to present information in engineering settings.

In order to pass this unit, the evidence you present for assessment needs to demonstrate that you can meet all of the above learning outcomes for this unit. The criteria below show the levels of achievement required to pass this unit.

To achieve a pass grade you need to:	To achieve a merit grade you also need to:	To achieve a distinction grade you also need to:
P1 interpret an engineering drawing/circuit/network diagram	**M1** evaluate a written communication method and identify ways in which it could be improved	**D1** justify the choice of a specific communication method and the reasons for not using a possible alternative
P2 produce an engineering sketch/circuit/network diagram	**M2** review the information sources obtained to solve an engineering task and explain why some sources have been used but others rejected	**D2** evaluate the use of an ICT presentation method and identify an alternative approach.
P3 use appropriate standards, symbols and conventions in an engineering sketch/circuit/network diagram	**M3** evaluate the effectiveness of an ICT software package and its tools for the preparation and presentation of information.	
P4 communicate information effectively in written work		
P5 communicate information effectively using verbal methods.		
P6 use appropriate information sources to solve an engineering task		
P7 use appropriate ICT software packages and hardware devices to present information.		

Introduction

Being able to communicate ideas and information in an effective manner is an essential requirement for engineers and working as an engineer requires you to understand and translate technical language. Engineers need to communicate engineering and technical information using sketches, circuit and network diagrams. They also need to be able to use effective verbal and written skills in an engineering context. The use of ICT to present technical information is, of course, an essential requirement for presenting information, and the ability to obtain information from appropriate sources and use that information to solve problems and complete tasks is a key requirement for technician engineers.

This section of the book is organised as follows; each of the subsections can be readily linked to the learning outcome (LO), pass (P), merit (M) and distinction (D) criteria.

Section/content	LO	P	M	D
2.1 Interpreting and using engineering sketches/circuit/network diagrams to communicate technical information	1	1, 2, 3		
2.2 Using verbal and written communication	2	4, 5	1	
2.3 Obtaining and using engineering Information	3	6	2	
2.4 Using Information and Communication Technology (ICT) to present information	4	7	3	1, 2

2.1 Interpreting and using engineering sketches/circuit/network diagrams to communicate technical information

This section will cover the following grading criteria:

P1 P2 P3

If you worked as an interpreter you would need to understand a foreign language and translate it into a language understood by others. All branches of technology have their own 'languages', and being able to interpret or understand technical information allows you to use it to work effectively. Many years ago car manufacturers realised that if they were to label the gauges on the dashboard, it would be helpful to have a common system to allow drivers in any part of the world to understand what each gauge means. To overcome language barriers, the symbol, shown in Figure 2.1, has become universally accepted and recognised.

Figure 2.1 Fuel tank symbols used in cars

Engineers frequently use graphical methods to produce diagrams and sketches; you will need to be able to understand what the symbols and conventions mean when viewing these drawings. Just like learning a language, the conventions can seem confusing. However, when you become familiar with these conventions, you will be able to interpret and use them with a degree of fluency.

Interpret

By the end of this section you should have developed a knowledge and understanding of interpreting an engineering drawing/circuit/network diagram.

Engineers need to interpret information effectively from a variety of sources. Diagrams, drawings and sketches are methods of graphical communication and there are conventions in the way that they are produced. It is important to present technical information graphically; this information should be in a form that is readily understood and can be interpreted by others.

> **Key Point**
>
> Interpreting drawings, sketches, circuits and network diagrams refers to the information available in a written or graphical form. It is how you interpret or translate this information to produce a final product that is of interest to engineers.

Looking at drawings, specifications, circuit diagrams, sketches, etc. is something that engineers have to do all the time. The information available, and how it is used, is known as interpretation, and what you use this for is crucial in engineering, manufacturing and related technologies.

For example:

- engineering drawings use dimensions to allow accurate measurements to be made
- circuit diagrams allow components to be selected and appropriate connections and settings to be made
- exploded/assembly diagrams show how complex components are put together and taken apart.

In the next section you will learn how to produce sketches, circuits and diagrams, including standard symbols and conventions.

39

Obtain information and describe features

You should be able to obtain information and describe key component features from any of the standard methods of communication used in engineering. Engineers often use sketches, engineering drawings, computer-aided design (CAD) and circuit diagrams to communicate information. Standard techniques are used and it is important that you can recognise these and use them to obtain relevant information and data.

You have a sketch of a broken machine part that you need to replace.

What sizes need to be used?

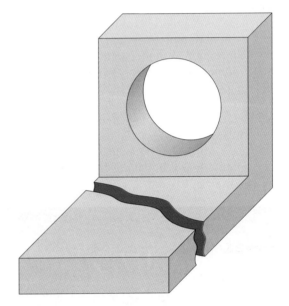

Figure 2.2 Sketch of a broken bracket

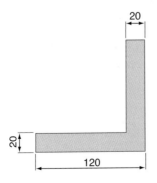
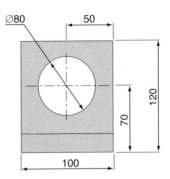

Figure 2.3 Orthogonal view of the bracket

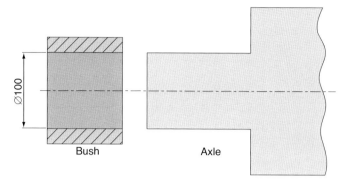

Figure 2.4 Nominal sizes of bush/axle

You need to press-fit a bush on to the axle of a skateboard.

For a press fit, which should be bigger, the hole or the axle?

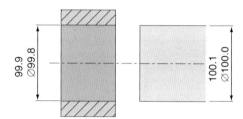

Figure 2.5 Bush/axle combination with tolerances applied

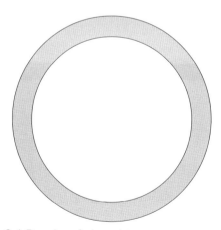

Figure 2.6 Drawing of a bracelet

You have a steel bracelet that needs to be very smooth and round.

How smooth and how round should the bracelet be?

These examples show how important information needs to be interpreted by the use of standard techniques that are easily recognisable. You are probably not very familiar with some of them; this is because they are part of the language of engineering and require translating. Later in this section you will learn more about how to use these and other conventions to produce sketches and diagrams containing all the key information that is required.

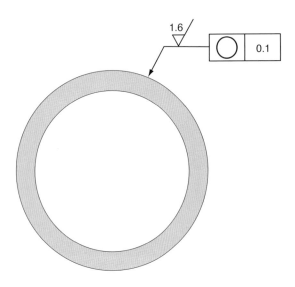

Figure 2.7 Drawing of bracelet with geometric/surface finish symbols added

Key Terms

Dimensions – The actual sizes of a component or feature. You should not use a rule to measure off a drawing, there should be a numerical value.

Tolerances – The maximum and minimum allowable sizes for a feature; the larger a tolerance is, the easier it is to reproduce.

Identify manufacturing/assembly/process instructions

You should be able to obtain information on how a product or group of products are used and develop further details from sketches and diagrams.

Assembly drawings

An assembly drawing is intended to allow the user to put together or assemble a group of components. The details of these parts will be represented in individual drawings or specifications; however, the assembly allows the parts to be put together in the right sequence and with the correct orientation. An assembly drawing should have a numbered parts list and the individual parts should be 'balloon' referenced, as shown in Figure 2.9.

Exploded views

Sometimes an assembly is represented by an exploded view. This technique is used to give a three-dimensional (3D) view of the assembly, as if it had been taken apart. Often, alignment lines are added to indicate which parts are connected directly to each other.

Activity 2.1

Study the drawing in Figure 2.8.

Why are there two views?

What does the series of diagonal lines through the right-hand view represent?

What dimension is missing from the drawing?

Can you guess what the ∅ symbol means?

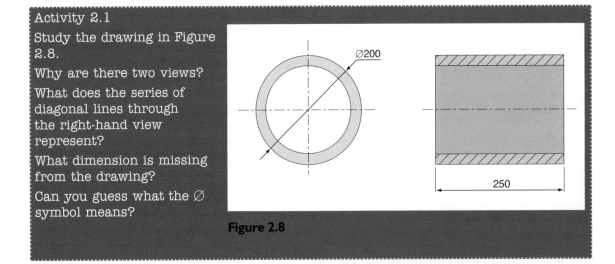

Figure 2.8

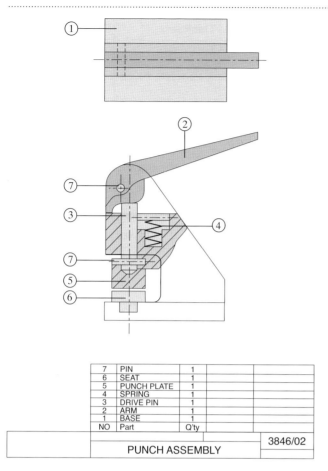

7	PIN	1	
6	SEAT	1	
5	PUNCH PLATE	1	
4	SPRING	1	
3	DRIVE PIN	1	
2	ARM	1	
1	BASE	1	
NO	Part	Q'ty	

PUNCH ASSEMBLY	3846/02

Figure 2.9 Assembly drawing and parts list

Figure 2.10 Exploded view drawing

Cutting lists

Making a full list of all the materials that will be needed to make a product is a very useful part of the planning process. If you have engineering drawings or sketches, it should be possible to extract the size and material information in order to generate a cutting list. Cutting lists are best presented in the form of a table. If you are preparing sheet material, it is usual to allow an extra 10 mm to the length and 5 mm to the width, to allow for final planing and trimming to length. Similarly, when cutting metal rod or similar sections of metal or plastic, 3 mm is usually added to allow for final trimming using a metal turning lathe or other similar techniques. The diameter column is useful for round sections such as a metal rod, dowel or plastic tube. N/A (not applicable) can be entered in the grid where no measurement needs to be listed.

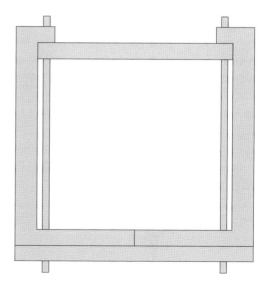

Figure 2.11 Drawing of a fabrication

43

Part	No. OFF	Material	Length	Width	Thickness	Diameter
Sides	2	Sheet steel	210	85	8	N/A
Top	1	MDF	110	85	6	N/A
Base	1	MDF	130	85	6	N/A
Pins	4	Steel rod	102	N/A	N/A	8

Table 2.1 A completed cutting list

Operating procedures

Standard operating procedures are used to ensure that all operators, particularly in a production environment, do their job in the same way. Essentially, a standard operating procedure is a visual guide, although photographs are often used, as well as graphic images, to ensure compliance.

Plant/process layout

Process layout drawings are frequently used in the chemical or process industries to show how large-scale pipework and machinery are put together. Although these are traditionally two-dimensional (2D), much more use is being made of 3D techniques. A 3D CAD drawing, for example, can show how a refrigeration plant will fit into a warehouse or how complex pipework is installed in an oil refinery.

Process layouts can be flow diagrams, showing how a process is carried out, and can be schematic, being used to represent how one operation follows another.

Alternatively, they can be a representation of how a process occurs in a factory. For example, a plan of a car manufacturing plant could show the flow of raw materials into the plant and the finished product, the car, as it moves down the production line.

Wet hands and forearms with warm running water and apply soap.

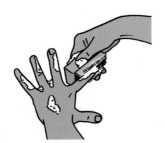

Scrub lathered hands and forearms, under fingernails and between fingers for 10–15 seconds. Rinse thoroughly under warm running water for 5–10 seconds.

Dry hands and forearms thoroughly with paper towels, or dry hands for at least 30 seconds if using a hand dryer.

Turn off water using paper towels.

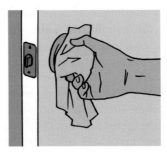

Use paper towel to open door when leaving.

Figure 2.12 Standard operating procedure

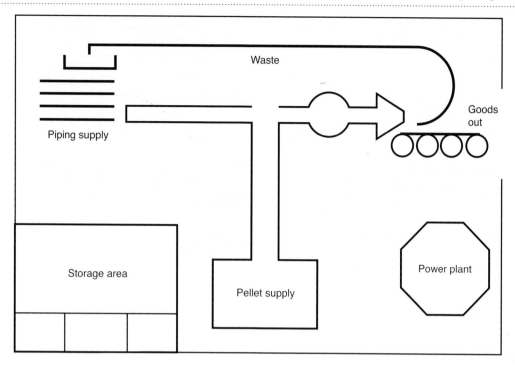

Figure 2.13 Plant layout

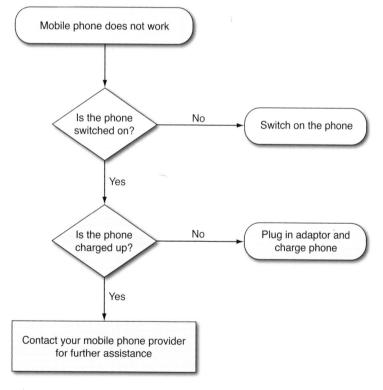

Figure 2.14 Process chart

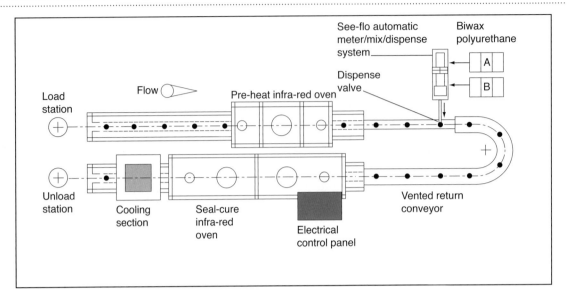

Figure 2.15 Production layout

Electrical/electronic communication circuit requirements

Circuit diagrams use symbols to represent components, parts or connections. They can be very complex and require an understanding of what each symbol represents. Fortunately, there are standard symbols that are used and many computer packages have libraries of these symbols, allowing you to develop and test circuits quickly and easily. Examples of the types of symbols used are shown in Figure 2.16.

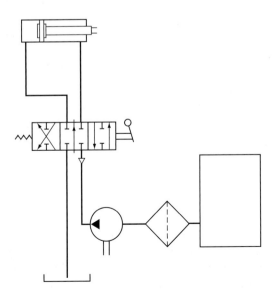

Figure 2.16 Circuit diagram

Activity 2.2

For the computer network shown, explain how each component interacts with its inputs and outputs. You should list each part and indicate its input(s) and output(s).

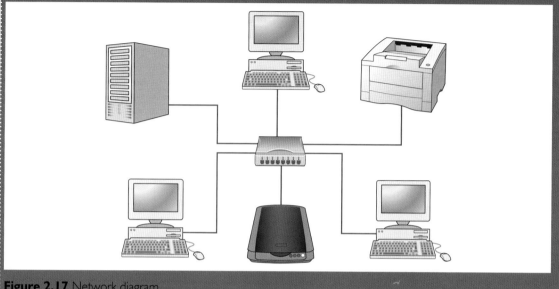

Figure 2.17 Network diagram

Activity 2.3

Generate a parts list for the exploded view shown in Figure 2.18.

You will need to check how many of each component are required and produce a grid, with labels, to identify each part.

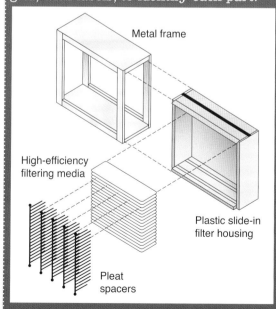

Figure 2.18

Graphical information used to aid understanding of written or verbal communication

When you read a technical manual, textbook or handbook, it is often difficult to understand if only text is used. Similarly, imagine you are being taught how to replace the wheel on a car, test an electric motor or build a computer network. You need to be able to visualise what the writer is trying to communicate or what the presenter is trying to explain to you.

The use of photographs, videos or access to components/assemblies would help, of course. However, graphical methods often allow us to clearly visualise what is being communicated to us.

Illustrations

2D technical illustrations are often used to represent 3D objects. The method involves graphic software or hand drawing of a product or assembly. Although many designs are represented by 3D CAD drawings, there is still much use of drawing techniques in engineering. Technical illustrations are intended to visually communicate technical information and should be an accurate representation in terms of size, showing the key features

47

and indicating what the object does or what it is used for.

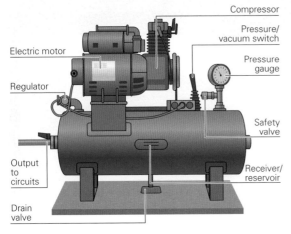

Figure 2.19 Technical illustration

Technical diagrams

Technical diagrams are similar to technical illustrations as they are constructed to convey information. However, they are predominantly 2D and, unlike the technique of technical illustration, are not intended for a non-engineering audience. The techniques are typically used by engineers to indicate how to build, construct, maintain or connect components and follow recognised formats.

Sketches

Sketches are a quickly produced drawing showing some details of a product. Sketching is used to develop ideas and identify areas for improvement. An engineering drawing/diagram is created from a sketch to provide a medium for allowing manufacture, assembly or construction. Sketches are often 3D representations of an object, using the technique of isometric projection. Alternatively, they can be a line drawing showing the arrangement of components in a circuit.

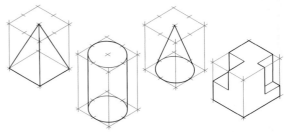

Figure 2.21 Isometric sketch

Key Terms

Illustrations – Drawings that give a visual representation of an object or assembly.

Technical diagrams – Diagrams that show how parts or components work or are put together.

Sketches – Freehand drawings that demonstrate the key features of a layout, circuit or component.

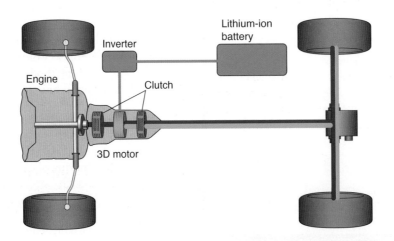

Figure 2.20 Technical layout

Activity 2.4

For the assembly shown in Figure 2.22, draw up a step-by-step operating procedure to allow consistent assembly of the parts.

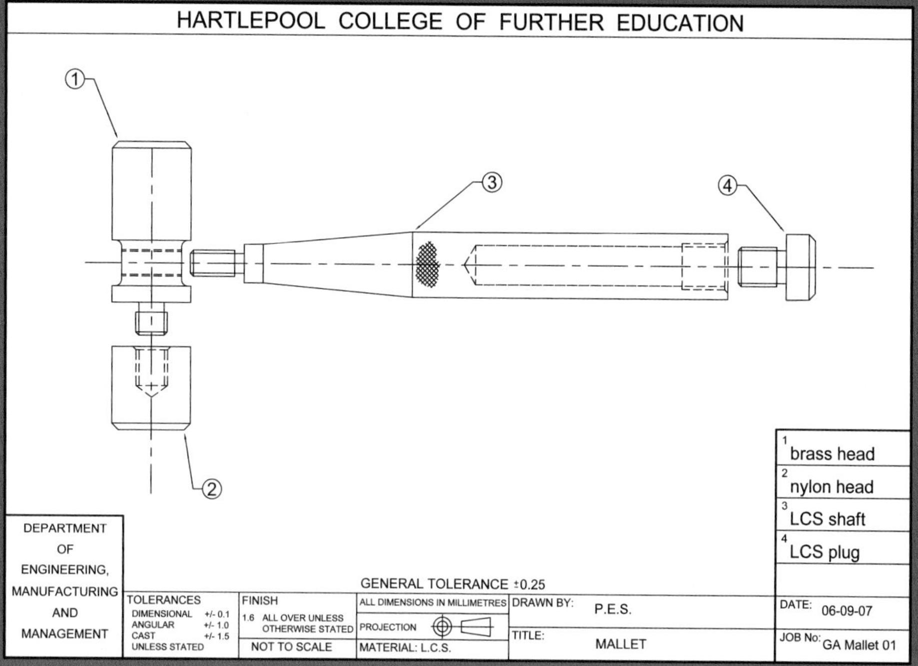

HARTLEPOOL COLLEGE OF FURTHER EDUCATION

| 1 brass head |
| 2 nylon head |
| 3 LCS shaft |
| 4 LCS plug |

DEPARTMENT OF ENGINEERING, MANUFACTURING AND MANAGEMENT

GENERAL TOLERANCE ±0.25

| TOLERANCES | FINISH | ALL DIMENSIONS IN MILLIMETRES | DRAWN BY: P.E.S. | DATE: 06-09-07 |
| DIMENSIONAL +/- 0.1 ANGULAR +/- 1.0 CAST +/- 1.5 UNLESS STATED | 1.6 ALL OVER UNLESS OTHERWISE STATED NOT TO SCALE | PROJECTION MATERIAL: L.C.S. | TITLE: MALLET | JOB No: GA Mallet 01 |

Figure 2.22 Assembly drawing

Engineering sketches/circuits/ network diagrams

By the end of this section you should have developed a knowledge and understanding of producing and using an engineering drawing/circuit/network diagram to communicate technical information.

Engineers need to use a variety of techniques to communicate information effectively. Different industries have different ways in which they communicate information, and there are conventions that must be followed to ensure understanding and consistency.

Key Points

Producing drawings, sketches, circuits and network diagrams refers to the layout and presentation of technical information in graphical form. It is how others interpret or translate this information to produce a final product that is of interest to engineers.

Getting ideas into a form that can be understood by others is not always easy. Preparing drawings, specifications, circuit diagrams and sketches, is something that engineers have to do all the time to allow communication of their ideas. The communication of this information in an accurate and consistent manner is crucial in engineering, manufacturing and related technologies.

Using different techniques is important to engineers, as the audience can be technical or non-technical. In addition, the techniques used can be specialist or non-specialist – for example, a hydraulic circuit diagram may be understood by a maintenance technician but not by an electrical engineer.

Sketching allows you to:

- create initial ideas
- develop equipment designs
- look at equipment performance
- communicate and demonstrate ideas
- speed up the implementation of engineering projects.

49

Make the Grade P2 P3

Grading criterion P2 requires you to produce an engineering sketch/circuit/network diagram. This does not mean that you will need to produce all three. The reason for the choice is to reflect the branch of engineering in which you are involved. A mechanical engineer might produce engineering sketches, while an electrical/electronic engineer might produce circuit diagrams.

Grading criterion P3 requires you to use appropriate standards, symbols and conventions in an engineering sketch/circuit/network diagram. This means that as well as producing a neat sketch, you will have to use the correct method of laying out your sketch and use symbols or standards.

Freehand sketches of engineering arrangements using 2D and 3D techniques

Freehand sketching is a very important technique for communicating technical information. Sketches can be used for communicating ideas, instructions and information.

Before formal drawings, models or CAD drawings are created, it is often useful to present detailed freehand sketches to analyse and discuss with colleagues.

Components

Sketching is a good way of recording ideas quickly, particularly in the design of a new or modified component.

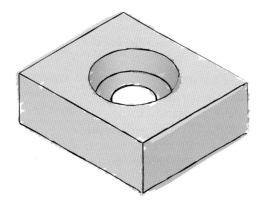

Figure 2.23 Sketched component

Sketches are often initial concept ideas, giving you a wide range of possible approaches to consider. Ideas should be completely different and not just variations on one idea. Freehand sketches can be drawn in either 2D or 3D. 2D sketches are a good method of showing the details of an idea. A 2D sketch will show what the product looks like from one particular direction, such as from the top or from the front. 3D sketches are more difficult to draw, but will give a better idea of how the whole product will look. A 3D sketch will usually show the top, front and one end of the product.

In addition, pencil crayons and marker pens can be used to produce different effects, depending on the shading techniques used. Colours and textures can be used to represent different materials and surfaces.

Adding notes to your sketches

You can draw attention to part of your idea or explain a particular feature by adding notes to your sketches. These should be placed around your drawings or on the drawing itself. Make sure your notes are concise and clearly written.

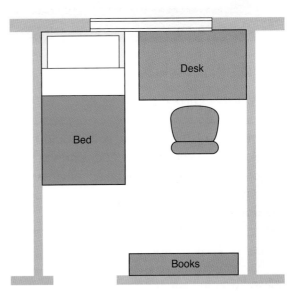

Figure 2.24 Sketch with notes

Sketching techniques

To produce neat engineering sketches, you will need a variety of tools, including:

- HB pencils
- eraser
- paper
- straight-edge or rule

- template for circles, blocks and symbols
- set squares and triangles.

It is useful to use squared or lined paper to keep diagrams neat and square. When designers use drawing film to create technical drawings, they often place a squared grid underneath to keep lines and text neatly arranged. In addition, you can use the grid lines to give you an appropriate scale, as it is unlikely that you will always be sketching components at their actual size.

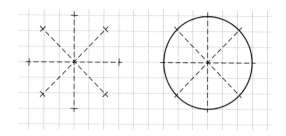

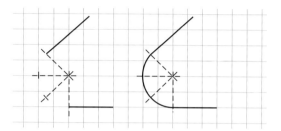

Figure 2.26 Shapes drawn on a grid with guidelines

You should always consider the use of guidelines when producing your sketches. This can allow you to place faint lines in position to help with arcs, circles, hexagons, and so on. You can place a series of guidelines on your squared paper to reduce the length of curved segments, show tangent points or help to reproduce polygonal shapes and objects.

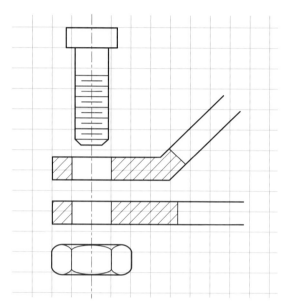

Figure 2.25 Sketch using grid lines

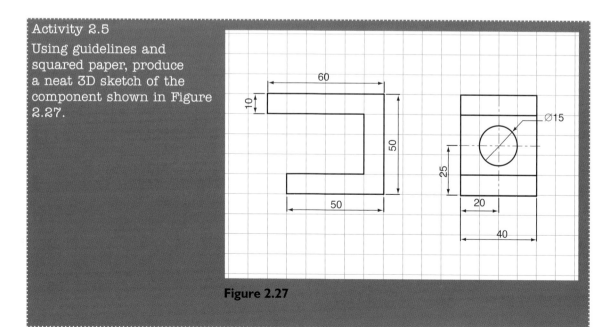

Activity 2.5

Using guidelines and squared paper, produce a neat 3D sketch of the component shown in Figure 2.27.

Figure 2.27

Isometric sketching

Isometric paper is used to create a representation of a 3D object. The guidelines are set at 30 degrees, 90 degrees and 120 degrees. This allows you to construct a sketch that can represent the front, side and top faces of an object.

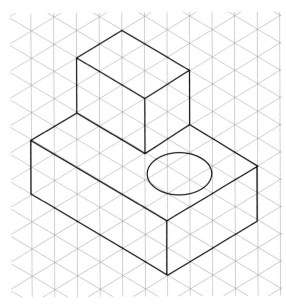

Figure 2.28 Sketch using isometric grid lines

If you want to represent a circular object (isometric circle), you have to construct it as an ellipse. Once again, the use of construction lines is advisable.

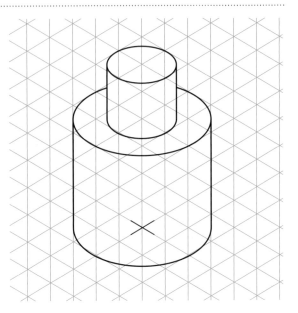

Figure 2.29 Sketch showing isometric circles

Key Points

Adding notes to your sketch really helps people to understand what you are trying to communicate. These can be neat labels explaining what each part is, a description of how to put the parts together or what you are trying to show in your sketch.

Activity 2.6

Using isometric paper, produce a 3D representation of the component shown in Figure 2.30.

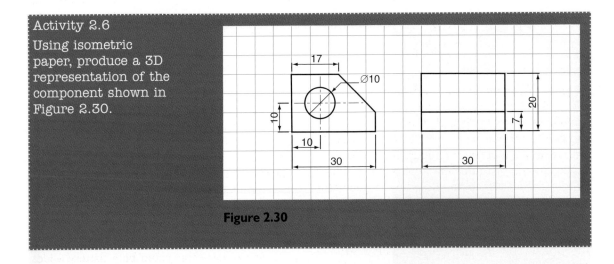

Figure 2.30

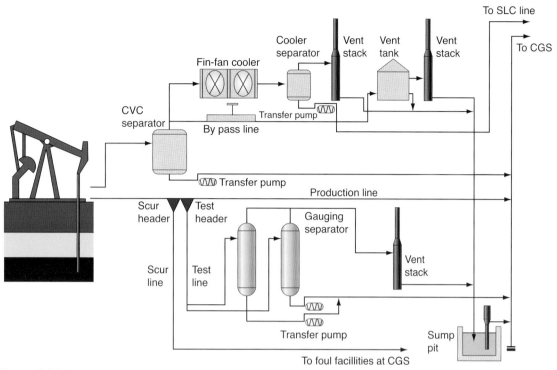

Figure 2.31 Plant layout sketch

Engineering plant or equipment layout

You can quickly sketch how plant is to be arranged or assembled; this type of sketch is often used to allow operators on a construction site or in a warehouse/factory to put together complex machinery in the right place.

You should use a rule and add labels to indicate the positioning of equipment or flow of materials. Although normally drawn in plan view, as a 2D sketch, this can also be a 3D representation.

Designs or installations

Sketches of installations are very similar to plant layout sketches; they are used to indicate how equipment is to be installed (or fitted) into a building. Alternatively, a sketch can be used to indicate how a piece of machinery is installed into another piece of equipment. You might produce an installation sketch to show how a generator is fitted inside a dam for a hydroelectric power scheme; you might then have a sketch to show how the turbine is fitted into the generator. Sketches are normally the first stage in producing a working drawing and would be used for generating ideas, not final requirements.

Electrical/electronic circuit diagrams

An electrical/electronic circuit can be constructed by using circuit symbols to produce a schematic diagram. Circuit diagrams show how a circuit is connected together; because it is purely schematic, the final circuit often looks quite different to the circuit diagram (see Unit 6 for more information).

An electric circuit allows electricity (current) to flow based on the potential difference (voltage) between components (resistors). An example of the symbols used in an electrical circuit diagram are shown in Figure 2.32 and electronic circuit symbols are shown in Figure 2.33.

Software packages can be used to produce circuit diagrams; the symbols can be placed on to a template using a 'drag and drop' technique, and connected by lines to represent connecting wires.

Electrical circuits need a power source, a path for electrons to flow, a load and a way of turning them on/off. If you consider a lighting circuit in a house, you would have to think of at least some of the following elements:

- power source – this could be a battery, which supplies direct current (d.c.); however, it is more likely to be mains electricity, which is alternating current (a.c.)
- path for electrons to flow – this is the wiring

53

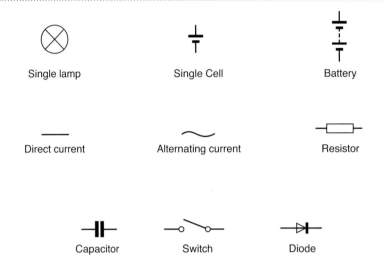

Figure 2.32 Electrical circuit symbols

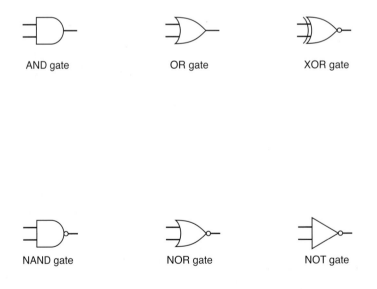

Figure 2.33 Electronic circuit symbols

in your house which connects the elements of a circuit

- load – a light bulb acts as a load; it is a resistor which glows because it is being heated
- a way of turning it on/off – this is a switch which is placed between the light bulb and the power supply and interrupts the wiring.

In addition to these components a household would have fuses or circuit breakers. If a fault in an electrical component or device causes too much current to flow in a circuit, you need a safety device to protect the circuit or appliance. The fuse is a thin piece of wire which overheats and melts, so you have to replace them when they fail. A circuit breaker is an electromagnet; excess current energises the electromagnet, pulling against a spring to 'open' the circuit. These can be reset; however, you should investigate the reason for the excess current first!

Tips for drawing circuit diagrams

- Ensure that you use the correct symbol for each of the components.
- Values should be added to resistors, capacitors, and so on.
- Use a ruler to draw straight lines, representing connecting wires, between components.
- A dot (●) is used to represent the point where wires are joined.
- The top of a power supply is normally positive (+).
- Signals normally flow from left to right – that is, inputs on the right and outputs on the left.

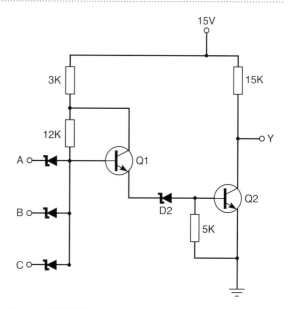

Figure 2.35 Electronic circuit diagram

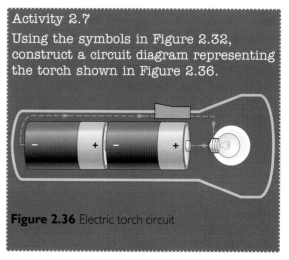

Activity 2.7
Using the symbols in Figure 2.32, construct a circuit diagram representing the torch shown in Figure 2.36.

Figure 2.36 Electric torch circuit

System/network diagrams

A system or network diagram is a graphical representation of an interconnected group or system. A tree diagram is one form of network diagram.

There are many different types of network diagram. For example, a computer network often shows the distribution from the main server to satellite computers and servers. They can often show printers, internet connections, wireless routers, etc.

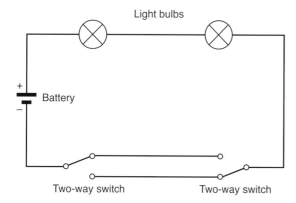

Figure 2.34 Electrical circuit diagram

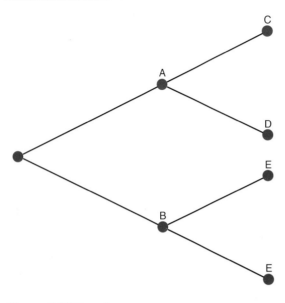

Figure 2.37 Tree diagram

Key Terms

2D sketch – A two-dimensional sketch shows separate views of the front, side or top of an object.

3D sketch – A three-dimensional sketch shows one view incorporating the front, side and top of an object.

A telecommunications network diagram might show the central telephone exchange for a building, with all the branch lines leading off to individual offices. A Programme Evaluation and Review Technique (PERT) diagram (see Figure 3.17 on page 113) is used by project managers to determine the length of time that different parts of a project will take and to find the critical path.

Use of common drawing/circuit/ network diagram conventions and standards

Whether you are creating an engineering drawing, a circuit diagram or a network diagram, there are certain standards and conventions you should follow to ensure that the information you are communicating is easy to understand.

Layout and presentation

Technical drawings should be drawn on a template. Templates are based on standard paper sizes.

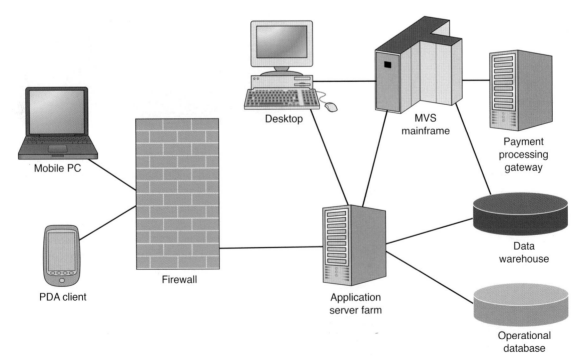

Figure 2.38 Computer network

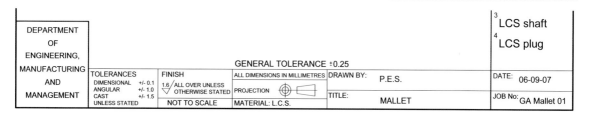

DEPARTMENT OF ENGINEERING, MANUFACTURING AND MANAGEMENT			GENERAL TOLERANCE ±0.25			[3] LCS shaft [4] LCS plug	
	TOLERANCES	FINISH	ALL DIMENSIONS IN MILLIMETRES	DRAWN BY:	P.E.S.	DATE:	06-09-07
	DIMENSIONAL +/- 0.1 ANGULAR +/- 1.0 CAST +/- 1.5 UNLESS STATED	1.6 ALL OVER UNLESS OTHERWISE STATED	PROJECTION				
		NOT TO SCALE	MATERIAL: L.C.S.	TITLE:	MALLET	JOB No:	GA Mallet 01

Figure 2.39 Title block

Drawing sheet sizes in millimetres

A0   1189 × 841
A1   841 × 591
A2   594 × 420
A3   420 × 297
A4   297 × 210
A5   210 × 148

The template is normally a rectangle drawn onto the paper sheet; it contains important key information, most of which is located within a title block. The title block is normally a smaller rectangle in the bottom right-hand corner of the drawing sheet. Drawing sheets and title boxes should conform to BS ISO 7200 technical drawings.

The title block could include:

- name – the name of the designer or originator of the drawing

- checked by – the person who checked the drawing prior to it being issued
- title – the name of the component/assembly/circuit/network, which should contain information to identify the type of drawing – for example, general arrangement or detail; it should also clearly describe accurately what the drawing portrays
- drawing number – a unique number allowing the drawing to be archived and catalogued
- issue number – as a drawing is changed or updated, the issue number increases to reflect the current status
- date – when the drawing was originally constructed
- scale – indicating whether the drawing is a full-size representation of the component or a scaled version

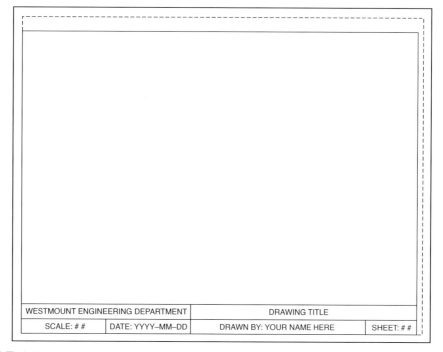

Figure 2.40 Typical template drawing

57

- projection symbol – indicating first- or third-angle projection.

Other information, such as tolerances, material specifications, surface texture, geometrical tolerances, etc., is often included.

The template should be a rectangle drawn inside the perimeter of the drawing sheet. The distance from the edge is normally 20 mm on large templates (A0, A1, A2) or 15 mm on smaller templates (A3, A4).

> **Activity 2.9**
> Create a drawing template using a blank A3 sheet of paper. You will need to include your company/college/school name somewhere on the template.

Engineering drawings are constructed to convey all the required information to allow a part to be manufactured. Because drawings are conventionally constructed in 2D, more than one view is required. The technique used to place views on a drawing is known as projection, and the two systems used are called first-angle and third-angle projection.

First-angle projection   Third-angle projection

Figure 2.41 Projection symbols

Orthographic projection is a method of producing a perfect 2D representation of an object. However, it is unlikely that you will be drawing them full size, so you will have to draw them to scale. This allows the object to be visualised accurately, with everything in proportion. Small objects are 'scaled up' and large objects are 'scaled down'. It is conventional to use standard scale factors or ratios, so something that is drawn half-size is at a scale of 1:2, while something that is drawn ten times bigger than its actual size is at a scale of 10:1.

Reduction scales		
1:1	1:2	1:5
1:10	1:20	1:50
1:100	1:200	1:500
1:1000	1:2000	1:5000
Enlargement scales		
1000:1	2000:1	5000:1
100:1	200:1	50:1
10:1	20:1	5:1
1:1	2:1	

Table 2.2 Standard drawing scales

You can tell whether a drawing is presented in first- or third-angle projection by referring to the symbols shown in Figure 2.41. However, the convention revolves around three views or elevations.

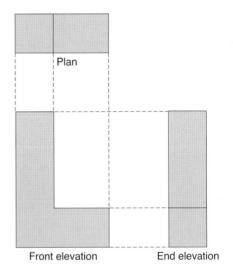

Front elevation   End elevation

Figure 2.42 Third-angle projection drawing

Third-angle projection has the three views arranged as shown in Figure 2.42.

1. Front elevation – the main or front view of an object. This is normally drawn in the centre of your template and is usually drawn first.
2. End elevation – this is a side view and you draw exactly what you would see if looking at the side of an object. You can construct with the left-hand or right-hand view (or both). The view is drawn at the same side as you would be looking at the drawing, so the right-hand view is drawn on the right.

3. Plan view – this is a view that represents what you would see if looking at an object from above. The view is drawn above as you would be looking from above.

All views will need to be at the same scale and aligned using guidelines, as shown in Figure 2.42. First-angle projection is very similar, in principle, to third-angle projection. However, once the front elevation has been constructed, the views are drawn in the opposite orientation, so the right-hand view is drawn from the left and the plan view is drawn from underneath.

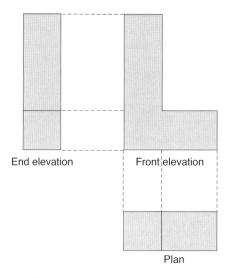

End elevation   Front elevation

Plan

Figure 2.43 First-angle projection drawing

Activity 2.10

Sketch three views of the component shown in Figure 2.44 using third-angle projection and an appropriate scale factor to allow all three views to fit on an A4 sheet of paper.

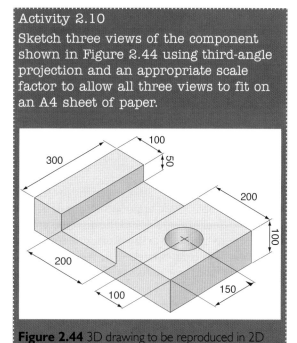

Figure 2.44 3D drawing to be reproduced in 2D

Dimensions

A dimensioned drawing should provide all the information necessary for a finished product or part to be manufactured. An example dimensioned drawing is shown below.

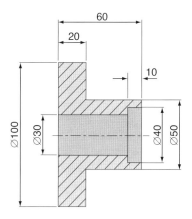
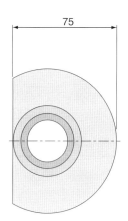

Figure 2.45 Dimension example

abcdefghijklmnopqrstuvwxyz1234567890
ABCDEFGHIJKLMNOPQRSTUVWXYZ

Figure 2.46 Standard lettering and numbering

Dimensions are added to a drawing using thin, continuous lines. Lines are projected from the start and finish points; these are known as projection lines.

A small gap should be left between the projection line and the object you are drawing; similarly, the projection lines extend beyond the dimension line, which is perpendicular to the projection lines. The dimension line is annotated with the dimension, which does not have units, as it is conventional to use millimetres in engineering drawings. All dimensions and additional notation should be the same size and, when using manual techniques, it is often useful to practise by drawing two parallel lines and trying to reproduce consistent lettering and numbering, as shown in Figure 2.46.

Tolerances

All manufactured products are subject to tolerances; you cannot make a part to the exact size on a drawing. For the most part, a general tolerance will be indicated in the drawing – for example, $\pm$ 0.5 mm. If a specific dimension is to have a tolerance, it is added as shown in Figure 2.47.

The largest size is placed above the smallest, and an appropriate and equal number of decimal places are used for both.

Geometric tolerances

Geometric tolerances are used to determine the allowable variation from true geometry. It is often visualised as the diameter or width of a zone within which the axis of a cylinder, surface or hole should fall.

The tolerance zone can be determined as one of the following:

- the area within a circle
- the area between two circles
- the area between two equidistant lines or between two parallel straight lines
- the space within a cylinder
- the space between two coaxial cylinders
- the space between two equidistant surfaces or two parallel planes
- the space within a bent pipe.

Line types

Lines on an engineering drawing have particular meanings; they are of two thicknesses, so you will need two pencils! Generally, the thick lines are twice the width of the thin lines.

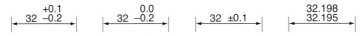

Figure 2.47 Tolerance dimension

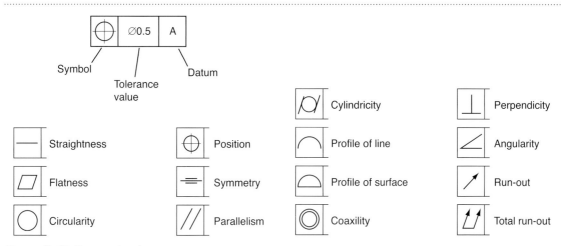

Figure 2.48 Geometric tolerances

Description and representation	Application
Continuous wide line	Visible edges and outlines
Continuous narrow line	1 Dimension extension and projection lines 2 Hatching lines for cross sections 3 Leader and reference lines 4 Outlines of revolved sections 5 Imaginary lines of intersection 6 Short centre lines 7 Diagonals indicating flat surfaces 8 Bending lines 9 Indication of repetitive features
Continuous narrow irregular line	Limits of partial views or sections provided the line is not an axis
Dashed narrow line	Hidden outlines and edges
Long dashed dotted narrow line	1 Centre lines 2 Lines of symmetry 3 Pitch circle for gears 4 Pitch circle for holes
Long dashed dotted wide line	Surfaces which have to meet special requirements
Long dashed dotted narrow line with wide lines at ends and at changes toindicate cutting planes	Note: BS EN ISO 128–24 shows a long dashed dotted wide line for this application
Long dashed double-dotted narrow line	1 Preformed outlines 2 Adjacent parts 3 Extreme positions of moveable parts 4 Initial outlines prior to forming 5 Outline of finished parts 6 Projected tolerance zones
Continuous straight narrow line with zigzags	Limits of partial or interrupted views Suitable for CAD drawings provided the line is not an axis

Figure 2.49 Line types

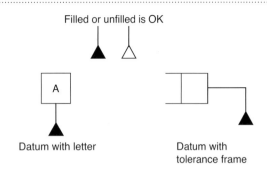

Filled or unfilled is OK

Datum with letter

Datum with tolerance frame

Figure 2.50 Geometric tolerance box

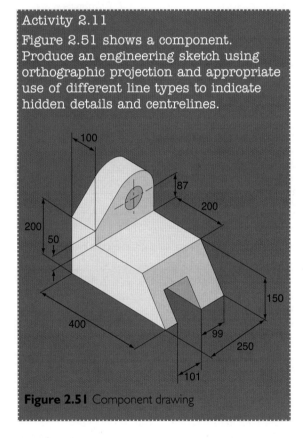
Hatching

Key Term

Hatching – a technique used to indicate where a part or component has been sectioned.

Sections are created to show hidden details. If you can imagine that a part has been sliced in two along a cutting plane, you have a section!

A section shows all visible elements that exist beyond the cutting plane, as indicated in Figure

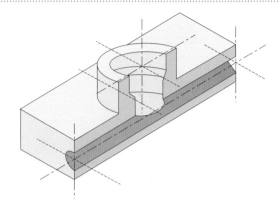

Figure 2.52 Sectioning

2.52. On an engineering drawing, the cutting plane is indicated by the centre line, with thick lines at the ends and arrows indicating the direction from which the sectioned view is being viewed. Letters are used for reference, as a drawing can have several sections, depending on its complexity.

Hatching is used to indicate where the cutting plane would encounter solid material. Hatching is normally constructed at 45 degrees, with a spacing of approximately 4 mm.

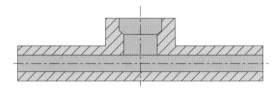

Figure 2.53 Hatching

All hatching for a single object should be at the same angle and spacing. However, when assemblies are to be hatched, the direction of hatching on adjacent parts is reversed. Similarly, if more than two parts are adjacent, hatching distances can be varied or the hatch staggered for clarity.

Surface finish

Engineering components invariably have to have a smooth surface; the measure of this smoothness is called surface finish. Surface finish indicates how deep the grooves or undulations are on a surface.

Otherwise known as roughness, it consists of surface irregularities, which often result from machining operations. It is the height of these irregularities with respect to a reference line that determines a value for surface finish. It is measured in millimetres or microns.

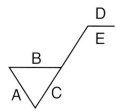

Figure 2.54 Surface finish symbol

Symbols

Symbols and abbreviations are often used in engineering drawings to represent features where space does not permit the use of text.

For example:

Across flats	AF
Assembly	ASSY
Centres	CRS
Centre line	In a note CL
Centre of gravity	CG
Chamfered	In a note CHAM
Cheese head	CH HD
Countersunk or countersink	CSK
Countersunk head	CSK HD
Counterbore	CBORE
Cylinder or cylindrical	CYL
Diameter	In a note DIA
Diameter	Preceding a dimension ∅
Drawing	DRG
Equally spaced	EQUI SP
External	EXT
Figure	Fig
Hexagon	HEX
Hexagon head	HEX HD
Internal	INT
Left hand	LH
Long	LG
Machine	M/C
Material	MAT
Maximum	MAX
Minimum	MIN
Not to scale	In a note and underlined NTS
Number	NO.
Pitch circle diameter	PCD
Radius	In a note RAD

Radius	Preceding a dimension R
Reference	REF
Required	REQD
Right hand	RH
Round head	RD HD
Screw or screwed	SCR
Sheet (drawing number)	SH
Sketch	SK
Specification	SPEC
Spherical radius	SR
Spotface	SFACE
Square	In a note SQ
Square	Preceding dimension □
Standard	STD
Thread	THD
Thick	THK
Tolerance	TOL
Typically or typical	TYP
Undercut	UCUT
Volume	VOL

Parts lists

When components are brought together, an assembly drawing is produced. As previously discussed, the purpose of this type of drawing is to indicate how parts are put together.

- Although detail dimensions are excluded from assembly drawings, overall dimensions or dimensions required for assembly can be included.
- Sectional views are often used to view internal parts.
- Each part is given a number indicated by a balloon reference – a number in a circle with a leader line pointing to the part.
- The leader line has either an arrow pointing to the edge of a component, or a dot if the line finishes within the outline of the part.
- A table of parts should be added to the drawing to identify each part.
- The parts list should have a description of the part, the quantity required and the part number.

Circuit/component symbols

As discussed previously, standard symbols are used to construct circuits. Although you have already looked at electronic and electrical symbols, there are other standard symbols, depending on your profession – for example, pneumatic and hydraulic symbols.

63

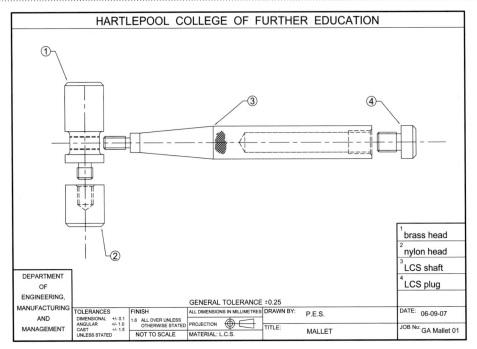

Figure 2.55 Assembly drawing with parts list

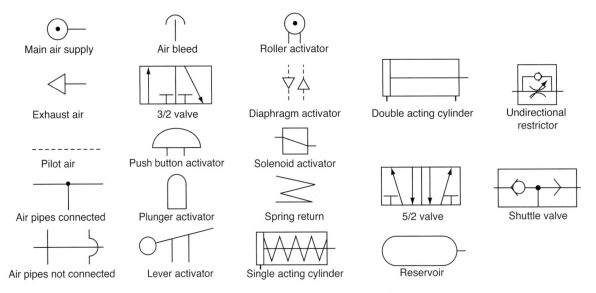

Figure 2.56 Example of typical pneumatic symbols

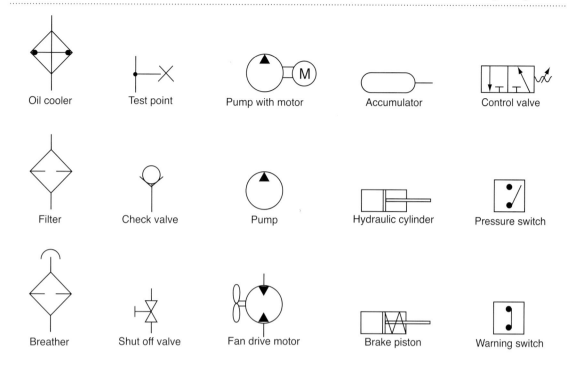

Figure 2.57 Example of typical hydraulic symbols

Use of appropriate standards

British Standards (BS) and the International Organization for Standardization (ISO) lay down strict requirements which have to be met when constructing drawings; this ensures standardisation and consistency. The use of technical standards allows the use of one reference source with all the information available.

Typical standards used by designers include:

BS/ISO 8888	Technical product specification
BS 3939	Graphical symbols for electrical power, telecommunications and electronics diagrams
BS 2917	Graphic symbols and circuit diagrams for fluid power systems and components
BS 5070	Engineering diagram drawing practice
BS ISO 14617	Graphical symbols for diagrams

Activity 2.12

Figure 2.58 shows a circuit. Redraw the circuit using conventional circuit symbols. You should refer to Figures 2.56 and 2.57 for the standard symbols.

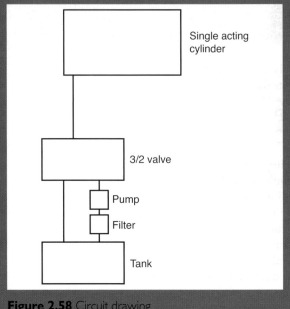

Figure 2.58 Circuit drawing

Activity 2.13

Figure 2.59 shows a component drawing. Copy the drawing on to an A4 sheet and add the following symbols, standards and conventions to show that:

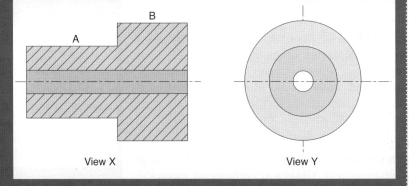

Figure 2.59 Component drawing

- face A is to have a surface finish of 3.2 microns

- face B should be parallel to Face A within 0.1 mm

- centre lines should be added where appropriate

- a cutting plane on view Y should be added.

You should also sketch a 3D model of the component using appropriate 3D sketching techniques.

2.2 Using verbal and written communication

This section will cover the following grading criteria:

Being able to communicate effectively is a key requirement for all engineers. Written communication can use a variety of methods, depending on the importance of what you are trying to communicate or on the audience. If you want to tell a friend you will be late for a social event, you might send a text message. However, if you are applying for a job, a formal letter might be a better form of communication.

The same is true of verbal communication. You could phone a friend to tell them you are going to be late; however, a formal interview would be more likely if you are applying for a job.

Written work

By the end of this section you should have developed the ability to use written methods to communicate information effectively.

Engineers need to be able to use a variety of written techniques to record and present information. There are prescribed formats and different styles, depending on the audience and use for the work being recorded/presented. In this section you are going to look at the different ways that engineers communicate and use different methods, depending on the situation and context.

> **Key Points**
> Communicating effectively in writing requires you to use a prescribed style, depending on who the audience will be.

Written work can vary from making simple handwritten notes for yourself, as a record of a meeting or discussion, to producing formal documents such as a business letter or memo.

When you are making handwritten notes, they are just for you – in other words, an audience of one person. Even if your handwriting is difficult to read, as long as you can understand what you have written, then it meets the need of the audience. But what if you are taking a message on the phone? If you are then going to leave a note for someone, it might be worth writing out the details carefully so that they can clearly understand the message.

> ## Make the Grade   P4
>
> Grading criterion P4 requires you to communicate information effectively in written work. In order to do this, you will need to present evidence of note-taking, the ability to use a specific writing style, proofread and amend text, use a diary/logbook and use a graphical presentation style. You may need to present several pieces of evidence to accomplish this; however, you will be demonstrating many of these elements within your project (Unit 3) and you may be able to use this as evidence.

Note-taking

Note-taking techniques depend upon personal preferences; notes can be handwritten using conventional text, but can also be in shorthand or diagrams. Shorthand is a form of writing which, typically, replaces words with abbreviated text or symbols. Although there are many formal shorthand systems, with names such as Pitman, Gregg and Teeline, many people develop their own methods of writing that they can use to write up more formally at a later date. Alternatively, you can use lists or bullet points to remind yourself of key points from a discussion.

Many people prefer a pictorial method to remind themselves of a discussion or to develop ideas. Flow charts are often very useful, as they can show a plan or process; another typical technique is the use of mind maps, which can be used to generate creative ideas or make plans (see Unit 3).

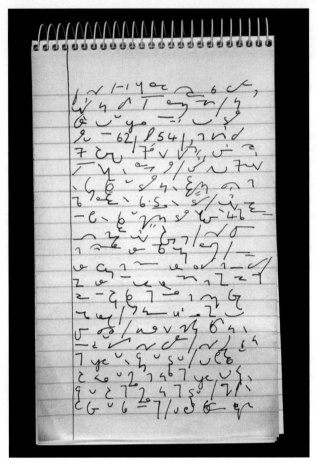

Figure 2.60 Shorthand example

Writing style

Different writing styles are used for technical or formal documents. There are conventions about the way you should present information that largely depend on who is reading what you have written and the message you are trying to get across. A business letter is a very formal document, with a recognised style and format. Similarly, a memo has a particular layout and content. However, emails can vary from being very formal and precise when used in business, to being informal when used among friends.

Business letter

A business letter should follow the format shown in Figure 2.61. This is a standard format which normally follows a pattern.

Est 1947 12 High Street Bristol BS1 4JB   Tel: 01234 572662   Fax:01234 572000

12 High Street
Bristol
BS1 4JB

Date: 14th April 2010

Mrs P Jones
41 Fern Close
Bristol
BS2 5HD

Dear Mrs Jones

Reference No: Order A751 - Hanley Bedside Cabinet

Thanks you for your order. I am pleased to confirm that delivery of the above
product will take place on 15th May 2010 between 8 a.m. and 6 p.m. Please
ensure that someone is available to accept delivery. We regret that we cannot
allocate specific time slots.

If you experience any problems with the delivery process, please do not hesitate
to contact me on 01234 572682.

Yours sincerely

K Adams

Ms K Adams
Delivery Manager

Figure 2.61 Business letter

Technical report

A technical report is used for a variety of purposes.
Typically, reports might be used for the following.

Technical background report

Used to give background information about a
specific subject – for example, how hybrid vehicles
work or why a particular computer game is popular.

Instruction booklet

This kind of report can be a simple procedure – for
example, how to build a circuit to control traffic
lights. They can also be very specific, such as how
to create a 3D model using a particular piece of
CAD software.

69

Research

This usually refers to work carried out in a laboratory, trial, field study, and so on. Usually, data are considered and analysed, and conclusions presented. It could be a report on whether a particular adhesive can be used to replace fasteners in an aircraft, for example. Alternatively, it might be used to indicate the voting preferences of certain groups of people in an election.

Feasibility

This kind of report is used to evaluate whether a particular project is worth undertaking. Often based around cost, it is used to determine whether to invest in a particular technology or facility – for example, drilling for oil in hazardous environments or upgrading the personal computers (PCs) used in a school or college.

Specifications

When a design is required, it is important to specify the performance, useability, materials, operating characteristics, etc. Specifications are frequently used by engineers to give them the requirements of a product they may be designing or building.

Reports are usually broken down into sections; if you are considering a substantial report, you would use the format explained in Unit 3 for a project report. However, for a small-scale report, such as a laboratory report following a test or investigation, the following format is recommended:

Title page – include your full name, the title of the investigation or test, the date and tutor name if applicable.

Objectives – explain why you are completing the experiment or test, what you hope to find out and/or what theory you are proving or testing.

Apparatus – list the equipment and materials you are using; a sketch or diagram is also helpful.

Method – outline the procedure carried out to complete the experiment/test in a logical, step-by-step manner.

Observations – record what you saw or measured.

Analysis – this section allows you to interpret the results of your observations. It could be carrying out calculations or plotting graphs.

Conclusions – what did you learn from doing the experiment/test? Did you prove or disprove a theory? Refer to your objectives and discuss how well they were accomplished.

Recommendations – were there any issues with the test equipment or materials? Is further study required? What would you suggest doing next?

Reports are normally written in the third person and past tense.

Memos

Memorandums, or memos for short, are usually short messages used within business, often to remind or inform colleagues of an event or policy/procedure.

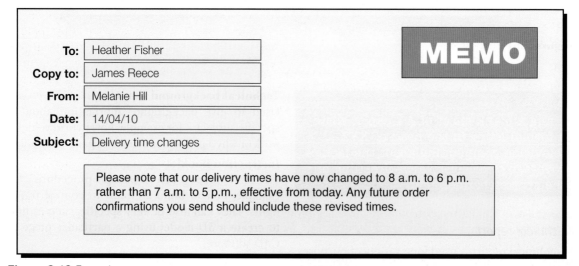

Figure 2.62 Example memo

A memo usually consists of:

Distribution list	Who the memo is being sent to and who it is from
Copy	Who is also being given a copy or whether the memo should be kept on file
Title	What the memo is about
Content	The information being communicated

It is worth noting that the use of memos has been replaced in many organisations by the use of emails, as they often perform the same function in a quick and efficient manner.

> **Activity 2.16**
>
> Write a memo to Mr Black, Ms Red and Mrs Blue (with copy to Dr White), reminding them of the need to attend the board meeting on 1 April at 9 a.m. and not 9.30 a.m.

Email

Using email is a quick and convenient method of transmitting information; it is used both to transmit formal information, such as quotes and requests, and as a quick and easy method of transmitting informal information. The ease of operation and speed of email communication has seen it become an essential business tool for all organisations, and the ability to attach drawings, documents and specifications to an email has revolutionised the way industry works. Documents/drawings not produced electronically can be scanned to produce an electronic file that can also be attached to an email. Many mobile phones have a facility to allow emails to be received and sent, in a similar manner to texting someone.

Fax

Before the development of email systems, many organisations used faxes to send information such as diagrams or specifications. Most faxes operate in a similar way to a photocopier, and allow A4 sheets of paper to be fed through a scanner, which sends the information via a phone line to a fax machine in another location, which prints a copy of the original document. Some organisations still require faxes to be used, as a signature is sometimes required – on a purchase order, for example. These are generally much poorer quality than the principle of attaching scanned or electronic images to an email, however.

> **Activity 2.17**
>
> Give a brief explanation of why and when you would use the following forms of communication:
> * memo
> * formal letter
> * lab report.

> **Key Terms**
>
> **First person** – Writing from your perspective – for example, I carried out the experiment.
>
> **Second person** – Writing from the audience's perspective – for example, you carried out the experiment.
>
> **Third person** – Writing from neither your perspective nor that of the audience – for example, the experiment was carried out.

Proofreading and amending text

Proofreading is a technique used to check that what you have written is really what you are trying to say. Proofreading gives you the opportunity to check for mistakes and ensure that the piece of work you have created is error-free. Professional proofreaders will often use a red pen and go through a document indicating errors and omissions, using conventions such as those shown in Figure 2.63.

Ideally, you should create your document or written work thinking of it as a first attempt or draft. You should then leave some time before finding somewhere quiet to look through what you have written. Many people find it difficult to read from a screen, so printing a hard copy can help. You can then annotate this document as you go through it.

Things to look for when proofreading:

* Words – Are they spelt correctly? Do they have the correct meaning? Are there any missing words, such as I, the, it and in.
* Grammar – Look for missing apostrophes and commas, and sentences starting without a capital letter. A common mistake is using i instead of I.
* Consistency of style and punctuation – For

71

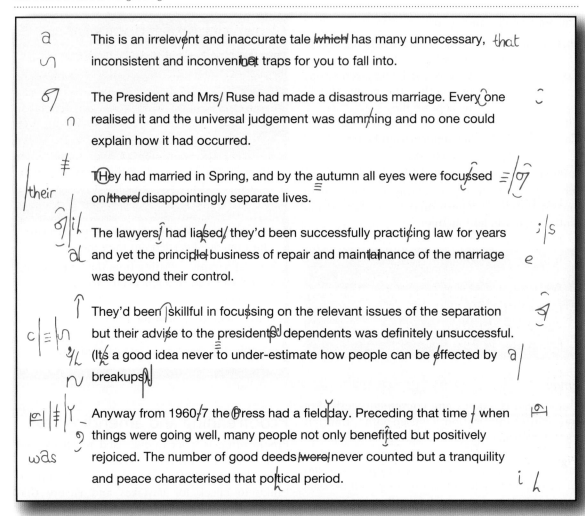

Figure 2.63 Example proofread script

example, have you used the first person in one sentence and third person in the next?

- Repetition – Two words repeated – for example, the the text is easy to read.
- Typos – Have you used the wrong word – for example, tree instead of three, or weather instead of whether?
- Ambiguity – When you read your work back, is it clear? Does it make sense? If you are unhappy with a particular section, you may need to rewrite it.

When using word-processing software, it is tempting to use the spelling and grammar-checking functions. While these are very useful in the first instance, they are not foolproof. The document shown in Figure 2.64 was produced using word-processing software and contains no spellchecking errors. Can you see what went wrong?

Key Points

Proofreading your own work is very difficult. It is surprising how you can reread something several times and still miss a glaring error. If you can persuade a tutor, teacher, friend or colleague to proofread your work for you, they will probably spot mistakes you would have missed.

Key Term

Editing – Amending a piece of written work after it has been produced, to remove errors and improve the style and content.

The spelling Chequer
(or poet tree without mist takes)

Eye have a spelling chequer
It came with my pea sea
It plainly marks four my revue
Miss steaks eye cannot see

Each thyme when I have struct the quays
Eye weight four it two say
If watt eye rote is wrong or rite
It shows me strait a weigh

As soon as a mist ache is maid
In nose bee fore too late
And eye can put the error rite
Eye really fined it grate

Ive run this poem threw it
I'm sure your policed to no
It's letter perfect in its weigh
My chequer tolled me sew

Figure 2.64 Poet tree without mist aches

Use of diaries/logbooks for prioritising work schedules

Diaries are often used by engineers in the same way as other people, to record dates, times, appointments, and so on. However, as an engineer, you are often conducting tests, being involved in design meetings and similar activities where record-keeping is important. A large-format diary is often useful as a logbook for recording data, discussions and logging completed jobs. An alternative to a diary is a logbook, which is designed to record information, results or settings/observations, as well as planning or prioritising work.

Key Terms

Diary – A personal daily record of activities, meetings or important events.

Logbook – An official record of events, with key information and data.

Activity 2.18
Look at the letter in Figure 2.65.
Can you spot any mistakes?
Rewrite the letter, correcting errors and clarifying any inconsistencies.

Hi Paul.

Can you help me. I want to application for a part-time job and need a reference from somebody what knows me at college.

I just wanted to to check it is ok to give you're contact details to thems. I'm sending the application toady so you should here from them sometime next week.

thanks for your help.

Regards,

Keisha

Figure 2.65 Formal letter

73

Graphical presentation techniques

Presenting complex information is often challenging. A large table of numbers is often not easy to interpret or understand. This is particularly true if you are giving a presentation and you need to communicate your results effectively. Graphs are useful to present discrete data to show a trend, as shown in Figure 2.66. Graph paper is used to make plotting of the data easier.

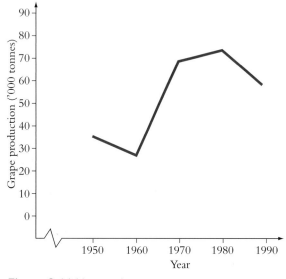

Figure 2.66 Line graph

> **Key Points**
>
> If you are producing a line graph, you should always label each axis; similarly, a chart should have each section or group clearly labelled.

When your data are grouped as shown in Figure 2.67, then pie charts, pictograms and histograms are effective ways to summarise information. Similarly, if you have produced a technical report which required a large amount of data to be collated, it is useful to present a graph, chart or diagram in the main part of your text and include the collated data in the appendices.

Number of people	Frequency
0–10	8
11–20	19
21–30	11
31–40	3
41–50	28

Figure 2.67 Frequency table

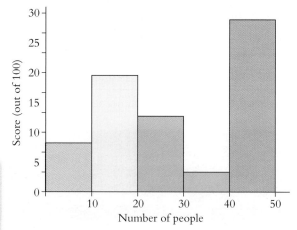

Figure 2.68 Histogram

Number of people	Frequency
0 – 10	ЖН IIII
11 – 20	ЖН ЖН ЖН IIII
21 – 30	ЖН ЖН I
31 –40	III
41 – 50	ЖН ЖН ЖН ЖН ЖН III

74   **Figure 2.69** Tally chart

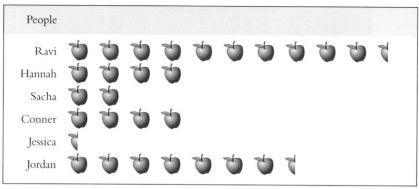

Pictogram
as of Aug 26, 2009

Figure 2.70 Pictogram

* each apple = 200000 units

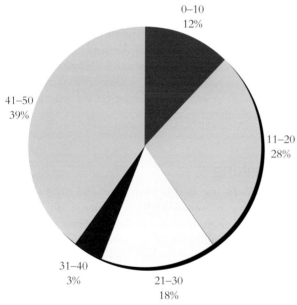

Figure 2.71 Pie chart

Key Terms

Bar chart – Another name for a histogram, named because of the vertical (or horizontal) bars.

Frequency table – A table with two columns: column one giving different values; column two giving how often each value occurs.

Make the Grade   M1

Grading criterion M1 requires you to evaluate a written communication method and identify ways in which it could be improved. This could be your work, someone else's or a combination of both. You will need to demonstrate that you understand different methods of communication, and show where changes can be made to improve the presentation and content. You might find it useful to review the reasons for proofreading and think about why we use graphical techniques.

Activity 2.19

Speed (r.p.m.)	0	10	20	30	40	50	60	70	80	90
Power (kW)	0	0	0.10	0.19	0.30	0.40	0.51	0.61	0.70	0.79

Plot a line graph of the information shown in the table. What can you say about the relationship between speed and power?

Mark		10	20	30	40	50	60
Number of students		5	7	12	11	6	2

Display the information about marks awarded to students using **one** of the charts shown in Figures 2.66 to 2.71.

Compare the method you selected with the other options. Which chart displays the information best? Why?

Verbal methods

By the end of this section you should have developed the ability to use verbal methods to communicate information effectively.

Communicating verbally depends, in a similar way to written communication, on the intended audience. However, there is more to verbal communication than what is said; the tone of voice, use of body language and other non-verbal communication is very important when trying to communicate.

Key Points

Communicating verbally is not just about speaking; you will need to demonstrate that you can adjust your style, depending on who the audience is.

People who speak different languages often manage to communicate information – for example, when they are seeking directions or buying goods. How do you think they do that?

Verbal communication is not always about speaking; listening is also important. Sometimes it is what *not* being said that is important.

Make the Grade — P5

Grading criterion P5 requires you to communicate information effectively, using verbal methods. You may need to do this over a series of tasks or activities; however, you will be demonstrating many of these elements within your project and you may be able to use this as evidence.

Speaking

You would use a different tone and style when making a presentation to customers or senior managers than when speaking with colleagues or friends. Speaking with customers or colleagues requires an appropriate level of technical language. However, if you are talking to someone who does not have an engineering background, you have to modify your language and avoid using complex engineering terminology, so that they can

Figure 2.72 Speaking

understand. Further guidance on giving formal presentations is given in Unit 3.4.

> **Key Points**
>
> When giving instructions, always:
> - tell people what you are *going* to tell them
> - *tell* them
> - tell them what you have *told* them.
>
> Finally, ask questions to check that people have understood.

Listening

Listening skills are as important as speaking skills, and it is often important to take notes in addition to listening carefully. If you are being given instructions on health and safety or guidance, or being shown how a particular machine or operation is maintained/performed, you will be expected to be able to remember and carry out instructions without supervision in future.

Figure 2.73 Body language when listening

> **Key Point**
>
> Always check that you have understood when you have been given instructions.

> **Activity 2.20**
>
> Listen to a recorded news story, by yourself, without taking notes. Explain the story to friends or colleagues, allowing them to take notes if they wish. Now have them listen to the recording. Did you get all the main points across to the group? Did they understand? Did they listen carefully?

Impact and use of body language in verbal communication

It is often said that the majority of communication is non-verbal. What this means is that the facial expressions, body language and gestures we make give clues to our audience about what we mean and our thoughts or opinions. When you are making a presentation to a group of colleagues, customers or members of a team, it is important to speak clearly. However, you also need to demonstrate positive body language.

Figure 2.74 Body language when speaking

Eyes

You should always look at your audience, but do not stare at individuals as this can be intimidating. You should also try to avoid rapid eye movements. People whose eyes move slowly and in a relaxed manner are perceived as being more sincere.

Tone

You should try to keep your tone of voice even. People whose voices get louder are often perceived as being angry.

Pace

Try to keep a steady pace when talking. People who frequently pause or hesitate give the impression of either not being prepared or of not knowing what they are talking about.

Stance

Try to stand tall and look confident; keep your feet together and your weight evenly distributed. Stooping or hunching over your notes implies that you lack conviction and have been placed in an embarrassing or awkward position.

Movement

Don't stay in one place. Be prepared to move, but in a natural way; however, walking around too much is distracting. Similarly, it is a good idea to use positive hand gestures and avoid crossing your arms or holding your hands behind your back.

Key Point

Body language can be as important as what you say in communicating information.

Activity 2.21
In a group of three or four people, ask each person in turn to walk up to each other and, without speaking, display either delight, confidence, anger, fear or nervousness. See how well the group members identify the emotions. What are the key signs of each emotion?

2.3 Obtaining and using engineering information

This section will cover the following grading criteria:

There are a variety of information sources that you will need to access in order to solve an engineering task. These can be computer and non-computer-based and could require you to prove facts, collect statistics, find data or research principles. Once you have obtained the relevant information, you will need to demonstrate that you can use it correctly to solve problems, prove theories and ideas or check your work.

Information sources

By the end of this section you should have developed the ability to obtain information from a variety of sources.

Many people obtain information from the World Wide Web (www). They think that by using a search engine and typing in a few keywords, the internet will provide all the answers. There is a great deal of information available, but is it always what we want, and is it reliable? In this section you will explore the use of a variety of information sources.

Key Points

Information sources are often described as primary or secondary. Primary sources are usually directly from a source or person, so they could be a journal article by the researcher, or statistics collected by you or a colleague, through a question-naire or interview, for example. Secondary sources analyse, summarise or comment on existing research; these could be a journal article reviewing other people's research, a review of a range of newspaper articles, or statistics summa-rised from a variety of surveys or tests.

Make the Grade — P6

Grading criterion P6 requires you to use appropriate information sources to solve an engineering task. You will need to use research skills to find information from a variety of sources (not just the internet). It is useful to produce annotated hard copies of your information sources where possible, identifying the key elements as appropriate.

Non-computer-based sources
Books

Books are often overlooked as we use the internet for so much of our information-gathering. However, they are a permanent and invaluable source of information that can be checked, and are often reviewed and checked for authenticity. They contain a more in-depth focus on a subject and, particularly with reference books, they have all the data and information you require in one place. One of the disadvantages of books is that they can quickly become out of date. For example, you might buy or refer to a book to learn how to use a particular piece of software. However, by the time the book has been researched, written and published, the software may have changed, as many companies release new versions every year.

Figure 2.75 Books

> **Key Point**
>
> Books have a unique referencing system called the International Standard Book Number (ISBN). The ISBN allows us to search catalogues and databases for particular books.

Technical reports

Many engineering companies carry out performance tests or reviews. This information is often contained within reports which will then be archived within the organisation. University researchers and students also write reports, which can be obtained from the appropriate library or learning resource centre.

Figure 2.76 Report

Institute and trade journals

These are publications which are published by professional bodies such as the Institution of Mechanical Engineers (IMechE) or the Institution of Engineering and Technology (IET).

Figure 2.77 Trade journal

Data sheets

Data sheets are often produced by companies to give size and performance specifications of their products. For example, a car manufacturer will produce data sheets for all the different models it produces, giving information such as fuel tank capacity, fuel economy figures, vehicle turning circle, and so on.

| DATA SHEET | PROJECT _____ UNIT _____ ITEM _____ | | | | DATA SHEET ____ of ____ SPEC _____ | | |

Fluid						Crit Press Pc	
		Units	Max Flow	Norm Flow	Min Flow	Units	Units

SERVICE CONDITIONS
- Flow Rate
- Inlet Pressure
- Outlet Pressure
- Inlet Temperature
- Spec Wt/Spec Grav/Mol Wt
- Viscosity/Spec Heats Ratio

LINE
- Pipe Line Size — In / Out _____
- Pipe Line Insulation _____

VALVE BODY/BONNET
- *Type _____
- *Size _____
- *Max Press/Temp _____
- *Mfr. & Model _____
- *Body/Bonnet Matl _____
- *Liner Material?ID _____
- End | In _____
- Connection | Out _____
- *Flow Direction _____
- *Packing Material _____
- *Packing Type _____

TRIM
- *Type _____
- *Size _____ Rated Travel _____
- *Characteristic _____
- *Balance/Unbalanced _____
- *Rated C_V _____ F_L _____ X_T _____
- *Seat Material _____
- *Stem Material _____

ACTUATOR
- *Type _____
- *Mfr. & Model _____
- *Size _____
- On/Off _____
- Spring Action Open/Close _____
- *Maximum Allowable Pressure _____
- *Min Required Pressure _____
- Available Air Supply Pressure:
- Max _____ Min _____
- Air Failure Valve _____

SWITCHES
- *Type _____ Quantity _____
- *Mfr. & Model _____
- Contacts/Rating _____
- Actuation Points _____

AIRSET
- *Mfr. & Model _____
- *Set Pressure _____
- Filter _____ Gauge _____

TESTS
- *Hydro Pressure _____

Rev	Date	Revision	Orig	App

SPECIALS/ACCESSORIES _____

Figure 2.78 Data sheet

Test/experimental results data

Many companies perform tests on their products and record the data. This can be very useful if engineers need to compare performance or test the suitability of a given product. For example, a company that produces lubricants and oil additives will test the performance of the oil at different temperatures and keep records of those tests. Consequently, if a customer requires a lubricating oil to operate effectively in a particularly hot or cold environment, the company will know which additives to blend with the lubricant in order to maximise performance.

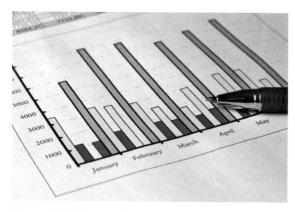

Figure 2.79 Results sheet

Manufacturers' catalogues

Many manufacturers produce comprehensive catalogues which list all the products they supply or manufacture. Typically, circuit designers or maintenance technicians will refer to these catalogues when selecting appropriate electronic, hydraulic or pneumatic components, depending upon the type of circuit in question.

Figure 2.80 Manufacturer's catalogue

Key Points

Libraries and learning resource centres are excellent repositories of information. The staff who work there are able to help you search for the information you want and can guide you to the correct source or obtain the information for you.

Activity 2.22

Can you complete the table and explain why you have selected the information source in each case?

Information required	Information source	Justification for using this source
Hardness value for mild steel		
Sizes of a given electric motor		
Capacity of a given model of air compressor		
Tensile strength of a copper sample you tested in the laboratory		
Subscription rate for a year's membership of an engineering institution		

Computer-based sources

Computers are an everyday part of modern life and we use them without thinking, particularly the internet, which is a valuable source of information.

Internet

The internet is a global system which links computer networks using Internet Protocol (IP) addresses. It is a network that connects millions of private, business and public systems. These systems are linked by fibre-optic cables, copper wires or wireless communication devices. Much of the information traditionally reproduced in manufacturers' catalogues is available online. The use of the internet as a repository for this kind of information is invaluable, as it is easily available and can be updated quickly and easily. Similarly, many manufacturers make data sheets available on their websites. The advantage for companies is that they have the extra opportunity to promote themselves and their products.

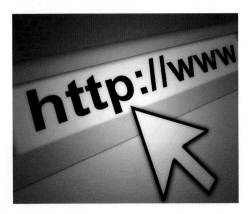

Figure 2.81 Internet

Intranet

An intranet is a computer network which operates in a similar way to the internet, but is a private computer network only available to employees of an organisation. An intranet is often used as a facility to store common documents, such as policies and procedures, or useful information, such as employee telephone extension numbers. However, it can also be used to store engineering documents, such as test results, or engineering drawings generated from a CAD system.

CD-ROM-based information

Although CD-ROMs or DVDs are frequently used to store software programmes, they are also used as a store of information. Many companies issue their catalogues as a CD-ROM.

CD-ROMs can also be used to store data files and analytical software, which allow you to perform simulations or engineering calculations.

Figure 2.82 CD-ROM

Spreadsheets

A spreadsheet is a computer application that allows a worksheet to be displayed on screen. Consisting of cells arranged in rows and columns, data and formulae can be included within the sheet, typically to allow experimental or financial information to be analysed quickly and easily. By entering different values, the formulae can automatically recalculate values, so the spreadsheet can be used to test theories or to plot charts or tables.

Figure 2.83 Spreadsheet page

Databases

A database is a computer application that allows the user to manage and organise data and information. The data can then be searched, based on different attributes. Manufacturers often store the details of their products in a database. This allows users to search for products by entering specific information. For example, a shop or online store selling or renting DVDs would store the information to allow customers to search the catalogue based on a variety of categories such as DVD type, year of release, popularity etc.

Figure 2.84 Database

> **Key Points**
>
> When researching information, particularly when using the internet, it is advisable to ensure that you have used at least two independent sources to verify that information.

> **Activity 2.23**
>
> Using the internet and a reference book, find the temperature at which copper wire melts. Do the values agree? Why do you think that finding this information might be useful?

Use of information

By the end of this section you should be able to use information to solve an engineering task effectively.

Whether information is obtained by research or data collection, it is often how this information is used that is important to engineers. Gathering information is only the first stage in the process.

> **Key Points**
>
> When using information sources, you should always make reference to them and never copy (plagiarise) other people's work. You can draw conclusions and write about what you have researched, but you must acknowledge source material and write things in your own words.

Make the Grade   P6

Grading criterion P6 requires you to use appropriate information sources to solve an engineering task. Once you have obtained the required information, you will need to show how you use it to solve the given task.

Solving problems

To solve problems in engineering, we often need specific information. For example, we might need to determine the electrical current in a circuit, and, by using a manufacturer's catalogue, the resistor values can be determined. However, once you have that value, you still need to use Ohm's law to determine the current. Finding the information is only part of the task.

Alternatively, to determine the direct strain in a steel bar, we might need to know the modulus of elasticity for the material. This information can be found from a reference book. Again, the modulus of elasticity is used in a formula, along with the direct stress, to determine the direct strain.

83

Product/service/topic research

Information about a product is important in order to determine its performance, size, etc. For example, a design team might use the size of a range of tyres if they are designing alloy wheels.

Similarly, service information is important and can often be found from technical literature or manufacturers' specifications. This information can be used to plan the amount of time a bearing can be used before it requires lubrication, or the time an electrical cable can withstand a given temperature before it melts.

Sometimes, as engineers, it is not a specific piece of information or data that is used, but more general information about a topic. An example of this could be the opinion of operators on the performance of a tool or component. Alternatively, you might have researched how long it takes to perform a service operation and use the data to plan maintenance procedures.

Gathering data or material to support your own work

As an engineer, you might be producing original work during your career or studies. This could include performing tests or designing products. To confirm or support your work, you will often need to use data or information you have collected. When new products are developed, a prototype is often manufactured and tested rigorously. This performance test data is important in order to assess whether the product meets the design specification. Another example would be using a computer programme to assess whether a design will carry a given load or whether a circuit will operate a device as expected.

Checking validity

Using information to check validity is important in order to ensure that the work you are engaged in meets set requirements. You would not use results from a tensile test to determine the electrical or thermal conductivity of a material, for example. However, if you gathered data from a tensile test and used this to determine the yield stress of a material, along with expected values for similar materials and data from previous tests, you could determine the validity of your test.

Figure 2.85 Checking in a book

2.4 Using information and communication technology (ICT) to present information

This section will cover the following grading criteria:

Information technology is something we often take for granted. However, we use it all the time for presenting information and it is important to use the most appropriate technology to present information correctly. For example, it is possible to use word-processing or desktop publishing (DTP) software to prepare drawings. However, it would be more sensible to use a dedicated CAD package to present engineering drawings in the most appropriate way.

Software packages

By the end of this section you should have developed the ability to use different software packages in an engineering setting.

> **Key Points**
>
> Using ICT to present information is an important technique to allow effective communication in engineering organisations. You might be presenting information formally to a large group of customers, clients or managers. Alternatively, you could be discussing a product online with a customer, supplier or colleague. Whatever the situation, you will need to feel comfortable using different software packages.

ICT refers to both hardware and software. In this section we are going to concentrate on computer software, which refers to the computer programmes used, such as word-processing, CAD or spreadsheets. Although there is a wide range of software packages available to the computer user, most engineering organisations use a relatively small range of software.

> **Make the Grade** **P7**
>
> Grading criterion P7 requires you to use appropriate ICT software packages and hardware devices to present information. You will need to demonstrate word-processing, data handling, drawing and communication. You may be able to demonstrate all of these in your project presentation.

> **Key Term**
>
> **Computer programmes** – Programmes are generally split into two categories: applications which allow us to produce drawings, documents, graphic images, and so on, and operating systems which allow the applications to run on the computer.

Word-processing

Word-processing software is widely used for the production of any kind of printable medium. It can be used to edit and compose documents and letters, which can include graphic images, tables, diagrams and charts. The software has a variety of additional features, depending upon the version and operating system, but usually including spell-checking and page numbering, for instance.

> **Activity 2.27**
>
> Using word-processing software, try to create the table shown here using the appropriate software tools. Make sure that you save this document, as you will be using it later.
>
> | Cost | Quantity |
> |------|----------|
> | £5 | 5 |
> | £7 | 3 |
> | £9 | 2 |

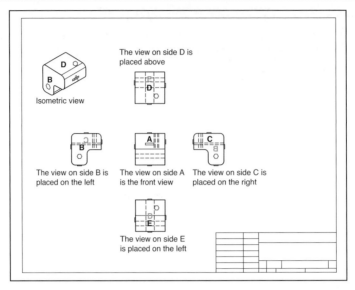

Figure 2.86 CAD drawing

Drawing

Many word-processing packages have a simple facility for drawing lines, circles and other simple geometry. However, to produce engineering drawings, circuit diagrams and other technical drawings and illustrations requires specific software.

2D CAD

CAD is a technology used to produce technical drawings, traditionally created on a drawing board, using specific software. The software can be linked to computer-aided manufacture (CAM) software and is in widespread use in engineering and manufacture.

Graphics software

Many applications are available to allow you to construct simple diagrams, flow charts, and so on.

However, to create impressive graphic images or manipulate digital images, you will need a dedicated graphics package. These packages often provide animation options, giving you the opportunity to produce videos or 'walk-throughs'. An engineer might want to produce a short video of how a mechanism such as a robot might operate. Architects might use graphic packages to simulate a person walking through a building they have designed.

Figure 2.87 Graphics software

> ### Activity 2.28
> Using an appropriate drawing package, draw the component shown in Figure 2.88.
>
>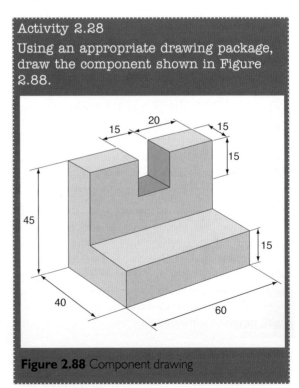
>
> **Figure 2.88** Component drawing

Graphics packages can often import 2D or 3D CAD images as a starting point.

Data handling and processing

Data-handling and processing software is used to organise and manipulate information. Usually, this refers to numerical data.

Key Term

Copy and paste – Highlighting information in a software application allows us to use the copy feature (often accessible by using the right mouse button). This copies information onto a 'clipboard' – in other words, it is temporarily stored. This information can be copied or pasted into another part of the current file. You can also copy information from one application and paste it into another.

Spreadsheets

In section 2.3 you learned how a spreadsheet is constructed and what a spreadsheet is used for. You will typically be expected to use a spreadsheet to collate and process data, such as test results, or to process information. The spreadsheet can be used to analyse the data and display it graphically.

Activity 2.29

Ensure that you have both spreadsheet and word-processing software applications active. Open the word-processing file in which you created the table earlier (see p. 85).

Copy and paste the table into a new worksheet and see if you can use formulae to display the totals shown here. Save the file; you will need it later.

(Hint: you may need to format the cells to get the £ to display.)

Cost	Quantity
£5	5
£7	3
£9	2
£21	10

Databases

In section 2.3 you learned how a database is constructed and what a database is used for. When working in many engineering organisations, it is important to review or check information. As an engineer, it is probable that you will be expected to use a database to retrieve the required information, and you may also be required to input data.

Presentation package

Presentation software is frequently used to generate slides, so that when you are giving a verbal presentation, you can produce slides with images, diagrams or key information to support your presentation.

Activity 2.30

Ensure that you have both spreadsheet and presentation software applications active. Open the spreadsheet file in which you pasted the table earlier (see opposite).

Copy and paste the table into a new slide, and resize it to fit onto one slide at a reasonably large scale, as shown in Figure 2.89.

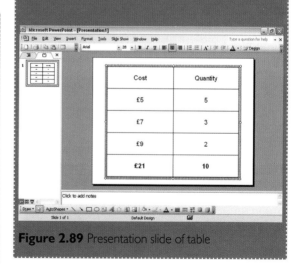

Figure 2.89 Presentation slide of table

Simulation package

Specific software such as simulation programmes are used to show the performance of different systems. Therefore, an engineer can use these specialised software applications to design, test and simulate electrical/electronic circuits or pneumatic/hydraulic circuits or plant/process systems.

> ### Key Points
>
> Although computers are generally reliable, occasionally the operating system can 'freeze' or 'crash'. Usually, you can simply restart the system and all is well; however, it is possible to lose work! It is good practice to save your work at regular intervals, and many companies back up or save their data, drawings and important documents on a secure server, or even on discs in a fireproof safe!

Communication

By the end of this section you should have developed the ability to use different communication technologies in an engineering setting.

Communication software is used to communicate over a computer network, using audio, video real time or text formats.

Email

We have already discussed the use of email in section 2.2. Email is one of the most widely used forms of communication in business and has the advantage of allowing attachments such as CAD drawings or word-processed documents. You will be expected to use email as a method of communication in the majority of engineering organisations.

> ### Activity 2.31
>
> Send an email to a colleague, friend or classmate, with an attachment of the presentation slide you prepared in Activity 2.30.

Fax

You learned a little about faxes in section 2.2. Although the advent of the email has revolutionised electronic communication, you may encounter a fax machine, which many businesses still use from time to time. Before the development of email, fax was widely used to transmit documents. Copies of letters, quotes, bills, technical diagrams, and so on, can be transmitted via a phone line and the image is reprinted at the receiving end. A special phone, known as a telecopier (which looks like a phone with a large keypad and printer combined), is required at the sending and receiving end.

Figure 2.90 Fax machine

Messaging

Software using real-time text formats refers to messaging software; this is less widely used in business as it is predominantly used for social networking. However, business organisations often have a presence on social networking sites in order to raise the profile of the company and its products.

Figure 2.91 Mobile phone text message

Videoconferencing

Video formats such as webcams are often used to talk to friends or colleagues; however, video conferencing is used by large companies to hold meetings, often with contributors from all over the world. The hardware required is sophisticated, often requiring remote-controlled cameras and multiple large-format screens.

Figure 2.92 Videoconferencing suite

Optical and speech recognition software

Optical character recognition (OCR) is a useful technique, working in conjunction with a scanner. The software 'reads' the letters and documents being scanned and recognises words; the result can be imported into word-processing software. While some editing is often necessary, it is considerably quicker than reproducing the whole document. Speech recognition software allows the user to dictate; using a microphone in a similar way to OCR, the software recognises what is being said and reproduces it on the page. Word-processing software is utilised once again, and some editing and proofreading is required.

> ### Activity 2.32
> If you have to give a presentation to five colleagues who are situated in different parts of the world, which communication technology would you use to allow interaction with all of them simultaneously?
>
> Prepare a list of benefits and drawbacks to the method you propose.
>
> What alternative technology could you use?

Hardware devices

By the end of this section you should have developed the ability to use different hardware devices in an engineering setting.

We use computers and associated systems in all walks of life. When we put fuel in our cars, the pumps, payment card readers and closed circuit television cameras (CCTV) to record registration details are all hardware devices that are computer-controlled. You can use a variety of hardware devices to help you communicate effectively.

> **Key Points**
> Hardware devices are those pieces of equipment that are used to facilitate communication using ICT. This is not only the computer system, but devices that can often be connected to a computer system.

Hardware devices used for communication are not only computer based. Telephones, fax machines and control systems are all ways of communicating; however, the majority of our communication systems have some computer involvement in them.

> # Make the Grade
>
> Grading criterion P7 requires you to use appropriate ICT software packages and hardware devices to present information. You will need to demonstrate the use of a computer system and input and output devices.

Computer system

A computer system can consist of a single PC or laptop; alternatively, it can refer to a network of interconnected computers, or a network of devices that are controlled by a computer.

PC

A PC will normally consist of a variety of components, as shown in Figure 2.93. The parts would normally consist of an outer case, power supply, motherboard, central processing unit (CPU), memory modules, hard disk drive, graphics card, sound card, network card, cooling fans, internal speaker, CD/DVD drive and various cables to link the components. Of course, to use a PC requires the addition of input and output devices, an operating system and application software.

89

Figure 2.93 Parts of a PC

A laptop is, in principle, a miniaturised version of a PC, but with input and output devices built in.

Network

A network is a system that usually connects computers across an organisation; this allows a server to hold records securely and allows multiple users to access the same files/folders. As well as for communicating and sharing information, a computer system is also used to control a process or production line. For example, as products are scanned at a supermarket checkout, the computer system registers the reduction in stock on the shelves. The system can then schedule restocking when a specified number of a product has been sold.

Input/output devices

Input devices are used to allow information to be input into the system. The most obvious example of this is the keyboard; however, scanners, drawing tablets and microphones would also be classified as input devices.

Output devices are quite the opposite and are used to retrieve information from a computer. An obvious example is a printer; however, the monitor or visual display unit (VDU) is also an output device, as is a speaker.

Typical input devices	Typical output devices
Keyboard	VDU
Mouse	Printer
Scanner	Plotter
Light pen and tablet	Speakers
Microphone	

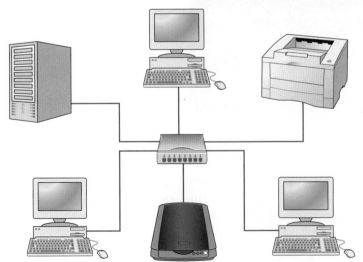

Figure 2.94 A computer network

Key Term

Hardware – A general term for any piece of computer equipment that can include the computer.

Peripherals – A computer is not a peripheral; however, peripherals are hardware devices which can be attached to and removed from a computer, allowing certain functions – for example, a digital camera can be seen as a peripheral as it can be used to transfer pictures from its internal storage to the computer or vice versa.

Present information

By the end of this section you should have developed the ability to present information using a variety of forms.

Technology is frequently used to facilitate presentation of information. The use of ICT is commonplace in engineering, and competent use is an expectation from employers and colleagues.

Key Points

Information can be presented in a variety of forms, something that we have been exploring throughout this unit. However, using hardware and software underpins all that we have learned, and the competent use of both is necessary for high-quality presentation.

You could spend a considerable amount of time generating a technical report, with tables, graphs and charts included. You might spend hours proofreading and editing it, ready to submit it to your tutor or employer. Finally, you come to print a hard copy to hand to them, and the ink runs out on your printer or the paper jams. The presentation of your work is compromised because of an issue with the hardware.

You could spend a considerable amount of time putting together a complex presentation on your laptop, checking that all the slides are correctly sequenced and the spelling and grammar are correct. You might embed video clips and plan smooth transitions between slides. Finally, when you save the presentation on to a memory stick and plug it into the host computer, you find that the software on that machine is a different version and is not compatible. The presentation is compromised because of an issue with the software.

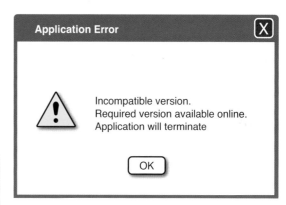

Figure 2.96 Software glitch

Make the Grade   P7

Grading criterion P7 requires you to use appropriate ICT software packages and hardware devices to present information. You will need to demonstrate the use of a computer system and appropriate software to generate a professionally presented report, and visual presentation material to support an oral presentation.

Figure 2.95 Printer jam

Report that includes written and technical information

Throughout this unit we have discussed the use of a variety of communication methods, including letters, memos, technical product/service specifications, emails, faxes, tables of data and graphical data. If you are producing any of this by hand, you would be expected to be as neat and careful as possible, and to make sure that all written work and diagrams are legible and easy to read. The same is true of any information generated using ICT. You should check each part of a report for clarity and legibility. Although this was discussed in the section on proofreading, if you have spent a great deal of time and effort producing a substantial report, you should make sure that the printed copy is of the highest standard.

> **Activity 2.35**
> Investigate the difference in terms of output quality between a dot matrix printer, laser printer and inkjet printer. Which would you prefer to use for your final project report?
> What other factors or settings might affect the quality of your printed work?

Visual presentation

Using ICT to present information is an important technique to allow effective communication in engineering organisations. You might be presenting information formally to a large group of customers, clients or managers. Alternatively, you could be discussing a product online with a customer, supplier or colleague. Whatever the situation, you will need to feel comfortable using presentation software. There is guidance on how to prepare an effective presentation in Unit 3.

Make the Grade M3

Grading criterion M3 requires you to evaluate the effectiveness of an ICT software package and its tools for the preparation and presentation of information. You will need to prepare a presentation and then reflect on its impact. Some questions you might ask yourself are:

- Is the text all in the same font and of the same size?
- Are the text and images aligned?
- How well are they positioned on the page?
- Are all labels, graphs and charts clearly visible?

It might be helpful to access a large screen or data projector to review your presentation, as sometimes what you see on a small screen does not look the same on a large one.

Make the Grade D1

Grading criterion D1 requires you to justify a specific communication method and explain the reasons you chose for using that method and not a possible alternative. Justification requires you to consider a range of different ways of communicating and to look at the pros and cons of each one. However, it is more than this in that you will need to give reasons why you chose the method you did. Consequently, you will need to demonstrate clearly that you considered more than one option and why you rejected alternative techniques.

Make the Grade D2

Grading criterion D2 requires you to evaluate the use of an ICT presentation method and to identify an alternative approach. It might be useful to video yourself giving a presentation or gather feedback from your audience. However, this criterion is not about the presentation, but the method used. If you used presentation software, for example, were the slides clear, and could you have used graphics instead of tables? Evidence could be a written evaluation of your work, or you might submit an improved version of your presentation, indicating where and why you have made improvements.

To achieve a pass grade you will have:	To achieve a merit grade you will also have:	To achieve a distinction grade you will also have:

P1 interpreted an engineering drawing/circuit/network diagram

P2 produced an engineering sketch/circuit/network diagram

P3 used appropriate standards, symbols and conventions in an engineering sketch/circuit/network diagram

P4 communicated information effectively in written work

P5 communicated information effectively using verbal methods

P6 used appropriate information sources to solve an engineering task

P7 used appropriate ICT software packages and hardware devices to present information.

M1 evaluated a written communication method and identified ways in which it could be improved

M2 reviewed the information sources obtained to solve an engineering task and explained why some sources have been used but others rejected

M3 evaluated the effectiveness of an ICT software package and its tools for the preparation and presentation of information.

D1 justified the choice of a specific communication method and the reasons for not using a possible alternative

D2 evaluated the use of an ICT presentation method and identified an alternative approach.

By the end of this unit you should be able to:

- keep records, specify a project, agree procedures and choose a solution
- plan and monitor a project
- implement the project plan within agreed procedures
- present the project outcome.

In order to pass this unit, the evidence you present for assessment needs to demonstrate that you can meet all of the above learning outcomes for this unit. The criteria below show the levels of achievement required to pass this unit.

To achieve a pass grade you need to:	To achieve a merit grade you also need to:	To achieve a distinction grade you also need to:
P1 prepare and maintain project records, from initial concepts through to solution, that take account of and record changing situations	**M1** maintain detailed, concurrent records throughout the project that clearly show progress made and difficulties experienced	**D1** independently manage the project development process, seeking support and guidance where necessary
P2 prepare a project specification	**M2** use a wide range of techniques and selection criteria to justify the chosen option	**D2** evaluate the whole project development process, making recommendations for improvements.
P3 agree and prepare the procedures that will be followed when implementing the project	**M3** evaluate the project solution and suggest improvements	
P4 use appropriate techniques to evaluate three potential solutions and select the best option for development	**M4** present coherent and well-structured development records and final project report.	
P5 outline the project solution and plan its implementation		
P6 monitor and record achievement over the life cycle of the project		
P7 implement the plan and produce the project solution		
P8 check that the solution conforms to the project specification		
P9 prepare and deliver a presentation to a small group, outlining the project specification and proposed solution.		
P10 present a written project report.		

Introduction

Solving problems and finding solutions is an integral part of what engineers do. Project work is a key element of many engineers' careers, and problem-solving is fundamental to many projects undertaken. Some problems are easily solved – replacing a fuse on a faulty circuit can be relatively simple – but of more importance to the engineer is the question, why does the fuse keep failing?

Large-scale projects require teams of engineers, often from different disciplines – for example, designing wind turbine generation equipment might involve electrical, mechanical, electronic, civil and aerospace engineers. However, smaller-scale projects can be undertaken by individuals, and this is the purpose of the engineering project unit.

There is an expectation that you will be able to apply the knowledge and skills gained from other units you may have studied, or from elsewhere in your programme of study, to produce a substantial piece of work that reflects the expectations of an engineering technician. You will need to demonstrate that you can plan effectively and follow this plan, and/or modify it if necessary, to achieve the desired outcomes. The result of your project could be a product or device; alternatively, it might be a design, process or service.

Some projects are modifications of existing products; a good example of this would be the modification of the simple vacuum cleaner, using the cyclone principle that made Sir James Dyson famous. Similarly, Trevor Baylis OBE used the energy stored in a spring to power a small-scale electrical generator, and consequently designed the wind-up radio. Both of these inventors would have had to achieve financial backing; to do this they would have had to present their proposals to financiers and colleagues.

Part of any project is its presentation, and you will be expected to prepare and deliver a presentation on your project to a given audience. In addition, you will be required to produce a report outlining your project and drawing appropriate conclusions, to allow recommendations to be made. There are conventions that should be followed in the presentation of written reports and the use of ICT will be required to demonstrate this.

It is likely that, as a project is a substantial piece of work, you will need to record or log your progress. Keeping a record of your work as you progress is an essential component of this unit, as is the production of a suitable project plan.

If, for any reason, the project you undertake does not go to plan or does not produce the desired result, it is not always a sign of failure. Many projects, although not initially successful, can lead to success in future developments. A good example of this is the ball barrow developed by Sir James Dyson. The principle behind this was to develop a wheelbarrow that replaced the wheel with a large ball to allow easy manoeuvring. Although this development was partially successful, the idea reappeared as a development on the Dyson vacuum cleaner and has proved to be a commercial success.

Although the project unit focuses on independent study, you should be allocated a project supervisor to guide you – this could be your tutor, manager, colleague or assessor. It is important that you keep in regular contact and review your project frequently, to ensure that things are going to plan.

This section of the book is organised as follows; each of the subsections can be readily linked to the learning outcome (LO), pass (P), merit (M) and distinction (D) criteria.

Section/content	LO	P	M	D
3.1 Specifying a project, agreeing procedures and choosing a solution	1	1, 2, 3, 4	1, 2	1
3.2 Planning and monitoring a project	2	5, 6		
3.3 Planning and monitoring a project	3	7, 8		
3.4 Presenting the project outcome	4	9, 10	3, 4	2

3.1 Specifying a project, agreeing procedures and choosing a solution

This section will cover the following grading criteria:

P1 P2 P3 **P4** **M1**
M2 D1

Being able to specify a project often depends upon the environment in which you are working. Many learners will be in employment, and negotiation with the employer can result in a project that benefits both the employer and the learner. When you have developed ideas for your project, you will need to produce a brief outline and consider the feasibility of what you are planning. Provided your project is achievable, you will need to agree procedures with all stakeholders, which might include tutors, employers, colleagues, and so on. You might have several proposals for the project you would like to undertake, and it is important to use appropriate techniques to determine which project you are going to undertake, how you allocate roles and responsibilities, and the fitness for purpose of your project proposal.

Project records

Project records are critically important in ensuring that a successful outcome is achieved. As you change your designs, test different theories, observe changing outcomes, and so on, you will need to make notes, take pictures, create drawings and plans, set targets and review progress.

By the end of this section you should have developed a knowledge and understanding that will allow you to prepare and maintain project records.

Key Points

Many large-scale projects run into problems with litigation and disputes, because records have not been properly kept and the outcome of the project work is not as anticipated.

Keeping a record of what you do is invaluable, not only to remember what you have done, but to allow you to monitor and plan your progress. When you come to writing a formal report and preparing a presentation, you will need to refer to these records. If you have incomplete or disorganised notes, it will be difficult and time-consuming to write your final report. A significant project will not be achievable if you rely on your memory, and the record of work you compile as you progress is invaluable to a successful project.

Make the Grade **P1**

Grading criterion P1 requires you to prepare and maintain project records, from initial concepts through to solution, that take account of and record changing situations. To achieve this, you will need to demonstrate that you can keep effective records of what you do during the project and keep accurate records of any changes to the plans you made at the start.

Written

Written records are invaluable in maintaining and organising your work. Many engineers use a large-format diary or notebook and, consequently, can make notes on the activities and records on a daily basis. A useful technique is to use numbered or dated sheets of paper, which are then placed in loose-leaf binders or lever-arch

Figure 3.1 Project portfolio

files. The advantage of this technique is that other documents can also be stored, in plastic wallets, for example. In addition to making notes, it is useful to keep a record of any sketches or drawings that are carried out, as well as plans and modified plans and drawings.

If you are undertaking a design project, you will be coming up with a variety of ideas and the final design will, very probably, be a combination of the design sketches and drawings produced over a period of time. Keeping all sketches is important, as ideas that are initially discarded can be resurrected at a later date. Similarly, if you are

designing a circuit to operate a particular system, you may produce a series of sketches and circuit diagrams, and keeping a record of all of these is of vital importance.

The plan you develop at the start of a project may vary as time progresses; this can be due to a variety of factors. It is very important to record why a plan has changed and, more importantly, what modifications are made to the plan to allow a successful conclusion.

Targets

Setting targets is important to allow a project to be undertaken successfully. If your project requires six components to be designed and made, over a period of six months, you might target one component per month. Setting targets is part of the planning process and, of course, you need to monitor your progress to ensure that you are meeting those targets.

Use of planning tools

Planning tools available to engineers include the more traditional paper-based plans, which are templates that allow you to fill in key dates and events. There is a range of software applications used for planning; these vary from spreadsheet applications to specific project-planning software. Software-based applications can, of course, be modified and updated on a continual basis.

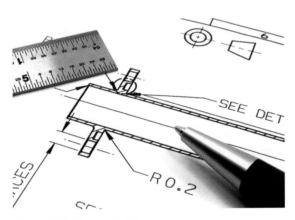

Figure 3.2 Design sketches

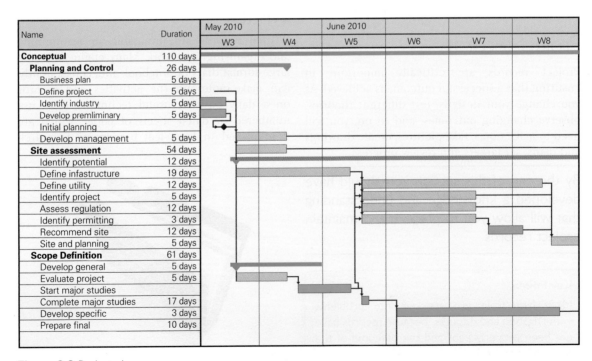

Name	Duration	May 2010		June 2010			
		W3	W4	W5	W6	W7	W8
Conceptual	110 days						
Planning and Control	26 days						
Business plan	5 days						
Define project	5 days						
Identify industry	5 days						
Develop premliminary	5 days						
Initial planning							
Develop management	5 days						
Site assessment	54 days						
Identify potential	12 days						
Define infastructure	19 days						
Define utility	12 days						
Identify project	5 days						
Assess regulation	12 days						
Identify permitting	3 days						
Recommend site	12 days						
Site and planning	5 days						
Scope Definition	61 days						
Develop general	5 days						
Evaluate project	5 days						
Start major studies							
Complete major studies	17 days						
Develop specific	3 days						
Prepare final	10 days						

Figure 3.3 Project plan

Figure 3.4 Project folder

Key Term

Project folder – It is useful to keep all your project work together in one folder. Drawings, designs, sketches, project records and plans can all be collected and subdivided using plastic wallets or coloured divider cards. This, along with your logbook/diary, should be maintained and updated on a regular basis. You might also find it useful to set up an electronic folder to store all your electronic files, project plans, CAD drawings, and so on.

Recording initial concepts

Keeping notes and records of the work you do as you progress your project is important, but it should not be forgotten that initial ideas and concepts can also be important and should be kept for reference. If you create mind maps, use brainstorming activities or draw flow charts, you should keep all of these records, as you may need to refer back to them later, and it may be difficult to remember what your thoughts were or what people said.

Activity 3.1

Refer to Figure 3.5 and compile a folder/file structure on an appropriate computer, either at work, home, college or school. It might be useful to save this electronic portfolio on a memory stick (USB flash drive), as you may be working in different locations and using different PCs or laptops.

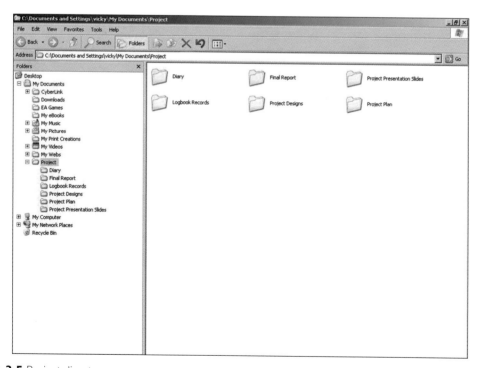

Figure 3.5 Project directory

Initial concepts

By the end of this section you should have developed a knowledge and understanding of how you will record your initial concepts.

When you consider undertaking a substantial project, it can seem quite daunting. To successfully achieve your objectives, you have to choose the right project to undertake, and this starts with recording your initial concepts. There are two strands to this, the most obvious being the generation of ideas, where you record the thought processes and creativity that go into a project. However, you also need to consider setting limits on cost, time and feasibility, and record your decisions based on these factors.

Many companies would like to undertake project work and research and development (R & D) to a greater extent than they currently do. The reasons for not being able to do more may be due to lack of resources, both physical and financial, or project work may simply not be feasible.

Key Term

Feasibility – Something that is feasible can be completed or carried out, so if you are determining the feasibility of a project, you are saying to yourself, can I accomplish this?

The first stages in developing a project are your initial concepts. These are generated by a variety of methods and techniques. An important part of developing these concepts is to decide whether your project meets the appropriate specifications and approaching ideas with an open mind. However, initial concepts and ideas are often used by an organisation to determine whether a project is worth investing in.

Make the Grade

Grading criterion P1 requires you to prepare and maintain project records, from initial concepts through to solution, that take account of and record changing situations. To achieve this, you will need to demonstrate that you have a record of your initial concepts or ideas, and that from these concepts you have developed the theme for your project.

Setting limits

When we refer to setting limits, we are deciding whether the project is feasible. Feasibility means can the project be carried out successfully? Cost is frequently a factor; many organisations have a specified R & D budget, and a viable project will not be undertaken if there is insufficient funding. Time is also a factor. If you have six months to complete a project, and your proposal will take two years to complete, it is obviously not going to be sanctioned. Many project proposals are rejected by organisations because they simply do not require the product being promoted. If something is not needed, then it will not be commissioned.

Value-cost-benefit analysis

Perhaps more important than the overall cost of a project is what the benefit to a company will be if the initial concept is implemented. A good example is whether it is worth a company investing £1,000,000 in a project to improve the manufacture of a product costing £10. If the saving is £1 per product and 10 million products are manufactured per year, then you make significant savings very quickly, but what if you make 500 000 products per year or only 100 000 per year? The payback on the investment is clearly an issue.

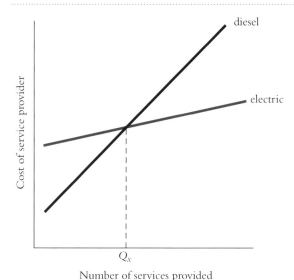

Figure 3.6 Cost-benefit chart

Activity 3.2

A company wants to replace the self-tapping screws it uses to secure the lid of a container. By replacing the screws with snap-fitting lids, it hopes to speed up production.

Write a list of reasons why this is a good idea.

Can you think of any disadvantages?

Generating ideas

There are a variety of methods of generating ideas to help you develop a given project. Making lists of ideas or notes on different themes is obviously useful, and it often helps to involve others in a group discussion.

Once you feel you have exhausted all the ideas, you can start to sort through them and carefully discard all unworkable options. This should leave you with a variety of ideas you can take forward to the next stage of your project planning.

Mind mapping

A mind map is a pictorial method of developing ideas or planning activities. Many people do not feel comfortable using conventional techniques, such as making notes or lists. The principle behind a mind map requires you to visualise a problem and produce a diagram, or map, which allows you to develop your ideas. Some students use mind-mapping techniques when making class notes! The structure of a mind map is thought to represent the brain's naturally fluid and random nature. Unlike notes and long lists, it is a form of presentation that is easily understood by your brain and will help you to be more creative in problem-solving.

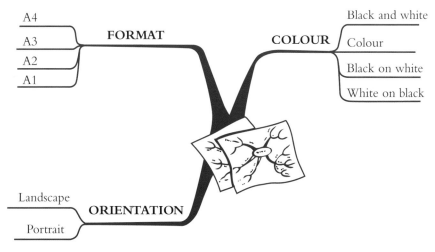

Figure 3.7 Mind map

101

To produce a mind map:

- place a blank sheet of paper in front of you in landscape format (sideways)
- draw a picture in the centre of the sheet – this represents your problem or idea
- draw a series of curves, evenly spaced, from around the picture
- at the end of each line, write keywords which represent each area you are going to investigate
- from each of these ideas, add more lines to represent each further thought or concern
- it is sometimes useful to add images instead of words if this is helpful
- you can use colour to highlight certain areas and connections.

Activity 3.3

Using an A3 sheet of paper, construct a mind map of how you might approach designing a new computer keyboard for a small child. Try to think of all the key features, issues and difficulties children might have, and use colours and shapes. Remember to start in the middle of your sheet. You may find this useful to do as a paired activity.

Brainstorming

Also known as a thought shower, the principle of a brainstorm is to generate a large number of ideas, particularly if you are trying to solve a problem. There are a set of ground rules which are often used to guide the process:

- quantity is best – although this might produce more ideas that are not going to be useful, the greater number of ideas is thought to encourage radical and effective solutions
- no criticism allowed – the purpose of this technique is to produce radical solutions and encourage all participants to contribute, without fear of ridicule or criticism, regardless of how unusual the idea might be
- unusual ideas – unusual ideas are welcomed in brainstorming sessions; this encouragement allows contributors to look at the problem from a new perspective and not to reproduce familiar and traditional solutions
- combine ideas – once the list has been generated, the combination of some of the ideas generated can produce a positive solution.

It is useful for one person to facilitate the session. This person should act as a leader and explain the problem to be solved, the rules for the session and who should participate. You will also need a

Figure 3.8 Brainstorming using sticky notes

note-taker, whose responsibility is to record all the ideas generated.

It is a good idea to gather people around a table and perhaps try a practice session with something quite unusual, such as ideas for the design of a solar-powered car or a clockwork television. The idea is to get your colleagues comfortable with the rules and concepts. Once you feel people are comfortable with the idea, you can start properly and introduce the session. The important part of the brainstorming activity is for you to encourage all ideas, however strange or unusual, and to encourage all participants to take part, build on the ideas being suggested and avoid any criticism of the ideas put forward.

Although paper, a whiteboard or flip charts can be used, it is sometimes useful to use sticky notes. The advantage of using these is that they can be moved around and grouped afterwards.

Activity 3.4

This activity requires you to work in a small group of three or four colleagues or friends/classmates.

Nominate one person to lead the group and record ideas. First, read out the rules on brainstorming, then spend five minutes brainstorming how you could redesign a computer keyboard for small children.

After the five minutes have elapsed, review your ideas. Can they be grouped into different categories or themes? Were any ideas generated that you would not have expected or anticipated? How does the process compare with the mind map you might have created for this?

Research techniques

By using brainstorming, mind mapping or similar techniques, you will hopefully have brought together everything you know, or need to know, about the project that you are about to undertake. Alternatively this could be a problem you are attempting.

You now have to find that information and you will use different research methods to achieve your goals. But where can you find what you are looking for? Many people use the internet as their primary research tool; however, while there is undoubtedly a great deal of useful information available via the World Wide Web, there is also a great deal of misleading and incorrect information too. Useful information can be found on 'wikis'; however, these are collaborative websites which can be directly edited by anyone with access. In other words, you should be very careful to ensure that what you read on a wiki is correct, as many mistakes are made which are not always corrected.

When researching, there are a variety of sources of information:

- textbooks
- product specifications
- manufacturers' brochures
- colleagues and family
- experts, such as engineers, architects, instructors
- teachers and lecturers
- trade journals
- government publications
- academic papers or papers from conferences and research papers.

Figure 3.9 Research

An important element of how you carry out your research is the methods you use. For example, you might undertake interviews with individuals or groups. This can be important – if you are proposing a new design for an existing product, you will need to obtain information from users of that product. You might carry out observations. This can be helpful if you are trying to provide a solution to a production line problem or to redesign a difficult assembly. The more information and background you have will help you to define your project and produce an appropriate solution.

Lines of communication

When working on a project, for the most part, you will be working independently. This means that you will need to set yourself goals and targets; in addition, you will need to keep a logbook/diary and record your progress. This is a significant challenge and it is important that you do not feel isolated or alone. When carrying out project work, you should ensure that frequent project reviews take place. If you are undertaking a group project this is particularly important, as you will be allocating different roles and responsibilities to group members, and you need to ensure that everybody is proceeding with their allocated tasks and decide how best to support members of the team who might be struggling.

Whether undertaking a group or individual project, it is important for you to have regular reviews with your tutor or supervisor, to ensure that you are proceeding correctly, keeping to your project plan and staying on the right track with your project. Similarly, if your project is work-related, you should be reviewing your progress with your employer at timely intervals.

Figure 3.10 Project review meeting

Make the Grade

Grading criterion M1 requires you to maintain detailed, concurrent records throughout the project that clearly show the progress made and difficulties experienced. You will have to do more than simply maintain a log of activities to achieve this. Think about how you might use the headings *Activity*, *Progress made* and *Difficulties experienced* as a prompt for comments on each logbook page.

Specification

By the end of this section you should have developed a knowledge and understanding of how to produce a project specification.

The project specification is an important document; it outlines the objectives, constraints and essential/technical requirements of the project.

Key Points
You will need to produce a project specification and have this approved by your project supervisor before you commence work on your project.

A project specification is a document that outlines what the project is required to deliver. A good example is an engineering drawing. For a simple component, the drawing should indicate the material, the dimensions and any additional information, such as heat treatment requirements or geometric tolerances. If the product has been manufactured correctly, and is as the drawing suggests, then it meets the product specification.

However, a substantial project is more comprehensive and will normally require a written document, which details all the information required to successfully produce a solution to the given problem.

Make the Grade — P2

Grading criterion P2 requires you to prepare a project specification. Your specification should reflect some of the ideas generated from the activities you carried out when generating initial concepts, and should include technical information as well as considering health and safety, resources, timescales, sustainability and quality issues.

Type of project

The type of project you undertake will largely reflect the nature of your course of study, your current employment or the industry in which you are interested. Typical engineering projects can include product design, plant layout/maintenance or production methods. Here are some examples of projects for different engineering pathways:

Aerospace

- Modification of an aeronautical product
- Specifying, evaluating and/or designing an aeronautical system or service
- Testing an aeronautical product

Electrical/electronic

- Modification of an existing electrical/electronic product
- Specifying, designing and building an integrated hardware/software system
- Testing and evaluating an electrical/electronic system or service
- Comparison and evaluation of a range of electrical/electronic CAD tools and systems

Manufacturing

- Modification of a manufactured product or service
- Designing and building a manufactured product or service
- Testing a manufactured product or service

Mechanical

- Modification of a mechanical product
- Specifying and designing a mechanical system
- Testing a mechanical product

Operations and maintenance

- Modification of a plant services
- Designing and building an inspection/ calibration test rig
- Testing plant service systems or subsystems

Key Points

Remember, you are looking for a problem or task to be solved, not necessarily for a finished item.

Activity 3.7

Consider the branch or area of engineering in which you currently (or intend to) work.

Make a list of the kinds of project work undertaken by engineers in that area. You may find it useful to do some research or to use a mind map.

Technical information

The project specification will be different, depending on the nature of what is being undertaken. Some ideas of what to consider for a project specification might include the following:

- Functionality – How does the product or service work? Is it overly complex? Can it be easily understood and used or will training be required?
- Aesthetics and ergonomics – Is it easy to use? Will special tools be required or can operators reach all the controls or parts? Does the product look fit for purpose, with elements of good design embedded?
- Operational conditions – Have you considered the environment you are working in? A product or service designed for arctic conditions will have different requirements to one used in the desert.
- Process capability – This is a measurable quantity which determines whether a product or process is fit for purpose. If you are designing a process to manufacture very accurate components, to within 0.001 mm, and your machinery is only capable of tolerances of 0.01 mm, it is

105

unlikely that the process is capable of meeting the standards.

- Size/scale of operation – The scale of an operation can refer to the physical size or the quantity required. It is unlikely that you will build a control system to fit inside a model aircraft if you use industrial-size electrical motors. Similarly, if you wish to modify a mass production system, it is unlikely that you should include a robot that requires maintenance after every 100 operations.
- Present and future trends – It is important to consider the 'future proofing' of your project. Designing a control system that uses a CD/DVD to store or relay instructions would be unlikely to be acceptable, as the quality of other devices is superior. Would it be better to take advantage of technology being used in mobile phones and MP3 players?
- Maintenance and likely product life issues – Is it a reliable product or service, or will a great deal of care be required in its operation? How long will the product last before it will need replacing? A light bulb might be expected to last 12 months, but with no maintenance, whereas a car would be expected to last 20 years, provided that it was regularly maintained.

often useful to indicate the standards expected in terms of the Control of Substances Hazardous to Health (COSSH), British Standards (BS) and the International Organization for Standardization (ISO).

Figure 3.11 Welder with protective equipment

Activity 3.8

For the reminder of this section, most of the activities are going to focus on the design of a computer keyboard specifically for children to use.

This is a group activity. In a group of three, think about the new children's computer keyboard. What technical elements do you think would be important to include in a product specification? See how many you can list; then divide them up between you and research each one, making notes as appropriate.

Health and safety issues

You should always specify that an awareness of health and safety is paramount, and ensure that all activities undertaken are properly assessed for risk, and that all components and equipment used and developed meet the relevant safety and operating standards. When creating a specification, it is

Environmental and sustainability issues

As we become more aware of the effect on the environment of what we do, as engineers, so it is important to consider environmental issues in all new products and services. For example, all major car manufacturers are investing in technology that reduces carbon emissions and improves fuel economy, and in designs that use

Figure 3.12 Consider recycling

alternative technology, such as electric vehicles or hydrogen power. However, environmental impact can be reduced if we use sustainable products and technologies – for example, when a new product reaches the end of its useful life, can it be recycled? Can a system that uses wood or paper ensure that only timber from sustainable forests is used?

> **Activity 3.9**
> Once again, we are going to consider the computer keyboard for children. List any environmental or sustainability issues that you think you will have to consider. You may need to do some research on materials.

Quality standards and legislation

There are a variety of standards used around the world. For example, in Britain, BS standards are used; in Germany, DIN (Deutsches Institute für Normung); in the United States of America, ANSI (American National Standards Institute). However, the international ISO standards are recognised throughout the world. These standards are important, as they set a benchmark for the quality and reliability of products and services. If you are building a test rig for your project, you should require that all the components you use meet the appropriate standards, to ensure that they are safe and fit for purpose. It is also important that standards are used to allow consistency – for example, BS8888.2006 is the British standard for writing technical product specifications. Legislation is also important. While all elements of a project might meet standards, there are still elements of legislation that need to be considered. An example of this would be an engine manufacturer who would carefully consider noise and emissions legislation if a design for a new aircraft engine was being considered.

> **Activity 3.10**
> This activity requires you to research standards that apply to computer keyboards and record the details.
> Do you think any other quality standards will apply?

Timescales

Clearly, timescales are important and form the key target behind developing project plans. In terms of a specification, it would be the usual procedure to specify a final date for either the complete project or each stage of a larger project. If a city was bidding to host a major international event, such as the Olympic Games, it is obvious that they have a target of the opening ceremony to work back from. In the case of a student project, it is likely that your tutor will specify the final completion date and you can plan your project from that date. If the project cannot be completed by that date, it is likely to be unfeasible.

Physical and human resource implications

Many large-scale projects are a team effort, and without the key personnel and resources, many projects are unworkable. If you were designing an innovative product, but the company you work for did not have sufficient resources to build or test a prototype, it is unlikely that the project would ever be successful. Similarly, a racing driver is unlikely to win a Grand Prix without a team of mechanics,

Figure 3.13 Quality standards

Figure 3.14 Teamwork

designers, fitness coaches, and so on, all working behind the scenes to prepare the car and driver for the race.

> **Activity 3.11**
> This is a group activity. Consider the computer keyboard for children. What physical and resource issues do you think you need to overcome to mass-produce the keyboard? You will need to think about the design and materials.

Procedures

By the end of this section you should have developed a knowledge and understanding of the procedures you will use to carry out your project.

When you undertake a project, it is important to ensure that everybody involved understands what part they have to play and what is required of them. It is also important that everybody understands who will report to whom and how decisions will be made. Finally, it is of vital importance that you know what resources are available to you and the consequences if you reach those resource limits.

> **Key Points**
> You often hear of projects going over budget. It is important that you understand what money you can spend or what other resources, such as materials, you have access to.

> ## Make the Grade   P3
>
> Grading criterion P3 requires you to agree and prepare the procedures that will be followed when implementing the project. It would be advisable to draw up an agreement between all the people who will be involved and present this as evidence. This is particularly important if you are engaged in a group project.

Roles and responsibilities

This is particularly important with a group project. Before the project work commences, each member of the team will need to know exactly what they are working on and what are their required tasks and targets. This can involve who is responsible for decision-making, holding meetings and keeping records. If you are undertaking an individual project, there are still responsibilities to be determined, although these are likely to be negotiated between you and the project supervisor and/or employer.

> ## Key Term
>
> **Role** – A person's role is the expected behaviour of that person.

Figure 3.15 Roles and responsibilities

Reporting methods

How you are going to report your findings should be established before you begin work on the project. Projects in industry often require frequent meetings and presentations, and you will need to establish when and how you will be expected to report your findings. You will probably agree with your project supervisor that regular meetings, possibly on a weekly basis, are necessary. You should ensure that you keep details of your discussions, possibly in your logbook or portfolio.

> **Key Points**
> You will be expected to deliver a verbal presentation using ICT, as well as writing a project report when you have completed your project.

Resource allocation and limits

Some projects require huge budgets – the design of a new aircraft, for example, or the building of a sports stadium. Other projects, such as the testing of different adhesives to assess performance, might be relatively small. Whatever the size of budget, it is important to establish what resources need to be allocated and an estimate of the required budget.

You will need to establish what resources you require in terms of:

- facilities – do you need access to a workshop, CAD facilities, a laboratory, and so on?
- cost – do you need to purchase any specialist equipment or consumables?
- support – will you need help from tutors, technicians, colleagues and so on – for example, in performing tests or manufacturing components?

Techniques

By the end of this section you should have developed a knowledge and understanding of how to use comparison methods and analysis to select the optimal solution.

Engineers often present a variety of solutions to a given problem. To determine which solution is the optimal one requires the use of a variety of analytical skills and techniques.

If you were given the task of determining which of two bags of identical coins you can keep, but were not allowed to count them, what would you do?

Hopefully you would think to weigh the bags and then select the heaviest!

You would be using a comparison method, and by analysing the results, you have selected the optimal solution.

Once you have developed a series of project proposals, you will need to decide which one of your proposals is the most appropriate project to undertake. You will clearly need to consider the feasibility of each proposal and whether it can be resourced.

Comparison methods

A useful method of analysing different project proposals is to compare them against set criteria and give each one a score. By totalling the scores, you can decide which project is the most likely to be successful. Of course, if any of the criteria make a project totally unsuitable, this will have to be indicated.

In the example shown in Table 3.1, Project 2 looks the most promising, while Project 1 is totally unsuitable, despite some high scores, as the process capability requirement is not met.

You can set whatever criteria are sensible for the project you are undertaking, and you should be consistent in how you apply your scores.

	Project 1	Project 2	Project 3	Project 4
Resource requirement	7	8	7	4
Cost	9	6	3	5
Fitness for purpose	8	7	5	6
Process capability	x	4	5	8
Total	x	25	20	23

Table 3.1

Activity 3.13

This is a group activity. Select three well-known brands of soft drink. Decide on key characteristics, such as taste, texture, price, and so on. Draw up a table similar to that shown in Table 3.1.

Each of you should score each drink individually, then compare results.

Do you all think the characteristics you selected should be weighted evenly or are some more important than others? What could you do to accommodate this?

Analysis

An alternative to using comparison methods to choose a project is to use criteria such as a cost-benefit analysis. An organisation is more likely to invest in a project that saves a small amount from the cost of each mass-produced item it produces than a one-off saving, even if this is considerably more. The analysis of long-term benefit is what is important.

Key Term

Optimal – Finding the optimal solution means selecting the solution that best meets the objectives (as a whole).

Make the Grade   M2

Grading criterion M2 requires you to use a wide range of techniques and selection criteria to justify the chosen option. This criterion is an extension of the P4 criterion. The key here is the words *a wide range*; criterion P4 suggests using *appropriate* techniques. By using a wide range and explaining which technique is the most appropriate one, you are demonstrating the higher-level thinking skills expected of merit-grade learners.

Make the Grade   D1

Grading criterion D1 requires you to manage the project development process, seeking support and guidance where necessary. The key to this is that you can demonstrate independence. You should not feel that you cannot ask for help; however, you should plan carefully and, if you do require assistance, try to ensure that you are presenting ideas and suggestions, rather than asking for help because you are unsure of what to do.

3.2 Planning and monitoring a project

This section will cover the following grading criteria:

If you have a project that will last a considerable time, you should break it down into manageable elements. This requires you to plan where you are going to start and what sequence you will follow. When planning a foreign holiday, you might draw a mind map of all the things that have to be organised (foreign currency, travel documents, accommodation, and so on), then plan and implement each one individually. You would, of course, keep checking or monitoring that everything is going to plan – is everyone's passport valid, have hotel reservations been confirmed, etc?

Planning

By the end of this section you should have developed a knowledge and understanding of how to plan a project.

Engineers use a variety of planning tools to decide how a project will run. These can be simple flow diagrams, sketched out with pen and paper, or complex computer software applications, full of key dates, milestones and review points. This choice is often dependent on the size and scope of a project.

Key Points

Projects are often broken down into five steps to help planning. You should ask yourself five questions:

1. What is the task?
2. Who is responsible for each activity?
3. How much time will each activity require?
4. What will the cost be for each activity?
5. What are the resource requirements?

Planning your project effectively will help you, as you will need to:

- identify key tasks
- demonstrate an understanding of the project requirements
- divide the key tasks into smaller elements
- organise tasks into a logical order
- plan enough time to complete each task
- include time for any emergencies or delays.

Once you have planned your project, you will need to ensure that it is on track by regularly checking your progress against the plan and recording how your plan has been adapted as you went along.

Make the Grade — P5

Grading criterion P5 requires you to outline the project solution and plan its implementation. Once you have determined the optimal solution to develop, you will need to provide evidence that you can plan effectively over a long period of time and set priorities. It is also important that your plan includes resource information.

Long-term planning

When undertaking a project which might last a considerable period of time, you will need to think about which planning technique you are going to use. There are a variety of specific planners available to you. For example, the wall planners you often see in offices can be used as a starting point; you can use coloured labels for clarity and different coloured pens. These charts are probably too simplistic, however, for the degree of planning required.

Flow charts

A flow chart is a useful tool for developing a programme or process. Different symbols are used to represent different actions, so a simple flow chart might use the following:

- circles – normally used to indicate the start or end of the process
- arrows – used to represent the direction or flow; this is the order decisions or operations follow

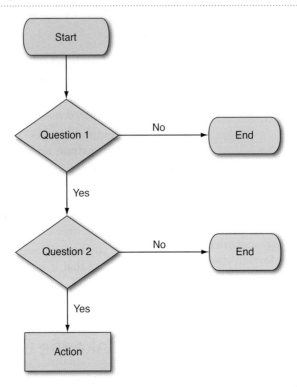

Figure 3.16 Flow chart

- parallelogram – used to indicate inputs/outputs
- rectangle – used to represent processes or actions
- diamond (rhombus) – used to indicate a true/false question.

Flow charts are often used by programmers or to show data flow. However, they can usefully be used to control systems or identify material-handling operations. If your project involves managing a process, this could be a very useful planning tool.

Gantt charts

A Gantt chart is a visual method of planning when each phase of a project starts and finishes. It looks like a bar chart that has been turned sideways, with the time taken measured on the horizontal axis. Although Gantt charts can be drawn by hand or generated using spreadsheet programmes, there are specific project-planning packages which can generate the charts for you.

When undertaking a project that has a series of smaller tasks, it is not always possible to complete each task one after another. The Gantt chart shows you where each task overlaps and when each task has its deadline for completion. The Gantt chart is constructed starting from the top left-hand corner and finishing in the bottom right-hand corner. From the chart, you can identify the critical path and add information such as monitoring dates, often called milestones, project meetings and formal reviews. Table 3.2 shows a plan for the following:

1. Agree specification
2. Design new components
3. Assembly drawing
4. Machine new component
5. Build prototype
6. Test prototype
7. Write report

Task	July									
	19	20	21	22	23	24	25	26	27	28
1. Agree specification	▓									
2. Design new components		▓	▓	▓						
3. Assembly drawing			▓	▓						
4. Machine new component			▓	▓	▓	▓	▓	▓		
5. Build prototype			▓	▓	▓	▓	▓	▓		
6. Test prototype								▓	▓	
7. Write report						▓	▓	▓	▓	▓

Table 3.2 Gantt chart

Critical path

Critical path analysis is a technique used for scheduling project activities and estimates the different routes, or paths, that a project can take. Like the Gantt chart, the key is to break down the project into tasks and use a diagram to map these tasks. The diagram is developed from a system known as Programme Evaluation and Review Technique (PERT). You can work out the earliest and latest times each task can start and finish, without affecting the overall project time. You can also determine which activities are on the longest path, and so critical, and any float or delays that will not impact on the overall project.

Individual tasks are joined by nodes, which are normally numbered in series of ten; this allows for extra nodes to be added in later, if the project changes or other activities need to be added.

In the example shown in Table 3.2 and Figure 3.17, the critical path is indicated by the longest time it takes to get from the start to the finish. You will notice that there is a link between 50 and 60; this is because, although there is no task

represented, the testing of the prototype cannot commence until the new component has been added to the prototype.

Although PERT diagrams can be drawn by hand or using graphics or CAD software, there are also specific software packages which can produce them for you.

Setting priorities

Although it sounds simple, setting priorities is a task that is often overlooked, with the consequence that you can spend a great deal of time focusing on small details and missing some of the key issues. If you are working on a design project that requires 3D CAD models to be created, your first priority might be to ensure that you have access to the appropriate software and can use it to the level of proficiency required. If you are using a Gantt chart or PERT diagram, you might think that you have organised your priorities, but think about the example given in Figure 3.17. You would need to ensure that when it came to testing the prototype, there was space and time in the test laboratory to arrange for the design to be tested and results produced.

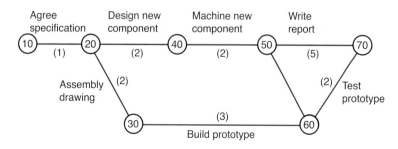

Figure 3.17 Network diagram

Useful resource information

Resource implications are important when considering all elements of your project. As previously discussed, if you require a 3D CAD model, is the software available to you or will it require purchasing? If you need tuition on the use of the software, is there a tutor available to help you or will you need to enrol on a course?

Figure 3.18 Using CAD

> **Activity 3.14**
>
> For the following example, construct a Gantt chart and network diagram for the following process:
> - agree specification – 2 days
> - design new components – 3 days
> - assembly drawing – 4 days
> - machine new component – 4 days
> - build prototype – 3 days
> - test prototype – 1 day
> - write report – 4 days
>
> You should assume the same constraints as in Table 3.2 and Figure 3.17, in terms of which processes can start before others and which processes are dependent upon others.

Monitoring

By the end of this section you should have developed a knowledge and understanding of how you will monitor and record achievement over the lifetime of your project.

To make sure that projects run to time and are completed safely and within budget requires the project manager to ensure that the project plan is feasible and that it is constantly monitored to ensure that if things begin to go wrong, they can be quickly rectified or the plan can be modified.

> **Key Points**
>
> Using a logbook to record everything you do is an invaluable source of evidence when you meet with your project supervisor, colleagues or your employer.

Having planned your project, you hope that everything will follow that plan and you will achieve a successful outcome. However, unforeseen events mean that it is unlikely that everything will go smoothly. How you deal with problems and what modifications you make will have a great effect on your overall project.

> ## Make the Grade   P6
>
> Grading criterion P6 requires you to monitor and record achievement over the life cycle of the project. To demonstrate this, you will need to keep a record of meetings and use your logbook/diary to record key events. If you need to change your plan, you should keep a record of your original, along with the new plan, for comparison and evidence purposes.

To ensure that you manage this properly, careful monitoring of your original plan is required. It is therefore important that you include some key project review dates at regular intervals. If you do have to update or change your project plan, you should keep a record of when and why you did so.

Monitor and record achievement

As discussed in Section 3.1 on record-keeping, it is important to use a logbook or diary to record all your project details and progress. This can also be useful for recording names, addresses, and telephone numbers, meeting dates, email and other correspondence lists.

Use of logbook

A logbook is particularly important if you are conducting a series of tests, for example. You can date and record the results to refer to when you later begin to write up your report, and if you are recording or analysing data, you can record the calculations at the same time. Logbooks are useful for:

- recording data
- analysing data
- keeping performance records
- modifying/updating charts and planners
- recording project goals and milestones
- initial concepts
- project solutions, technical decisions and information.

Key Points

It is likely that your project supervisor or assessors will want to review and sign off your logbook and/or diary on a weekly/regular basis. Consequently, you should have it available for all project sessions.

Make the Grade

Grading criterion P1 requires that when you construct your project plan, you include regular review sessions. At each of these sessions, make records of how your project is proceeding and how your plan will require modifying if necessary.

This evidence should be kept in your logbook or portfolio for future reference.

3.3 Implementing the project plan within agreed procedures

This section will cover the following grading criteria:

If you have carefully planned your project and put milestones in place, when you are prepared to monitor and evaluate your progress, you should be well placed to implement your project.

Implement

By the end of this section you should have developed a knowledge and understanding of what resources you need to implement your project.

Before engineers start work on a project, it is usual for them to gather together and plan what resources will be required to carry out the various activities required by that project.

Key Points

If your project requires practical activities, such as in an engineering workshop, laboratory or in a manufacturing/production environment, you should ensure that you have received appropriate training, are wearing the correct personal protective equipment (PPE) and have completed a risk assessment of the activity.

To ensure that you can complete all the stages of your project, you will need to allocate time carefully. This part of the project process may be where you will work independently and with less guidance and support than at the planning or presentation stages. Consequently, you will have to be disciplined and record everything you do in your logbook. If you come across any problems or issues, you should immediately seek advice from tutors, employers or colleagues, as it is vitally important that you keep to your plan.

Make the Grade — P7

Grading criterion P7 requires you to implement the plan and produce the project solution.
You will be following the plan you have created and monitoring it as you progress. You will need to maintain records and use resources carefully in order to carry out the project. Whether it involves manufacture, design, building, testing or a related engineering activity, there will need to be a specific solution to measure and assess at the end of your project.

Most projects require the use of resources, whether it is test samples being used in a laboratory, components being used to build circuits, or material being used to machine artefacts. In order to ensure that tools, equipment, facilities and materials are available, it is important that you liaise with your project supervisor to ensure that you can carry out the desired activities. Remember that part of your project plan should have involved an investigation of the feasibility and cost of carrying out the project.

Make the Grade — P1

Grading criterion P1 requires that before you commence working on your project, you should already have a significant amount of written material collected. This could include notes, sketches, drawings, plans, mind maps, lists, project specifications, comparison charts, and so on. It is useful to collect this evidence together into a portfolio that could constitute a large file or folder, which should also include your logbook/diary. This could be an e-portfolio, if you have managed to record all your data electronically. You will continue to add to this portfolio, with calculations, observations, photographs and data, during the implementation of your project.

Proper use of resources

Often, project work requires the use of tools and equipment. Many engineering technicians rely on the use of specialist tools and equipment, and take great care to ensure that they are clean, calibrated, well maintained and fit for purpose. However, every year in the workplace there are accidents caused by the incorrect use of work equipment, including machinery. The Health and Safety at Work Act 1974 outlines the responsibilities of employers, managers and supervisors, but safety is everybody's responsibility and you should seek guidance if you are using tools or equipment with which you are unfamiliar or have not had proper training in how to operate safely.

Tools and equipment refer to any equipment used in a workplace, including:

● machines such as photocopiers, drills, lathes, compressors, hydraulic presses and computers
● tools such as soldering irons, screwdrivers, knives, hand saws and torque wrenches
● lifting equipment such as hydraulic jacks, vehicle hoists and lifting slings
● ancillary equipment such as stepladders, test equipment and weight scales

Before using any equipment, you should check to ensure that it is in working order. If it is an electrical device, does it have a Portable Appliance Testing (PAT) notification? If so, is it up to date? You should assess the risks inherent in the use of tools and equipment and consider what can be done to prevent or minimise these risks, including checking if these measures are already in place.

Figure 3.19 CE label

Make sure that machinery is safe. All new machinery should have a CE mark; however, you should not rely on this and should carry out your own safety checks. Make sure you have been given instruction on how to operate any machinery or equipment before you use it.

Controlling risk often means ensuring that dangerous machinery and equipment are guarded. Check all fixed guards before you use any machines, and if guards are removable, check that they are interlocked to prevent accidental operation while the guard is not in place. In addition to the use of guards, ensure that you use jigs, holders and other appropriate work-holding devices. Make sure you know where start/stop and emergency stops are, and never work unsupervised in any kind of workshop or potentially dangerous environment.

PPE ensures that you are always supplied with the correct protective clothing and equipment before undertaking practical activities.

You should ensure that you are aware of fire safety and use of hazardous material when in unfamiliar surroundings.

For more guidance on health and safety, please refer to Unit 1.

> ### Key Points
>
> If you are carrying out any activities in the workshop, laboratory or anywhere you might use PPE, you should record your activities using a digital camera to show you working safely. These pictures can be used later in your presentation or project report.

Time management

Once you have had appropriate training and organised all the necessary tools, equipment, materials and other resources you need, you are ready to carry out your project.

Remember:

● keep daily or weekly records as the project progresses – these records will be an important part of your final report
● take pictures of practical activities
● keep your notebook up to date
● observe all health and safety requirements
● wear appropriate PPE
● document all changes to your project plan and keep before and after records.

Some activities may require careful time management, as you will have agreed timescales you need to work to. It may be that you need to access the laboratory to complete some testing; you may have limited time and will need to plan carefully to ensure that you have sufficient time to complete your activity, record your results and tidy up.

Figure 3.20 Time management

Reviews

Having previously discussed the importance of reviewing your plan, you should arrange with your project supervisor to meet on a regular basis. At these meetings, you can review recent activities, discuss any issues or problems and agree actions. This could, of course, include any amendments to your plan and revised schedules. It is often useful to complete a record of your discussions, which you can sign and date. This is further useful evidence to add to your portfolio or e-portfolio.

Checking solutions

By the end of this section you should have developed a knowledge and understanding of using previously learned and new techniques to check solutions.

Many projects can be broadly separated into two stages. When you are engaged in a design project, the first stage is gathering information and developing your design, and the second phase is manufacturing/building and testing. Alternatively, your project might involve measurement, testing and/or gathering information, then analysing the information and data. Whether or not your project follows this pattern, it is likely that after a period of time you will need to review the progress. In many large-scale industrial projects, engineers will present interim reports. The purpose of these is to indicate what early findings they have made and where the focus of the project will be in the future.

Key Points

The evaluative and analytical techniques you used in Section 3.1 to determine an option for development might be useful when you are comparing different solutions during the implementation of your project.

When you are gathering data or information in a practical setting, it is very important that you record your findings or data carefully, as when you come to analyse it, probably later, you will need to understand what you recorded.

Make the Grade   P8

Grading criterion P8 requires you to check that the solution conforms to the project specification. You should keep referring to the specification when you explore possible solutions, outcomes, designs, and so on.

Use of evaluative and analytical techniques

You will have used a range of planning techniques, such as Gantt charts and critical path methods, during the planning stage of the project. If you have set milestones or review points, it is important that you use these to take an overview of your work. Has it followed the plan? Do you need to make modifications? Will new activities need to be planned? If this is the case, you should review the plan with your tutor and adjust your plan accordingly.

If you have been carrying out tests, it might be appropriate to use graphics techniques to compare the relative performance of different materials or components.

Refer to Section 2.2 of Unit 2 for examples of graphs and charts you could use.

If you are designing a product, circuit or installation, you may be at the stage when you have several potential design solutions. Looking back to when you were deciding on which project to undertake, you might have used the matrix method outlined in Table 3.1 on page 110. Here is another opportunity to use this technique to determine your optimum design.

Statistics are often represented in graphs, but can also be tabulated or analysed to determine trends or key information, such as whether a process is capable of producing expected results reliably (process capability).

If you are designing a pneumatic, hydraulic or electronic/electrical circuit, there are software packages which allow you to simulate the performance of the circuit before you construct it. Similarly, if you have produced a design using CAD that you intend to manufacture, there are software applications which allow simulation of the computer-aided manufacture (CAM) process. Alternatively, you can use rapid prototyping techniques (3D printing) to build a model of your design. Engineers use this kind of software to save time and money, as building real products and circuits and testing them costs significantly more money, as well as being more time-consuming and requiring greater resources.

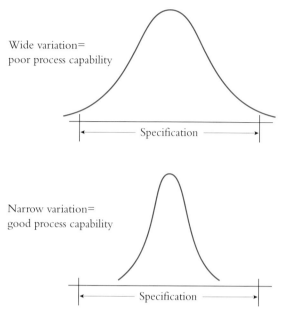

Wide variation= poor process capability

Narrow variation= good process capability

Figure 3.21 Process capability

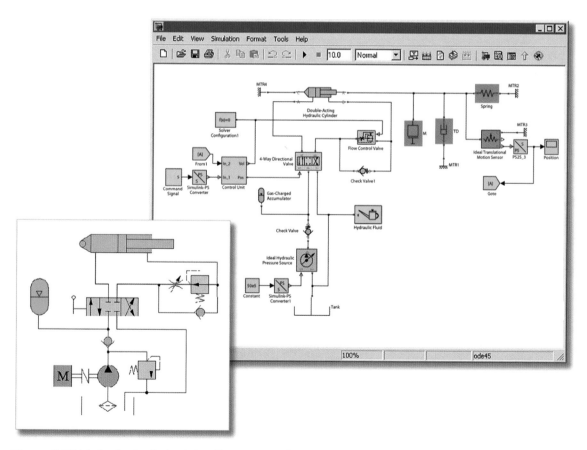

Figure 3.22 Hydraulic circuit simulation (Simttydraulics® from The Mathworks extends Simulink® to enable Multi Domain Modelling of Hydraulic Systems)

3.4 Presenting the project outcome

This section will cover the following grading criteria:

 P9 **P10** **M3** **M4** **D2**

As you come to the end of your project, you will need to consider how you are going to present your findings. You should start working on this, even though you may not have completed building, testing, evaluating, and so on. If you have a look at the Gantt chart on page 113, you will notice that, although this project only took ten days, the final report stage was planned over the last four of these days. If you leave it until the last few days to prepare your report and presentation, you will inevitably not have enough time to produce a piece of work that reflects all your hard work and effort.

> **Key Points**
>
> Presenting the project outcome means two things:
> - delivering a verbal presentation to an audience
> - writing a formal written project report.

Presentation

By the end of this section you should have developed a knowledge and understanding of how to prepare and deliver a presentation to a small group.

Engineers often have to present information to groups of people. These can be familiar groups, such as colleagues, or they can be unfamiliar groups, such as customers.

> **Key Points**
>
> A well-rehearsed and planned presentation is much easier to deliver if you remember that the audience is supportive and may have to do a presentation themselves at a later date.

Giving a presentation is often the most challenging aspect of project work. Oral presentations are increasingly important, as they are often necessary to communicate important information to the team you are working with; increasingly, they are also used in selecting candidates for jobs.

> ## Make the Grade P9
>
> Grading criterion P9 requires you to prepare and deliver a presentation to a small group, outlining the project specification and proposed solution. Remember, the presentation is not about using elaborate tricks with presentation software, nor is it an opportunity to describe everything you did during the project. You have to focus on the project specification you created at the start of your project, and demonstrate to the audience how you met the criteria laid down in that specification with your solution.

Deliver a presentation to a small group

Although standing in front of an audience may seem overwhelming, with a little preparation and practice you can deliver an effective and coherent presentation. There are four key elements to giving a presentation and you should consider each one separately.

1. The presenter

Most people are quite comfortable talking to individuals such as tutors, colleagues, friends or other students. Put yourself in front of a group, with all eyes on you, and things can suddenly change. Effects can include:

- nausea
- feeling faint
- sweaty palms
- tension
- breathing difficulties
- butterflies.

Even the most experienced presenters suffer from these symptoms from time to time. Experience helps, but there are several 'tricks of the trade' which can help calm the nerves and reduce the panic.

Relax

Tension can cause your voice to become constricted and will become evident in your body language. Deep breathing before you speak helps to slow the heart and relax the body. You should breathe in through your nose and out through your mouth, perhaps holding your breath for five to ten seconds. Make sure you breathe from the abdomen – this is what singers are taught to do, and it helps to control your breathing.

Tension builds up the neck and shoulders, and gently exercising this area can relieve tension and stress. To relax the neck and shoulders try the following:

- Rotate the neck slowly round to one side, until it cannot comfortably go any further; hold that position for five seconds, then return to looking forward. Repeat this exercise five times for one side, then five times for the other.
- Keep your head forward and tip your ear towards one shoulder – again, only go as far as you feel comfortable; pause for five seconds, then return to the upright position. Again, five times on one side, then five on the other.
- Facing forward, take a deep breath in; while breathing out, slowly roll your shoulders backwards. Repeat this five times backwards, then five times forwards.

If you feel pain or discomfort, you should not carry on. Always perform each exercise slowly and carefully.

Audience

Although you may feel very nervous, it is not always apparent to the audience. If you feel intimidated, it is often a good idea to imagine them sitting in fancy dress! It is likely that some of your audience will also have to give a presentation, or have done so in the past; consequently, they will be sympathetic and supportive. You should never admit to being nervous.

Speaking

Singers and actors do exercises to help them project their voice and to avoid damaging their vocal chords. Practice is important here and you should practise with some simple exercises. Try to recite a poem or limerick with a friend or colleague; you can position them at varying distances and see how loudly you need to speak to be heard. Mumbling or quiet speaking are signs that you need to practise projecting your voice (without shouting).

Body language

Another good exercise is to practise your presentation with a friend or colleague. You should try to be as natural as possible and do not be afraid of moving as you speak; most importantly, maintain eye contact with your audience. You can try videoing yourself, or your colleagues, and looking for any bad habits.

2. Content

Despite being nervous, the better prepared you are, the easier your presentation will be. Presenters who know their subject and have prepared a coherent and well-structured presentation will give a confident appearance and attitude.

Teachers and presenters are often told to:

- tell the audience what you are going to tell them (introduction)
- tell them (content)
- tell them what you have told them (recap).

There are a variety of key elements to a successful presentation:

Preparation

Having notes can help; however, if you are simply reading these, you will not be engaging the audience or maintaining eye contact, so cue cards may be better. These are small cards with a series of bullet points that act as reminders of the sequence of your presentation.

Figure 3.23 Student presentation

Objective

At the beginning, explain the purpose of your project, what you are going to talk about and why.

Content

Try to keep to three or four key themes/topics and tell the audience what they are at the start. During the presentation, you should indicate when each theme or topic starts and finishes. This breaks down your presentation into manageable chunks.

Language

You should keep to short sentences and avoid using overly complex technical language. Many in your audience may be familiar with engineering terminology, but there may be audience members who are unfamiliar to you and it is best to assume that they are relative novices.

Practice

You should make sure that you have practised giving your presentation with family/friends/ colleagues or that you can record yourself. This will give you an idea of timing and pace, as well as rehearsing your content, so that you become more familiar and comfortable with the content.

3. Audience

Although it is likely that most of your audience will be familiar to you, there may be people with whom you are unfamiliar, such as employers, different students or tutors. There are several things you can do to prepare the audience:

- organise the room beforehand; ensure that all members can see you and any presentation materials you are using; check for adequate heating and ventilation
- introduce yourself at the start and stand facing the audience, maintaining eye contact; try to appear relaxed
- explain to the audience how the presentation will flow; you may want people to keep questions until the end – if so, leave sufficient time at the end.

4. Preparation and use of visual aids

It is likely that you will use some sort of visual aids during your presentation. These could involve using a whiteboard/flip chart, handouts, demonstrations, overhead transparencies (OHTs), models or presentation software.

Using a whiteboard/flip chart

If you have never used a flip chart or whiteboard before, you should ensure that you are familiar with how they work and practise writing notes on them. You can then seat yourself at the back of the room and check that you can read what you have written.

Flip chart tips

- Prepare flip chart pages in advance; you will be able to check spelling and legibility, and your audience will not have to wait while you write or draw on the paper.
- If you want to draw diagrams, graphs or charts, you can create outlines in pencil that will not be visible to the audience, but will allow you to create neat, professional-looking images.
- Remember to have your objectives on the first page and return to this page to check them at the end.
- Leave blank pages between pre-prepared sheets; you might want to add extra notes or diagrams.
- Do not try to write too much on each page; two or three key points, with room for further notes, is ample.
- Stand to one side as you write, to allow the audience to see what you are writing/drawing.

Whiteboard tips

- Practise drawing your diagrams, charts and graphs beforehand, using the pens and whiteboard that will be available for your presentation.
- Check that there is sufficient light, but also that no glare or reflection is on the board.
- Use strong colours; you should avoid orange or yellow, as they are faint. Green is good for positive points, and red is useful for danger or very important points.
- Keep it simple. Do not try to create over-elaborate diagrams, as the audience may become bored and your presentation will lose its momentum.
- If using an electronic whiteboard, practise with all the tools and applications carefully. It is worth having a back-up in case the technology fails for any reason – handouts or a flip chart, for example.

Overhead projector (OHP)

You can create OHTs to illustrate points, display diagrams or write notes. When you place acetate sheets on to the plate of an OHP and switch it on, a light bulb illuminates the OHT and a lens projects the image onto a wall or screen. Before the use of presentation software and data projectors, OHTs were the most common form of supporting presentations. Laser transparencies can be used to allow information to be photocopied or printed onto the film. If you are planning to use OHTs, you should practise with the film and projectors, as care is needed to ensure that slides are visible and in focus. Many of the techniques mentioned in the section on the use of presentation software apply to the use of OHTs as well.

Demonstrations/models

If you plan to include a demonstration as part of your project, you will have to practise and rehearse carefully. Demonstrations, however, are very powerful, showing what you have developed in achieving your objectives, to show how a circuit operates, how a mechanism operates or how a tool is used, for example. You may wish to consider a video clip if you are concerned that a demonstration might not go to plan or if your project is based in the workplace or is not portable. Video clips with accompanying audio can be embedded into presentation software or delivered via a DVD or digital camera/camcorder, provided that a data projector or large enough screen is available, with speakers, of course.

Displaying a model or passing it round the audience is another useful technique which allows the audience, literally, to get a feel for what you have been doing on your project. Do be careful to ensure that there are no health and safety issues, such as sharp edges or trailing leads, if you do use models. Remember that any delicate or sensitive parts risk being damaged if mishandled.

Presentation software

Presentation software is now widely used to help support presentations and introduce key points, illustrations and diagrams. In addition, video and audio clips can be embedded, as can links to other software, such as graphics and simulation software. The principle behind the software is the preparation of slides, which can be activated in sequence and have largely replaced photographic slides used with projectors or OHTs used with OHPs. A data projector and PC loaded with appropriate software are required, however.

Figure 3.24 Presentation slide

Here are some key points when preparing presentation slides:

1. Think of your slides as a sequence. Do not try to get too much information on each slide.
2. Use landscape, rather than portrait format; keep wide margins, as it is difficult for the audience to follow a long line of text.
3. Use a large text font for legibility, say 24 point; this will also prevent you from trying to put too much on each slide.
4. Use a sans serif font, such as Verdana or Arial.
5. Do not use capital letters throughout, making it easier for the audience to read, as the shape of the word aids comprehension.
6. Three or four points are generally sufficient; remember that the audience cannot comfortably read a lot of text and listen to you at the same time.
7. Keep to key points. The slides should help you and the audience focus on issues you are discussing, but you are giving a verbal, not written, presentation.
8. Do not rely on the slide show. You are the person giving the presentation, and remember, the audience may have seen several presentations already.
9. Use bullet points and paragraph spacing to emphasise your points.
10. Use graphic images where appropriate; use diagrams and charts to emphasise points, but not clip art.
11. Do not introduce sophisticated transitions between your slides; these can distract the audience. It is better to 'wipe' from left to right, allowing words to be revealed without the text itself moving on the screen.
12. Use a pastel colour for the background; black text on a white background is hard on the eyes.
13. Ensure that the version of the software and operating system of the computer are compatible with your presentation. If the file you have worked on fails to load properly or the data projector fails to display, can you still give your presentation? Have a hard copy of the slides available, just in case.

Make the Grade

It is useful to video your presentation, as this will allow assessors, examiners and external verifiers to assess your performance.

Project report

By the end of this section you should have developed a knowledge and understanding of how to construct your project report.

When a project has been completed, it is customary to present a report of your findings, to allow whoever has supported you in the work, whether it is employers or other stakeholders, to judge the project solution reached.

> ### Key Points
> A project report is a formal document that follows rules and conventions in its style and layout.

The final report you submit is a substantial report which should document your project. It is a formal document and this section gives you guidelines on preparing and presenting the report.

Make the Grade   P10

Grading criterion P10 requires you to present a written project report. Although there are standards and conventions you should follow in the presentation style of your report, you should not lose sight of the reason for your final report. Just as with your presentation, you should focus on the project specification, how it was or was not met, and what conclusions you can draw from this.

Logbook/diary

Although the project report is the piece of work you will be assessed on, you might want to submit your logbook/diary as supplementary evidence. Provided that you have kept a record as you went along, your logbook/diary will be an invaluable resource as you prepare your report.

Written technical report

A technical report is a formal report that presents information in an understandable and logical order. There are conventions on how you construct your report, the sections you should use and the information that the report should contain.

Structure

Title page

This should include the title of your project, your name and the date the project was submitted.

Acknowledgements

If you have had particular help or support, this is your opportunity to acknowledge key people, whether it be colleagues at work, other members of your group or friends. The list should be small, however, and should only be included if you have had exceptional help or support.

Summary

This section is one that confuses many people; it is not an introduction to what you have done; rather, it is a brief overview of what you learned in your project. If somebody wanted a summary of everything you achieved in one paragraph, this is where they would look.

So there should be three things in your summary:

- why you did the report
- what you found out
- what conclusions/recommendations you would make.

All of this should be in less than 200 words, if possible!

Contents

You should record all sections and subsections with the appropriate page number against each.

Introduction

Here, you will state the objectives of your project and explain the background to why you are undertaking this piece of work, what you hope to achieve and how you plan to do it.

Major sections

You should divide your report into sections, depending on how your project is constructed. It is often useful to begin with how you developed your project ideas, the process that was required to select a suitable project and how you evaluated the different options. This should demonstrate how you arrived at the project specification. You might follow this with a discussion of your planning process and your project plan. There should be a section which refers to the process undertaken to complete your project, the recording of information, and how you monitored your progress and recorded your achievements.

Conclusions

This section is your opportunity to discuss what you learned from your project. You should compare what your objectives were with what you achieved, and compare your results, findings and experiences with published data or perceived wisdom.

Recommendations

Having stated your conclusions, you should consider what recommendations you would make. This could be further work to be carried out, whether it is more research, an improved design, further testing or a change in process. You can also include any issues with the process. You might recommend using different tools and equipment, improving the method of recording data or carrying out test processes. Recommendations are one of the most important elements of project work and should not be overlooked.

References

If you have quoted directly from books, websites or journals, you should list them here, using an acknowledged system such as Harvard referencing.

> **Key Points**
>
> There are a variety of different referencing systems, such as Harvard, APA and numeric. You should check with your project supervisor which is their preferred system and seek guidance on its use.

Bibliography

Any books, websites, journals and tutor notes that you might have used for background reading should be included here and listed in a similar format to your references.

Appendices

This section is for additional material that you have discussed or mentioned in your report, but which would be too bulky, or would disturb the flow, if included in the main body of your work. Examples of the information might include technical drawings you have produced or used, printouts from test equipment or raw data, specifications and large-scale diagrams.

> **Activity 3.15**
> Take time to go to your nearest library/learning resource centre and see if they have handouts or notes on how to use Harvard, numeric or APA referencing systems.

> **Key Term**
>
> **Plagiarism** – This is the use of other people's work without proper referencing. You should never 'copy and paste' information from a report, website or book and suggest that it is your work. Any content which is a direct copy should be placed in quotation marks, with a reference to the source. If you have used a reference source as background reading, it is acceptable to mention the source in a bibliography.

Use of information and communication technology (ICT)

Technical reports should be presented on single-sided A4 paper; handwritten work is not acceptable, so you will need to use word-processing software. This has the advantage of having spelling and grammar-checking facilities, and you can include digital images, spreadsheets, CAD drawings and other graphics to enhance your work and reinforce any points you are trying to make.

You should keep diagrams simple and create them specifically for the report. Remember to reference and label diagrams; the second and third diagrams in the first section might be titled:

- Figure 1.2 Test pieces under load
- Figure 1.3 Test pieces after fracture

You would refer to diagrams in the text in the form, 'see Figure 3.2', for example. Remember that diagrams need to be inserted in the document as close to the relevant part of the report as possible. If you are referring to data or documents, however, it is appropriate to include a note, 'see Appendix 1', for example. If you have tables of information, you can use the tools available in spreadsheet software to produce the information as a graph or chart, as this is often easier to follow for the reader.

Proofreading

Once you have completed the report, you will need to consider this as a first draft and carefully check through the work to ensure that everything is correct and that what you have written makes sense to the reader. The process of checking written documents is called proofreading and it

125

is useful if you can persuade a second person, preferably somebody who is not familiar with your project, to read through your work and help you to rewrite any confusing sections and correct any mistakes, to help produce a report in an understandable form.

Make the Grade — M3

Grading criterion M3 requires you to evaluate the project solution and suggest improvements. You should therefore include a discussion in your final report that critically assesses how well your project solution met the project specification. When suggesting improvements, the recommendations section of your report should be used to explain these.

Make the Grade — D2

Grading criterion D2 requires you to evaluate the whole project development process, making recommendations for improvements. Although the evidence may come from your logbook/diary, it would be useful to include, in the recommendations section of your report, suggestions as to how you could have managed the initial development of your project. This might be to do with time management, teamwork, lack of planning or other issues.

Make the Grade — M4

Grading criterion M4 requires you to present coherent and well-structured development records and final project report.

To achieve a pass grade you will have:

P1 prepared and maintained project records, from initial concepts through to solutions, that take account of and record changing situations

P2 prepared a project specification

P3 agreed and prepared the procedures that will be followed when implementing the project

P4 used appropriate techniques to evaluate three potential solutions and selected the best option for development

P5 outlined the project solution and planned its implementation

P6 monitored and recorded achievement over the life cycle of the project

P7 implemented the plan and produced the project solution

P8 checked that the solution conformed to the project specification

P9 prepared and delivered a presentation to a small group, outlining the project specification and proposed solution

P10 presented a written project report.

To achieve a merit grade you will also have:

M1 maintained detailed, concurrent records throughout the project that clearly show progress made and difficulties experienced

M2 used a wide range of techniques and selection criteria to justify the chosen option

M3 evaluated the project solution and suggested improvements

M4 presented coherent and well-structured development records and final project report.

To achieve a distinction grade you will also have:

D1 independently managed the project development process, seeking support and guidance where necessary

D2 evaluated the whole project development process, making recommendations for improvements.

By the end of this unit you should be able to:

- use algebraic methods
- use trigonometric methods and standard formulae to determine areas
- use statistical methods to display data
- use elementary calculus techniques.

In order to pass this unit, the evidence you present for assessment needs to demonstrate that you can meet all of the above learning outcomes for this unit. The criteria below show the levels of achievement required to pass this unit.

To achieve a pass grade you need to:	To achieve a merit grade you also need to:	To achieve a distinction grade you also need to:
P1 manipulate and simplify three algebraic expressions using the laws of indices and two using the laws of logarithms	**M1** solve a pair of simultaneous linear equations in two unknowns	**D1** apply graphical methods to the solution of two engineering problems involving exponential growth and decay, analysing the solutions using calculus
P2 solve a linear equation by plotting a straight-line graph using experimental data and use it to deduce the gradient, intercept and equation of the line	**M2** solve one quadratic equation by factorisation and one by the formula method.	**D2** apply the rules for definite integration to two engineering problems that involve summation.
P3 factorise by extraction and grouping of a common factor from expressions with two, three and four terms respectively		
P4 solve circular and triangular measurement problems involving the use of radian, sine, cosine and tangent functions		
P5 sketch each of the three trigonometric functions over a complete cycle		
P6 produce answers to two practical engineering problems involving the sine and cosine rule		
P7 use standard formulae to find surface areas and volumes of regular solids for three different examples respectively		
P8 collect data and produce statistical diagrams, histograms and frequency curves		

To achieve a pass grade you need to:	To achieve a merit grade you also need to:	To achieve a distinction grade you need also to:
P9 determine the mean, median and mode for two statistical problems and explain the relevance of each average as a measure of central tendency		
P10 apply the basic rules of calculus arithmetic to solve three different types of function by differentiation and two different types of function by integration.		

Introduction

One of the main responsibilities of engineers is to solve problems quickly and effectively. This unit will enable learners to solve mathematical, scientific and associated engineering problems at technician level. It will also act as a basis for progression to study other units, both within the BTEC National Engineering qualification, such as *Unit 28: Further Mathematics for Technicians*, and at BTEC Higher National level, or as part of a degree qualification.

This unit enables learners to build on knowledge gained at GCSE or BTEC First Diploma level and use it in a more practical context for their chosen discipline. The first learning outcome will develop learners' knowledge and understanding of algebraic methods, from a look at the use of indices in engineering to the use of the algebraic formula for solving quadratic equations. Learning outcome 2 involves the introduction of the radian as another method of angle measurement, the shape of the trigonometric ratios and the use of standard formulae to solve surface areas and volumes of regular solids. Learning outcome 3 requires learners to be able to represent statistical data in a variety of ways and calculate the mean, median and mode. Finally, learning outcome 4 is intended as a basic introduction to the arithmetic of elementary calculus.

This section of the book is organised as follows; each of the subsections can be readily linked to the learning outcome (LO), pass (P), merit (M) and distinction (D) criteria.

Section/content	LO	P	M	D
4.1 Using algebraic methods	1	1, 2, 3	1, 2	1
4.2 Using trigonometric methods and standard formulae to determine areas and volumes	2	4, 5, 6, 7		
4.3 Using statistical methods to display data	3	8, 9		
4.4 Using elementary calculus	4	5, 10		2

4.1 Using algebraic methods

This section will cover the following grading criteria:

P1 P2 P3 **M1** M2

D1

Indices and logarithms

By the end of this section you will have developed a knowledge and understanding of the laws of indices and logarithms, and how to use them to solve engineering problems.

First of all, what do we mean by a law? Generally, we think of a law as a code of conduct by which members of a society or group are expected to carry out their activities. Break the law, and you get punished.

In maths, science and engineering, we have many laws to work to. In a way, they are similar to the laws of society, because if you do not follow the law, or you try to do things in an unacceptable way, you end up with the wrong answer, then you fail and probably have to do the work again.

Figure 4.1

Introduction to indices

In this section you will learn about the following laws of indices:

$$a^m \times a^n = a^{m+n}$$
$$\frac{a^m}{a^n} = a^{m-n}$$
$$(a^m)^n = a^{mn}$$

You will understand them and be able to use them to solve engineering problems.

The language of maths

Learning any new language is strange at first, but just like learning a foreign language, learning the language of maths comes with practice and understanding.

Exercise 4.1

Remember the laws.

In the table below, some common mathematical rules have been completed for you. See if you know these, and think of any others to complete the table (for example, any rules you remember from school or any other previous learning you have done).

The law or rule	Explanation
BODMAS	Brackets, Order, Division, Multiplication, Addition, Subtraction – for the correct order for doing sums
SOHCAHTOA	sine = opposite over hypotenuse cosine = adjacent over hypotenuse tangent = opposite over adjacent These are the relationships between the sides and angles of a right-angled triangle.
Pythagoras' theorem	$a^2 = b^2 + c^2$ the relationship between the sides of a right-angled triangle

Symbol shock

This is a natural feeling when we see a new equation or set of symbols for the first time and we feel overwhelmed. After some practice and a few examples things seem less confusing.

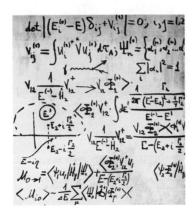

Figure 4.2

We already know that an index is the part of a reference book or textbook where all the main topics or words are listed, with the page number alongside each entry allowing easy location of the topic.

In maths, an index means something completely different to what we normally think of when we hear the word 'index'. In maths, the word 'index' means more or less the same as 'power', or more specifically, 'raised to the power'.

For example, 3^2 can be read as 'three squared' or 'three to the power of two', but it could also be read as 'three index two'.

The number written in a smaller typeface and raised up to the top (called a superscript) is also called the 'exponent', so '3 exponent 2' is another way of reading 3^2.

Finding your way around your scientific calculator

Have a look at your calculator and you should see a key labelled EE, EXP, or E. This is used to enter the exponents or powers of 10 – for example, 253 can be entered as 2.53×10^2. The keys you need to press are

$$2.53 \text{ EE } 2$$

or

$$2.53 \text{ EXP } 2$$

depending on the type of calculator you are using.

If you enter a multiplication sign and 10 before EE, you get 10 times the required answer.

Figure 4.3

This is because the way numbers are entered into the calculator is slightly different to the way we normally read or say the number.

For example 2.53×10^2 would normally be read as: 'two point five three multiplied by ten to the power of two'.

Pressing the 'EE' key to represent 'to the power' causes all sorts of problems.

What your calculator sees:

$$2.53 \times 10 \ (\times 10 \text{ to the power of) } 2.$$

Try this on your calculator.

$$[2.53 \times 10 \text{ EE } 2 =]$$

You probably got 2,530, which is not what you wanted.

The EE button of your calculator should be pressed to mean 'multiplied by ten to the power of'.

Using the computer

Microsoft® Word has a tile or button which can be used. It is the X^2 key (found in the Format menu).
Type the number 3, then select the X^2 tile.
Type the number 4, and you should see 3^4.
You may need to press the X^2 tile again, to stop it making everything into a superscript as you continue typing. This is useful for typing Pythagoras ($A^2 + B^2 = C^2$), which looks much more professional (and correct) than A2 + B2 = C2 when you are typing up assignments or notes.

More computer tips

Alternatively, there is a keyboard short cut. Select the characters you wish to make into super-scripts, and hold the shift key and 'Ctrl' together, then press the = key (to the left of the backspace key.

Repeating this toggles the superscript off again. When you repeat this without holding the shift key (that is, just the 'Ctrl' and = keys), you get a subscript – useful for electronics $R_A + R_B$ or chemistry H_2O, and so on.

The index or exponent is the power to which the number is raised, and in this case, all the following mean the same:

- 3^2
- 3 squared
- 3 to the power 2
- 3 index 2
- 3 exponent 2
- 3×3
- two threes multiplied together.

If multiplication was invented to save us having to add up many similar or identical numbers, then the use of indices can be seen as having been invented to save us writing out long strings of numbers multiplied together, and adding the powers.

Exercise 4.2

This is an exercise to practise rearranging indices and numbers. Look at the following strings of numbers. Copy and complete the indices column. The first one has been done for you.

String of numbers	Simplified using indices (or index notation)
$2 \times 2 \times 2$	2^3 Effectively, there are three twos multiplied together
$3 \times 3 \times 3 \times 3$	In this case, there are four threes multiplied together
$4 \times 4 \times 4 \times 4 \times 4 \times 4$	
$2 \times 2 \times 2 \times 3 \times 3 \times 3 \times 3$ (think carefully about this one)	Note that you cannot abbreviate different numbers in a shorthand form.
(now carry on) $3 \times 3 \times 4 \times 4 \times 2 \times 3 \times 4 \times 2 \times 2 \times 4 \times 3$	The numbers now need sorting into value order before you start; otherwise, it is exactly the same process.
$7 \times 7 \times 4 \times 5 \times 6 \times 4 \times 5 \times 3 \times 9 \times 8 \times 9 \times 3 \times 2 \times 9$	
Can you see any relationship between the indices and the frequency with which each number occurs? Write the relationship in your own words in the box to the right.	
Can you see any advantage in using the index notation instead of the string of numbers in the left-hand column? Write your comments in the box on the right.	

Exercise 4.3

Remember the basic laws of addition, subtraction, multiplication and division? Copy and complete the following table.

String of numbers	Simplified using indices (or index notation)
$2 \times 2 \times 3 \times 3 \times 4 \times 4 + 2 \times 2 + 3 \times 3 \times 3 + 4 \times 4$	$2^2 \times 3^2 \times 4^2 + 2^2 + 3^3 + 4^2$
$3 \times 3 \times 3 - 2 \times 2 \times 2 \times 2 - 4 \times 4 \times 4 \times 4 - 5 \times 2$	
$\dfrac{6 \times 3 \times 6 \times 6 \times 3}{2 \times 3 \times 2 \times 6 \times 3 \times 6}$	

Before we start talking about laws of indices, have a look at your answers in the right-hand column.

Can you see any ways to simplify what you have written down?

Worked Example

$$\frac{6 \times 3 \times 6 \times 6 \times 3}{2 \times 3 \times 2 \times 6 \times 3 \times 6}$$

becomes

can also be written as

$$6^1$$

$$\frac{6 \times 3 \times 6 \times 6 \times 3}{2 \times 3 \times 2 \times 6 \times 3 \times 6}$$

Look at the indices and describe what has happened.

On the top and bottom, there are some common numbers, such as $6 \times 6 \times 6$ on the top, and 6×6 on the bottom.

It looks like you have done $3 - 2 = 1$, which is exactly what has happened.

$$\frac{6 \times 6 \times 6}{6 \times 6}$$

This simplifies by cancelling to 6×1 or just 6.

This is one of the advantages of using indices and logarithms. They allow multiplication and division to become addition and subtraction.

Written in terms of indices:

$$\frac{6^3}{6^2}$$

Exercise 4.4

Simplify the following:

1 $2^3 \times 2^4 \times 2^5 \times 2^7$

2 $\dfrac{2^7 \times 2^8 \times 2^5 \times 2^6}{2^6 \times 2^4 \times 2^2 \times 2^3}$

3 $\dfrac{x^3 + x^6 + x^2}{x^2 + x^4 + x^1}$

When you see a negative index, what does it mean? For example, x^{-2}.

Put simply, the negative power inverts the expression, or the part which it is in.

For example, $x^{-2} = \dfrac{1}{x^2}$, so a negative index means 'one over'.

Make the Grade P1

Grading criterion P1 requires that you manipulate and simplify three algebraic expressions using the laws of indices and two using the laws of logarithms.

Activity 4.1

Class activity

Now that you are becoming proficient at working with indices, try the following:

1. All the class splits into pairs.
2. Each of you write a problem, similar to those given above.
3. Swap your problem with your partner, and the first one to solve the problem moves into the next round.
4. In the next round (and all subsequent rounds) the winner goes into the next round.
5. In the final, the two remaining members of the class carry out the same activity, and the faster of the two wins.

What is the prize for the winner? He or she chooses the pairing of students for the next activity.

Exercise 4.5

Use your calculator to solve the following:

1. 3^{-2}
2. $\frac{1}{3^2}$
3. 6^{-3}
4. $\frac{1}{6^3}$

Spend a few moments trying a few others by writing your own numbers in place of the 2, 3, and 6. You will see that a negative index means 'one over' or 'the inverse of'.

Developing the laws of indices

Activity 4.2

In small groups or pairs research the laws of indices and produce a handout or A3 poster (or flip chart) and a problem sheet to give to the rest of the class as part of a short talk.

After about 20 minutes, present what you now understand about the laws of indices to the rest of the class. This can be a poster, or just writing the details on the board for all to see.

Compare what you have all done with the next few pages, and correct any errors as you go along.

Laws of indices

Key Points

These are usually quoted, or listed, as:

$$a^m \times a^n = a^{m+n}$$
$$\frac{a^m}{a^n} = a^{m-n}$$
$$(a^m)^n = a^{mn}$$

Laws are generally easier to understand if they are written in words.

Exercise 4.6

The following expressions have question marks in place of some numbers where indices should be. Rewrite the expressions and insert the correct numbers in place of the question marks.

1. $3^6 \times 3^? = 3^9$
2. $6^4 \times 6^? = 6^7$
3. $24^6 \times 24^8 = 24^?$
4. $y^5 \times y^7 = y^?$
5. $(p^2)^? = p^6$
6. $(4^3)^? \times 4^8 = 4^{17}$
7. $\frac{5^4}{5^2} = 5^?$
8. $\frac{6^3}{6^5} = 6^?$

Activity 4.3

From the research you carried out into the laws of indices for Activity 4.2, and from listening to the explanations given by others, copy and complete the following table, writing a summary definition and explanation alongside each of the following laws.

Laws of indices	Write your own explanation of this in words	Give an example or two
$a^m \times a^n = a^{m+n}$		
$\frac{am}{an} = am^{-}n$		
$(a^m)^n = a^{mn}$		

Laws of logarithms

Logarithms were invented by John Napier in 1614. According to the Oxford English Dictionary, the word logarithm is derived from the Greek 'logos' = reckoning and 'arithmos' = number.

What is a log?

Again, it is a simple everyday word. It has several meanings already, but in maths it means something totally different again.

Key Term

The logarithm (or log) of a number, to any base – The power to which that base has to be risen to equal the number.

Base – The number b in the expression of the form b^n. For example, in the expression 2^3, the number 2 is the base.

Remember the indices?

$$2^3 = 8$$

$$2 \text{ index } 3 = 8$$

Alternatively, we could say this differently, by referring to the definition of a logarithm.

The logarithm of a number (8), to any base (2), is the power (3) to which that base (2) has to be risen to equal the number (8).

This would be written as $\log_2 8 = 3$, which says 'log to the base two of eight is three' or 'log to base 2 of 8 equals 3'.

Take another look at the definition above, then try another example.

Worked Example

Take the number 6 to the base 10 (because that is the base of our normal or common number system).

Raised to the power? We did that earlier (indices) and this is the link between logs and indices, hence, log of 6 to the base 10 is written as

$$\log_{10} 6$$

which also reads as the log, to base 10, of 6.

The $\log_{10} 6$ is 0.778151250383644.

You may be thinking, where did that come from? It came out of a look-up table, which can be found in all scientific calculators, as well as the calculator on a computer, and it is a function in Microsoft Excel®). You could also find it in log tables.

Look back at the definition. The logarithm of 6 to the base 10 is 0.778151250383644. The definition says that **the log of any number (6) to any base (10) is the power to which the base (10) has to be risen to equal 6.**

Hence $10^{0.778151250383644} = 6$

Try it on your calculator. Depending on the number of digits you can enter or have displayed on your calculator, you should get an answer somewhere around 6.

You may have noticed a problem, here. Where did the number 0.778151250383644 come from? The numbers were all worked out and put into the tables. They were then put into calculators when the latter were invented.

To help you remember the pattern and the process, think of some simple or more manageable numbers.

$\log_{10} 100 = ?$

Look it up in your calculator, and you will get 2.

Think back, again, to the definition of the logarithm, which says that **the log of any number (100) to any base (10) is the power to which that base (10) has to be risen to equal 100.**

We all know that 10 has to be risen to the power ^{2} to give an answer of 100.

Exercise 4.7

1. What is the log to the base 10 of 1 000?

2. What is the log to the base 10 of 100 000?

3. What is the log to the base 10 of 10 000 000 000 000?

Multiples of 10 are easy, but what about other numbers, such as 4.7? Look up 4.7 in your log tables, or enter 4.7 log in your calculator, or enter log 4.7 if it prefers it that way round. (Different calculators require different sequences of operation. Get to know your calculator.)

Working out these logs took a long time, but the benefits were well worth the time. As people had started trading products, such as spices, cloth, grain, food, fish, slaves and weapons, in very large amounts, they were having to multiply some extremely large numbers together, and multiplication was a difficult skill to gain. Addition was easier, so the conversion was welcomed.

Figure 4.4 Logs helped when trading products such as spices in large quantities

A logarithm is made up of a **mantissa** and a **characteristic**.

For example, if 2.7654 is the logarithm of a number, the characteristic is 2 and the mantissa is .7654.

Key Terms

Mantissa – The decimal part of a logarithm (for example, for 2.7654, the mantissa is .7654).

Characteristic – The digit to the left of the decimal point (for example, 2 in the number 2.7654).

If calculators had been invented hundreds of years ago, logarithms may never have been discovered. The mathematicians who developed the logarithmic tables helped immensely, as the following example indicates.

Worked Example

432 × 57 295

Convert to logs, then add the logs, then 'un-log' the answer.

'Un-logging' is actually called antilogging, and that is what the anti-log button is for.

Only work to four figures; this is accurate enough for now.

Number	Logarithm	
432	2.6355	
57,295	4.7581	antilog
Sum →	7.3936	24 751 413.18

To find the antilog on some calculators, you may have to press 'inv' or 'shift', then 'log', or you may have 10^x as a second function on the log key.

Having the key called '10^x' is very helpful, because that is the antilog of 'x', to base 10.

Check the multiplication on your calculator.

$$432 \times 57\,295 = 24\,751\,440$$

The 0.18 is due to rounding the logs back to four figures. The more digits we use in logs, the more accurate the answer, in the same way as rounding values at each step as you work through any problem reduces accuracy.

137

Pressing the correct keys should result in the answer being shown in the bottom right-hand box of the table.

10^x ? We have met that before in indices, and it is part of the definition of a logarithm, so there is no escaping the link between logarithms and indices.

Get to know your calculator. Experiment with the keys. Enter the number 2, then antilog (or 10^x) and you should get 100.

Exercise 4.8

Copy and complete the following statement by using the words and numbers given beneath the statement.

Since ____ is the _____ of 100, it follows that _____ is the _____ of 2.

- 2
- 2
- 100
- logarithm
- antilogarithm

Exercise 4.9

1. Take the logs of the following numbers to four decimal places:

Number	Log to base 10
2.3	
23	
2 300	

2. Complete the following:

(a) The characteristic is _____

(b) The mantissa is _____

Can you see another link between the characteristic and indices? (Hint – think of the link between the powers of 10 and the characteristics.)

3. Copy and complete the following:

Number	Log to base 10 (to 4 d.p.)
2 300 000	
23 000 000 000	
2 300 000 000 000 000 000 000	

4. After doing the above exercise, write down what you have observed about the characteristic and mantissa of the logs.

Your summary should have said something about the mantissa being four or more numbers that follow a decimal point, representing the logarithm of 2.3.

The characteristic tells us how many times the original number 2.3 is multiplied by 10.

Hence, the log of 2.3 is 0.3612 (and the zero at the beginning tells us that it is multiplied by *no* tens or by 10^0).

For 23, the mantissa is the same, but to get from 2.3 to 23, we multiply by one ten or 10^1.

For 2 300 000, we multiply 2.3 by ten, six times, or by 10^6.

Exercise 4.10

Here is a simple extension to the work covered so far.

1. What are the logs to base 10 of 0.23 and 0.0023? Try it on your calculator.

2. What can you say about the mantissa?

3. What can you say about the characteristics?

4. Can you explain this, thinking about indices and powers of 10?

   Hint: Think about how many times you have to multiply 2.3 by 10 to get to 0.23 or to get to 0.0023. It may help to think about positive and negative indices.

5. Complete the following to summarise what you have found out. Use the answers supplied below.

When the _____ is a decimal fraction _____ than 1, the _____ has the same _____, but the _____ becomes _____.

**characteristic   number   negative
logarithm   less   mantissa**

Now that you have learned what mantissa means, or used to mean when we worked with log tables, it is important to note that it also means something else. You will have come across this already when working in standard form.

Key Term

Standard form – Where every number is written in a format which has a number that is greater than 0 but less than 10, multiplied by 10 to a power.

For example, 245 written in standard form is 2.45×10^2.
10^2 is called the exponent of the number.
2.45 is called the mantissa.

The numbers for this example were simple, but with numbers like 432 and 2756, the multiplication would be more complicated. The alternative would have been to write out 2756 four hundred and thirty-two times and then add them up.

Making use of the laws of logs

The 'laws' of logs are used to find a solution without working out all the maths, apart from simple addition and subtraction, with the odd multiplication.

Worked Example

Let us look more closely at how logs could be used when they were devised.

We have 400 boxes of ammunition, each containing 2000 rounds. Using logs to base 10, calculate the total number of rounds.

Number	Log to base 10	
400	2.602059991	
2000	3.301029996	**Antilog**
Sum →	5.903089987	= 799 999.9994 or approx. 800 000, depending on the number of decimal places used

Exercise 4.11

$logA + logB = logAB$

Write this in words.

You should have written something like, the log of two numbers multiplied together gives the same value as taking the logs of the individual numbers and adding them together.

There is one mistake to avoid:

$$\frac{logA}{A}$$

From your basic memory of algebra, it appears that there is an A on the top and bottom, and they should cancel, leaving 'log' as the answer.

This is wrong.

logA actually says 'the logarithm to the base 10 of A', and cancelling top and bottom of an equation can only be done when multiplication or division are implied.

Logs involving division

Exercise 4.12

Again, by investigation, work out the following solutions to verify another rule or law of logs:

1. $logA - logB = log(\frac{A}{B})$
2. $log33\ 138 - log42 = log(\frac{33\ 138}{42})$

Both sides should give the same answer. Work them out, then discuss your answers with someone else. Did you identify this answer? Do you get the same? Since you have worked with the logs of 33 138 and 42, the answer must be the log of the answer $\frac{33\ 138}{42}$. What did you get?

If you need more practice, try using logs, and subtract the log of 42 from the log of 33 138, then take the antilog of your result to get (approximately) 789.

Now for a slightly more complex law of logs. You will remember that earlier, we wrote $2 \times 2 \times 2 \times 2$, and that it could be written as 2^4.

Key Point

Using logs changes multiplication into addition, and division into subtraction.

Exercise 4.13

Look at the following expression. Write down exactly what this means to you and give an example in simple numbers, such as 2^4, where $A = 2$ and $n = 4$.

$logA^n = nlogA$

Key Points

Logs reduce multiplication to addition and they reduce powers to multiplication.

Exercise 4.14

Can we then reduce that multiplication to addition? For example, try to show that $2^3 = 8$ by using logs and addition.

Common logs

So far we have only considered logs to the base 10, or **common logarithms**, as they are sometimes known. You may have heard about **natural logs**, which are also called logs to the base e, where e is a natural number.

Key Terms

Common logarithm – A log to the base 10.

Natural log – A log to the base e, where e is a natural number.

It might seem strange that a letter can represent a number, but e is used to represent what is known as a natural number (more on this later). Another natural number is the ratio of a circle's circumference (C) divided by its diameter (D), and we use the Greek letter π to represent this ratio.

Remember $C = \pi D$, from $\pi = \frac{C}{D}$.

As mathematicians will also tell you, e and 8 are also **transcendental numbers**, which means their value is not subject to a limited value, being an approximation which gets closer and closer to the value, but never actually gets there (more on this later).

Key Term

Transcendental number – A number whose value is not subject to a limited value, but is an approximation that gets closer and closer to the value, never actually reaching it.

Natural logarithms (base e)

The value which is normally used for *e* depends on where we are using it and why, and the level of accuracy we require, in the same way as pi (ϖ) can be $\frac{22}{7}$ or 3.142 or 3.14159, and so on.

As mentioned earlier, logarithms were invented by John Napier in 1614. They were also called **Naperian logs**, because he did not work to the common base of 10, but to e.

The interesting thing about the value *e* is that if you plot a graph of *x* against values of e^x, you obtain a curve, and at any point on that curve, the gradient is equal to the corresponding value of *x*.

The curve should start to look a bit like the sketch shown in Figure 4.5. Draw a tangent at any point on the curve and work out the gradient at that point. It should equal the *x* value at that point.

For example, if you draw a tangent at *x* = 0.5, the gradient you get should be about 0.5. The gradient is the 'rise' (*y* value) divided by the corresponding 'run' (*x* value) (see pages 148–9 for more information on this).

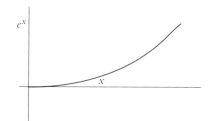

Figure 4.5 Graph

How to find the value of e on your calculator

For a chance to apply your new knowledge about logs and antilogs, using your calculator, what is the log of 10? Hopefully, your calculator should provide a value of 1.

Think of this in terms of the definition of logs: the log to base 10 of the number 10 is the power to which 10 has to be risen to equal 10.

In other words, how many 10s do you need to multiply together to make 10? Answer = 1. By putting 1 into your calculator, and asking it for the antilog base 10, it should give you 10 as the answer.

Therefore, if $\log_{10}10 = 1$, and $antilog_{10}$ of 1 gives us a value of 10 as the answer, this is telling us the numerical value of the base, which is 10.

To find a value for *e*, we carry out exactly the same process:

Enter the number 1 into the calculator and take the natural antilog of 1 (by using INV ln or SHIFT ln, or e^x as the second function).

Did you get 2.718281828?

Exercise 4.15

Plot the following values on a graph, using graph paper.

x	0	0.1	0.2	0.3	0.4	0.5	0.6	0.7	0.8	0.9	1.0
$y = e^x$	1	1.105	1.221						2.226		2.718

The missing values of e^x can be obtained from your calculator. If you have a graphical calculator, or have access to maths software, the graph will be drawn for you.

Comparing logs – common versus natural

Number	Logs to base 10	
8	0.903089987	
6	0.77815125	**Antilog**
Sum →	1.681241237	48
Number	Logs to base e	
8	2.07944542	
6	1.791759469	**Antilog**
Sum →	3.871201011	48

This shows that using logs, to any base, provides a means of turning multiplication into addition, and so on, and getting the same answer.

The purpose of the previous few pages has been to generate some understanding of logs as being more than just a mysterious button on your calculator. Whichever base of logs you use – 10 or e – the laws are exactly the same, so unless you are converting the logs into numbers, the base tends to be irrelevant.

Using the laws of logs

You have learned about logarithms and their laws. You have learned how to use logarithms to convert multiplication into addition and even how to convert powers into multiplication, or how to use logs twice to convert them into addition as well.

Now that you are familiar with logs, you are going to use the laws of logs to solve problems, without actually converting them into numbers or carrying out all the working out, leaving that until the end, if relevant.

Worked Example

Using the laws of logs, solve the following expression for x.

$$\log 12 = \log x + 2\log 2$$

This means, find a value for x from the expression.

Using the laws of logs, you can do this without using any difficult calculations.

If we are looking for x, we need to get everything else onto the other side (by using standard techniques of rearranging equations).

First step:

$$\log 12 - 2\log 2 = \log x$$

Key Points

When you are asked to show all your working out, this is what it means – clearly show the steps in your working out, not just the answer.

Key term

By inspection – A mathematical term meaning 'by looking at it'.

What is another way to write 2log2?

It can be written as $\log 2^2$, which also means $\log(2 \times 2)$ or $\log 4$.

So the equation now becomes

$$\text{Log}12 - \log 2^2 = \log x$$

Then

$$\log 12 - \log 4 = \log x$$

By inspection, and remembering that when taking logs, division becomes subtraction, the equation can then be rewritten as:

$$\frac{\log 12}{4} = \log x$$

which simplifies to

$$\log 3 = \log x$$

which must mean that

$$x = 3$$

Now without the words, solve the following to find x, using the laws of logs and showing all the working out.

$$\log 12 = \log x + 2\log 2$$
$$\log 12 - 2\log 2 = \log x$$
$$\log 12 - \log 4 = \log x$$
$$\frac{\log 12}{4} = \log x$$
$$\log 3 = \log x$$
hence $x = 3$

You could try it for yourself by using logs, but only if you wish to convince yourself that $x = 3$.

Exercise 4.16

Solve the following to find the unknown; show all your working.

1. $\log 25 = \log x + \log 5$

2. $2\log 3 + \log x = \log 36$

3. $\log 3 + \log 4 = \log x + 2\log 2$

4. If the equation for the instantaneous value of current, i, at any instant in time, t, for a charging capacitor is

$$i = Ie^{-t/CR}$$

where I is the current $(\frac{V}{R})$ when the capacitor is initially discharged, C and R are the capacitance and resistance values in the circuit, rearrange this equation, making t the subject.

5. The expression used to evaluate the tension of a drive belt is given by:

$$T = T_0 e^{\mu e}$$

Rearrange this equation to make T_0 the subject.

Exponential growth and decay

Some of the examples you have just been solving and rearranging are exponential decays and growths. In this section you will develop a better understanding of them.

As an introduction to this topic, consider a problem, the answer to which may not be as straightforward as it first seems.

Worked Example

Tea at Lords

Imagine that the England cricket squad have just won the Ashes, and by way of celebration, they are given a pot of tea. All the team stand waiting with their cups, and the captain, having the largest cup, pours half the contents of the teapot into it. The man of the match is the star bowler, and he pours half of the remaining tea into his cup. The top batsman is next, and pours half of the remaining tea into his cup, and so on.

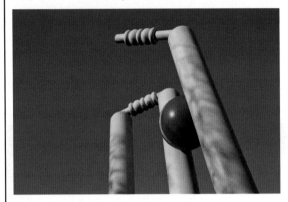

Figure 4.6

How many people would be able to pour an amount of tea from the pot, ignoring evaporation and other physical phenomena?

The answer is that it would go on forever, because the amount of tea left in the pot after each pouring would be halved.

Assume that they started with a litre of tea in the pot.

0.5 litres for the captain
0.25 litres for the star bowler
0.125 litres for the top batsman
0.0625 for the next
0.03125 for the next
And so on.

Putting this into a spreadsheet, which carries out multiple activities very quickly, the first 30 or so would receive the following amount of tea, in litres:

0.500000000000000000000000000000
0.250000000000000000000000000000
0.125000000000000000000000000000
0.062500000000000000000000000000
0.031250000000000000000000000000
0.015625000000000000000000000000
0.007812500000000000000000000000
0.003906250000000000000000000000
0.001953125000000000000000000000
0.000976562500000000000000000000
0.000488281250000000000000000000
0.000244140625000000000000000000
0.000122070312500000000000000000
0.000061035156250000000000000000
0.000030517578125000000000000000
0.000015258789062500000000000000
0.000007629394531250000000000000
0.000003814697265625000000000000
0.000001907348632812500000000000
0.000000953674316406250000000000
0.000000476837158203125000000000
0.000000238418579101562000000000
0.000000119209289550781000000000
0.000000059604644775390600000000
0.000000029802322387695300000000
0.000000014901161193847700000000
0.000000007450580596923830000000
0.000000003725290298461910000000
0.000000001862645149230960000000
0.000000000931322574615479000000

Hence, the next one gets half of that, and so on.

Although the proportion being removed each time (50 per cent) remains constant, the actual amount of tea received is reducing. This is an example of an exponential decay, and each subsequent reduction gets less and less, but never becomes zero.

(That is the theory – but in reality the tea is made up of a finite number of molecules which cannot be divided in two.)

Exponential growth

Exercise 4.17

Think about this in a small group and produce a similar result for a person who puts £100 in the bank and accumulates 3 per cent compound interest per year.

For the first 10 years the account looks like this:

£100.00
£103.00
£106.09
£109.27
£112.55
£115.93
£119.41
£122.99
£126.68
£130.48

Using what you already know about antilogs, can you devise an expression which will tell you how many years it will take for the sum in the account to pass £1,000?

(It is actually about 78 years, if that helps; now devise the expression.)

An engineering example of exponential growth – by calculation

Capacitor and resistor (CR) charging and discharging circuits form the basis of many electronic timers. The rate at which a capacitor charges is exponential – in other words, when it is empty (discharged), the charge flows into it very rapidly. As it fills up, the charge that has gone in already opposes the new charge that is trying to get in, so it slows down. The rate of charging becomes slower and slower as the capacitor approaches full charge.

Consider the following series CR circuit.

$R = 20\,k\Omega$ (kilo ohms)
$C = 33\,\mu F$ (micro farads)
The time constant for this circuit is $C \times R = \tau$ (tau).

When connected to a d.c. voltage of 100 V, charge starts to flow into the capacitor.

How long does it take for the charge in the capacitor to reach 40 V?

Using the exponential growth formula,
$$v = V(1 - e^{-t/\tau})$$
By convention, the use of V represents the maximum value of voltage = supply voltage. The use of v represents the instantaneous value of voltage.

We are trying to find the time taken, so we need to rearrange this for t.

Exercise 4.18

Have a go at rearranging this equation to find t.

Radioactive decay

Radioactive materials emit particles, and although the emissions are random quantum events – that is, each individual emission cannot be predicted exactly – such a material has what is known as a half-life.

Make the Grade — D1

Grading criterion D1 requires that you apply graphical methods to the solution of two engineering problems including exponential growth and decay, analysing the solutions using calculus.

Key Term

Half-life – The time it takes for half of the radioactivity to have dispersed.

Exercise 4.19

Using the equation $N = Ni\,(e^{-0.1386t})$

where

N = the number of particles remaining at time t (in years)
Ni = the initial number of radioactive particles
rearrange the equation for t and find the half life of the material.

(Hint, start by letting $N = 0.5\,Ni$.)

Other examples of exponential growth and decay can be located in a range of textbooks which relate to a wide variety of subjects, including biology, natural occurrences, engineering and manufacturing.

One such application is the Taylor's tool life equation $C = VT^n$. This allows manufacturing engineers to calculate, with some accuracy, when a machine cutting tool may need replacing, resharpening or changing.

Exponential growth and decay will be covered later, near the end of this unit, when the solutions to growth and decay problems will be analysed using the calculus.

Make the Grade   D1

For the D1 criterion, differentiation and exponentials and so on, if you have time, or want to start looking, go to http://mathworld.wolfram.com/ExponentialDecay.html

Linear equations and straight-line graphs

In this section you will develop a knowledge and understanding of linear equations, straight line graphs and experimental data.

Linear equations

Mention a straight-line graph to a mathematician or an experienced engineer and immediately they will think of the equation $y = mx + c$.

Until you develop familiarity with this, it can be difficult to see how a graph and an equation can be linked together, but the following section should help.

Worked Example

A customer requires a telephone and looks through all the brochures and adverts. He has found a company that will provide the service he requires (a simple line rental and telephone), which will cost £9 per month, and the calls will cost 3p per unit. He is confused by all the words and numbers, and asks for help.

Figure 4.7

He asks, 'If I have the phone and do not use it, it costs me £9 per month. If I make 20 units of calls per month, it will cost me £9.60. What will the total bill come to if I use 230 units in a month?'

Each month the bill contains:

- **the standing charge or line and phone rental**
- **a variable amount, depending on how many calls he makes, or the number of units of calls.**

This could be solved by creating a table which shows the cost of the monthly bill for any number of calls, but that would be a very large table.

An equation might help:

Bill = cost per Unit × Number of units + Standing charge

146

Some of the words have capital letters, and we'll look at those, ignoring all the rest:

Bill = Unit cost × Number + Standing charge

or

B = UN + S

Back into words – the total Bill is the cost per Unit times the Number of units plus the Standing charge.

Compare this equation with the straight-line equation identified earlier:

$y = mx + c$

and

B = UN + S

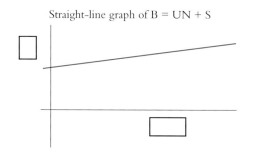

Figure 4.8 Straight-line graph for $y = mx + c$

Now look at the following straight-line graph for

$y = mx + c$

The straight-line graph shows two things: one is very obvious – a straight line; the other, less obvious thing is the intercept.

Key Term

Intercept – The point of the graph where the line intercepts (or crosses) the y axis.

Exercise 4.20

Copy the graph shown in Figure 4.8 and mark the point where the line intercepts the y axis.

Now let's draw a graph for the phone bill.

Straight-line graph of B = UN + S

Figure 4.9 B = UN + S graph

Exercise 4.21

Copy and complete the graph shown in Figure 4.9, filling in the two small boxes to label the axes. Write the symbols into the boxes in place of y and x for this graph

B = UN + S

Which axis is which?

When drawing graphs, there is always the decision to be made which variable (y or x, or in this case, B or N) is represented by which axis.

> **Key Points**
>
> Generally, the independent variable becomes the horizontal axis and the dependent variable becomes the vertical one.

If something is dependent on another factor, such as the bill (or B) being dependent on some aspect of the use of a telephone, then B must become the vertical axis (the y axis).

On the other hand, the number of units of calls does not depend on any other aspect of the cost, so this is called the independent variable and goes along the horizontal axis.

In summary, the variable which determines whether the bill increases or decreases each month is the number of units of calls that are made.

Exercise 4.22

1. Using a page of A4 graph paper, select a suitable scale, and plot the graph of the telephone charges that were discussed above. Place the graph paper in landscape and use a scale from zero to about £16 or £18, depending where you start, for the vertical axis. Across the horizontal axis, label a scale which allows up to about 200 to 260 call units.

> **Key Points**
>
> When plotting graphs, always try to make the most effective use of the graph paper to allow accurate plotting of the points, and accurate reading of values.

Because we know this is a straight line, we only need two points to draw the line, but using three points will help check that we have not got one of the points wrong.

2. Plot the following points, and draw the straight-line graph of B = UN + S.

Number of calls	Bill or total cost
none	£9.00
100	£12.00
200	£15.00

3. Now join up the points using a rule. The straight line should go through all three points if you have plotted them accurately. If not, check to see where you went wrong.

4. One item of a graph is called the intercept. Write down in your own words what the intercept of a graph means.

5. For the B = UN + S graph, what is the value of B at the point where the line of B = UN + S crosses it?

   If you have drawn the graph accurately, you should have obtained £9.00, which is the standing charge or monthly line and phone rental cost.

6. Remember that the customer had been told that if he used 20 call units a month, the bill would come to £9.60. How much is that for each call unit?

Gradient of the slope of line

You will be familiar with the terms 'slope' or 'gradient', and these are often defined as the 'rise' over the 'run', or

$$\text{Gradient} = \frac{\text{rise}}{\text{run}}$$

For example, the cyclist in Figure 4.10 is shown cycling up a sloping road, or a gradient. After travelling the length of this slope, indicated by the term 'run', the cyclist will have increased in altitude slightly. The distance by which the position is higher than it was previously is referred to here as the rise.

If the cyclist has risen by 1 metre after travelling 10 metres away from his starting point, the slope or gradient which he has travelled along is given by:

$$\text{Gradient} = \frac{\text{rise}}{\text{run}} = \frac{1\,\text{m}}{10\,\text{m}} = 1 \text{ in } 10 \text{ or } 0.1$$

Note, as the units are both distance $\left(\frac{\text{m}}{\text{m}}\right)$, they disappear, and the resulting gradient, in this instance, has no units. It is just a ratio of two values.

When straight-line graphs are used to represent equations and varying quantities, the gradient of the slope tells us the relationship between both variables.

Select two points on the graph you plotted in the previous exercise where you can clearly read the vertical and horizontal values. This may be two of the points which you plotted, and these would be:

Number of units	Total bill
100	£12.00
200	£15.00

The gradient or slope is the $\frac{\text{rise}}{\text{run}}$.

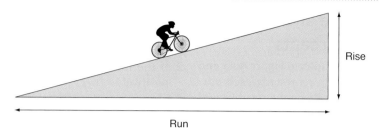

Figure 4.10 Rise over run

The rise is the change in total cost – from £12.00 to £15.00.

The run is the change in the number of units – from 100 to 200.

Using these data:

gradient

$$= \frac{£15.00 - £12.00}{200 - 100}$$

$$= \frac{£3.00}{100}$$

$$= 0.03$$

What are the units of measurement for this value?

£/number of call units gives us '£ per call unit'.

Or in this case, £0.03 is better written as 3p; hence the gradient of the graph is 3p per unit.

Using the units makes the answer more correct, and they help us to check that we have not made a mistake by having top and bottom values mixed up.

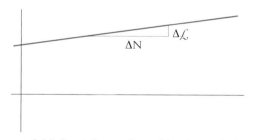

Figure 4.11 Graph for gradient of the slope of a line

Key Term

The capital Greek letter delta (Δ) – Used to mean 'a change in or a difference in'.

The gradient is thus $\frac{\Delta £}{\Delta N}$ or the change in cost divided by the change in the number of units.

Positive or negative gradient

The graph that you have just worked with increases in value as we follow the line to the right. It is said to have a positive gradient.

The following graph shows a line which reduces in height as we travel along it.

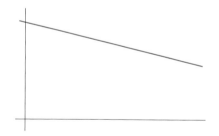

Figure 4.12 Graph with negative gradient

It is also possible to have a negative intercept, with either a positive or a negative gradient.

To simplify the equations, we will use the symbols that are conventionally used in a straight-line equation, *y, m, x* and *c*.

Identifying positive and negative gradients and intercepts

Look at the straight-line graphs in Figure 4.13 and identify the equation which represents each one. The graphs are drawn to indicate the positive and negative sections in all cases.

Basic rules:

1. If the line slopes downwards, the gradient (m) is negative.
2. If the line slopes upwards, the gradient (m) is positive.
3. The intercept is always where the line crosses the y axis, at either $+ c$ or $- c$.

Graph	Description of the line graph	Equations for the graphs
A	Negative gradient, positive intercept	$y = -mx + c$
B	Positive gradient, negative intercept	$y = mx - c$
C	Negative gradient, negative intercept	$y = -mx - c$
D	Positive gradient, positive intercept	$y = mx + c$

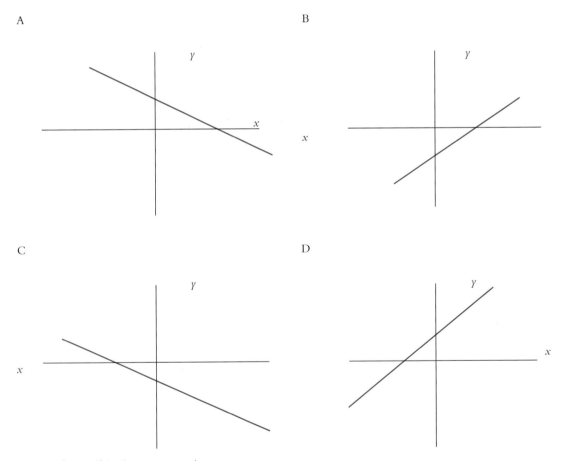

Figure 4.13 Identifying intercepts graphs

Exercise 4.23

1. Once you have studied the table and the sample straight-line graphs opposite, cover the page and sketch them, writing their equation alongside. Check your answers.

   The following problems involve the use of data that were obtained experimentally. In each case, use the equation and the data to plot a straight-line graph, and from the graph, determine the gradient, the intercept and the equation of the line.

2. The cable drum of a winch is designed to hold a cable which has a mass (m) of 2.5 kg for each metre of its length (l). As the cable is wound onto the drum, the total mass (M_t) increases. The equation for the total mass of the cable drum is a straight-line equation and the following data are known: the total mass of the drum and cable, when holding 25 m, 60 m and 80 m of cable is 912.5 kg, 1,000 kg and 1,050 kg, respectively.

   Using these data, plot the straight-line graph and obtain the intercept (mass of empty drum) and the gradient of the line (mass per metre for the cable). Also, write out the equation for this relationship.

3. The resistance of the coil in a hydraulic solenoid varies with temperature. The following maintenance records indicate the temperature over a day of operation.

Time	Temperature (°C)	Resistance
8 a.m.	20	22.5 Ω
1 p.m.	60	32.5 Ω
6 p.m.	40	27.5 Ω

   Plot a straight-line graph of the resistance (dependent variable) against temperature and obtain the intercept and gradient of the line, and the equation for this straight-line relationship.

4. An aircraft makes a flight with two stops in the journey. After two hours of flying, the aircraft lands and has a total mass of 942 tonnes. Without refuelling or changing any passengers, the aircraft then flies on to the next destination. Arriving three hours later, it is found to have a mass of 931 tonnes. At the final touchdown, three hours later, the total mass is found to be 919 tonnes.

   Ignoring extra use of fuel during take-off, and assuming that the fuel consumption follows a linear relationship, plot the data onto a graph and draw a straight line.

   Figure 4.14

   Use this to obtain the fuel consumption in tonnes per hour and determine the original take-off mass (m) of the fully fuelled aircraft.

5. The value of a resistor changes with temperature, and the following experimental data were obtained across a range of temperatures.

Resistance (R)	40 Ω	41 Ω	42 Ω	44 Ω	46 Ω
Temperature (T) (°C)	6	9	12	19	25

   Plot the graph of R against T and determine the straight-line law for this relationship.

Use of simultaneous equations to solve two unknowns in two equations

You may have seen two equations being solved using graphs, and that method is acceptable if you only need an approximation to the correct answer, or if the values are not very complicated, allowing easy reading of the scales of the graph.

This process is known as solving simultaneous equations. It is actually solving for two unknowns using two equations, simultaneously.

Make the Grade   M1

Grading criterion M1 requires that you solve a pair of simultaneous linear equations in two unknowns.

Worked Example

A motor dealer supplies seven identical cars and 13 identical vans to two engineering companies.

Company A buys four cars and six vans, paying a total of £72 000.
Company B buys three identical cars and seven identical vans, for a total cost of £69 000.

If both companies paid the same price for each car and each van, what was the price of each car and each van?

Step 1 – create an equation for each company.

Company A $4C + 6V = 72$ (working in £k)
Company B $3C + 7V = 69$

Now, label the equations 1 and 2 for identification and manipulation processes:

$$4C + 6V = 72 \qquad (1)$$

$$3C + 7V = 69 \qquad (2)$$

Step 2 – make the coefficients of one variable the same (that means change the numbers which multiply the Cs and Vs to get the same number in front of either the C or the V).

For example, if we multiply equation (1) by 3 and equation (2) by 4, both of them will have 12C in them.

(1) $\times 3$ gives $12C + 18V = 216$ (3)
(2) $\times 4$ gives $12C + 28V = 276$ (4)

Now, subtracting equation (3) from equation (4) will get rid of the Cs.

(4) $-$ (3) gives

$$10V = 60$$

Hence,

$$V = \frac{60}{10} = £6{,}000$$

which is the price of each van.

Substituting the value for V back into any of the equations (1) to (4) will lead us to a value for C, the price of each car.

Substitute $V = 6$ in (1) (that is, in equation (1))

To get
$$4C + 6(6) = 72$$
or
$$4C + 36 = 72$$
then
$$4C = 72 - 36$$
or
$$4C = 36$$
so
$$C = \frac{36}{4} = £9{,}000 \text{ for each car.}$$

This can then be checked by substituting the values £6,000 and £9,000 in equation (2).

Exercise 4.24

Set your own problems for each other to solve.

Writing problems such as the ones with the cars and vans either needs experimental data to start with, or the two prices, and a problem is then developed around it.

For example: the car was £9k and the van was £6k.
Selecting any number of cars and vans, work out the total cost and use your numbers of cars and vans to create equations similar to the ones given at the start of the worked example opposite.
For example, company A buys two cars and four vans.
Company B buys three cars and three vans.
Complete the equations and give them to someone else to solve.

Change the prices of the vans and cars and create some more.
Note, you do not always have to work with cars and vans!

Exercise 4.25

Solve the following engineering problems using simultaneous equations.

1. An electronics company buys many components and parts from a wholesaler. Two orders have been made for resistors and capacitors, and the original price for each individual component has been lost. Create a pair of equations from the following data and solve to find the individual component costs.

Batch 1 contains 68 resistors and 92 capacitors, and the total cost is 323.2p.
Batch 2 contains 18 capacitors and 37 resistors, and the total price is 89.3p.

2. A small engineering company manufactures trailers to order. One customer wants a small box trailer and a trailer for a quad bike so that he can take his daughter racing. To hold them both together, identical nuts and bolts are to be used.

The box trailer will contain 53 bolts and 96 nuts, which will cost £36.99.
The quad bike trailer will use 29 bolts and 42 nuts, which will cost £18.45.

What is the individual price of each nut and each bolt?

Having worked through these problems, spend a few minutes thinking about similar types of problems of your own, and make up a few simultaneous equation problems to share with others in your group.

Always start with the answers – for example, a tin of paint costs £5.94 and a paintbrush costs £2.33.

Six tins of paint cost £35.64 and four paintbrushes cost £9.32. Total = £44.96.
Four tins of paint cost £23.76 and eight brushes cost £18.64. Total cost = £42.40.

Figure 4.15

Then make up the problem:

If six tins of paint and four paintbrushes cost £44.96, and four tins of paint and eight brushes cost £42.40, work out the cost of one tin of paint and one brush, and so on.

Make the problems relevant to your chosen trade or a hobby, or relate them to the work or hobby of the person who will have to solve the problem.

Factors and factorising in mathematics

In this section you will develop a knowledge and understanding of the various methods of factorisation.

You may remember from previous studies when you were introduced to the term 'factor'.

Key Term

The factor of a number – Any other number that can divide into a number exactly (leaving no remainder).

Any number can be divided by the number 1 and itself.

Hence, the factors of 6 are 1, 2, 3 and 6.

The factors of 12 are 1, 2, 3, 4, 6 and 12.

What are the factors of 24?

Which has the greater number of factors, 24 or 30?

How many factors has the number 100?

Factorising and algebra

When we move from arithmetic to algebra, the terminology is still the same.

In arithmetic, we factorise the number 12 and obtain 1, 2, 3, 4, 6 and 12, because all these numbers will divide into 12 and leave no remainder.

What are the factors of x^2? They are 1, x and x^2.

As things become more complicated, they become less obvious and certain techniques can be used to factorise an expression.

We know that 3 and 4 are factors of 12 because we know that $3 \times 4 = 12$.

Have another look at the factors of 12.

If a factor is any number that will divide into another number and leave no remainder, are the following also factors of 12?

$$-1, -2, -3, -4, -6, -12$$

Since -3 multiplied by $-4 = 12$, they must be factors, and so must the other negative numbers. (More on this later.)

In algebra, it helps to learn factorisation by multiplying out some factors in the first place.

Multiplying expressions in brackets

Recognising factors is a skill that develops with practice.

Exercise 4.26

Work through the following to see how factors multiply together to create complex terms.

1. $2(3 + x) =$

2. $3(x - 4) =$

3. $x(6 - x) =$

4. $x(x + x) =$

5. $3(3 + 3) =$

The final problem in Exercise 4.26 contains only numbers, allowing a single number solution to be found. All the others have variable terms in them, and although they can be combined, they cannot be replaced with a single value until we give the variables a value, in which case they stop being variables and become constants.

Exercise 4.27

Now try the following:

1. $x(14x - 2y) =$

2. $3(y + z) =$

3. $14(f - \frac{R}{L}) =$

4. $8(x + 0.5) =$

If any of your answers are different to those given at the back of the book, have another look at your solution or discuss it with someone else, perhaps a tutor.

Let us take this a stage further.

$$3(x + 3)$$

This means multiply all the terms in the bracket by 3.

It follows, then, that

$$(x + 2)(x + 5)$$

must mean, multiply every term in the first bracket by every term in the second bracket.

Worked Example

Have a go at multiplying out this expression.

$$(x + 2)(x + 5)$$

Taking the first bracket apart, we have x plus 2. Then work out x times the second bracket. Then work out 2 times the second bracket. *Add* the two results together.

$$x(x + 5) = x^2 + 5x$$

$$2(x + 5) = 2x + 10$$

$$\text{SUM} \rightarrow x^2 + 5x + 2x + 10$$

Combining like terms $(5x + 2x)$

$$\rightarrow \underline{x^2 + 7x + 10}$$

Another way to view this kind of problem is to draw little curved arrows to remind you that each element in each bracket must be multiplied by each element in the other bracket.

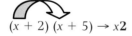

$$(x + 2)(x + 5) \rightarrow x2$$

$$(x + 2)(x + 5) \rightarrow 5x$$

$$(x + 2)(x + 5) \rightarrow 2x$$

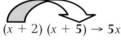

$$(x + 2)(x + 5) \rightarrow 10$$

Add all the individual results up, combine the like terms, and get $\underline{x^2 + 7x + \mathbf{10}}$.

> **Key Points**
>
> You may also find problems asking you to expand the brackets, which means just the same as multiply out the brackets.

You may have heard the term 'quadratic' equation. This means an equation in which a squared term is the highest order.

For example, $x^2 + 7x + 10$ is a quadratic equation. An equation like $x^3 - x^2 + 7x + 10$ is called a cubic equation.

> ### Activity 4.4
> See if you can find the names of equations which contain fourth powers, fifth powers and others.

Exercise 4.28

Have a go at a few of these to get some practice.

1. $(x + 4)(x - 8) =$

2. $(R - 5)(x + 7) =$

3. $(n - 3)(n + 2) =$

4. $(v + 3)^2 =$

There are a few formats of brackets which are worth remembering because they crop up time and time again, and being able to recognise them makes factorisation, or expanding, simpler.

$$(x + 2)(x + 2) = x^2 + 4x + 4$$
$$(x - 2)(x - 2) = x^2 - 4x + 4$$
$$(x + 2)(x - 2) = x^2 - 4$$

The same expression without any constants (numbers):

$$(a + b)^2 = a^2 + 2ab + b^2$$
$$(a - b)^2 = a^2 - 2ab + b^2$$
$$(a + b)(a - b) = a^2 - b^2$$

If we multiply $(x + 3)$ by $(x + 2)$ we obtain $x^2 + (3x) + (2x) + (6)$.

Pulling these together gives us $x^2 + 5x + 6$.

Factorising equations

Key Term

Factorisation – As the expansion of bracketed terms by multiplication is the step which we have just performed, then factorisation is the reverse process.

155

Worked Example

Factorising $x^2 + 5x + 6$ would now be easy because you have seen the factors already, so what about the following:

Factorise $x^2 + 6x + 8$.

Looking at the first example:

When we multiplied $(x + 3)$ by $(x + 3)$ and obtained $x^2 + 5x + 6$

We have an x^2, which obviously came from x multiplied by x.

Hence, we have made a start $(x + \ )(x + \ )$, but what goes after the x's?

How do we know the mathematical operators after the x's are $+$ and not $-$?

The other values, $5x$ and 6, do not appear to be too obvious, but look again at the intermediate step we used:

multiplying $(x + 3)$ by $(x + 2)$, gave $x^2 + (3x) + (2x) + (3 \times 2)$

The (3×2) is the product of the two separate digits in the original expressions.
The $(3x)$ and the $(2x)$ are the intermediate values of multiplying the first item in the first bracket by the second item in the second bracket $= (3 \times x)$, and the second item in the first bracket by the first item in the second bracket to obtain $(2x)$. We then added these together to get $5x$.

Hence, we want two numbers which add together to make 5 and multiply together to make 6.

List the factors of 6	Which of these add together to make 5?
I and 6	$1 + 6 = 7$
2 and 3	$2 + 3 = 5$
also	
-1 and -6	$-1 + -6 = -7$
-2 and -3	$-2 + -3 = -5$

The sum of factors which $= 5$ indicates that 2 and 3 are the missing factors, and we can complete the following:

Factorising $x^2 + 5x + 6$ gives us $(x + 2)(x + 3)$.

If you write $(2 + x)(3 + x)$, it means exactly the same as $(x + 2)(x + 3)$, but by convention, we generally write orders of x (from highest order or power to lowest, followed by numbers). This can help us to recognise patterns and factors as well.

Key Term

In the term $3x$ — the number 3 is called the coefficient of x.

Look again at the second problem:

factorise $x^2 + 6x + 8$.

Making a start

$$(x + \ )(x + \ )$$

As before, the $6x$ is made up of two lots of x's added together, and the coefficients of these x's must multiply together to make 8. To find which numbers add up to make 6 and multiply to make 8, a table provides a well-organised method – and it allows for error checking.

List the factors of 8	Which of these add together to make 6?
I and 8	9
2 and 4	6
also	
-1 and -8	-9
-2 and -4	-6

Therefore, the factors are $(x + 2)(x + 4)$.

Key Points

When multiplying out brackets, we multiply every term in each bracket by every term in the other bracket(s) (three, four or more sets of brackets can all be multiplied together, with no limit).

List the factors of -10	Which of these add together to make -3?
-1 and 10	9
1 and -10	-9
-2 and 5	3
2 and -5	-3

Try this one:

$(x + 4) (x - 2)$ expands to make

$$x^2 - 2x + 4x - 8$$

or

$$x^2 + 2x - 8$$

So factorise the following:

$$x^2 - 3x - 10$$

Hence

$$x^2 - 3x - 10$$

factorises to

$$(x + 2) (x - 5)$$

or

$$(x - 5) (x + 2)$$

It is always good practice to check your answer.

Expanding the brackets should take us back to the original quadratic equation.

Exercise 4.29

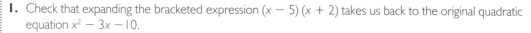

1. Check that expanding the bracketed expression $(x - 5) (x + 2)$ takes us back to the original quadratic equation $x^2 - 3x - 10$.

Work through the following examples, and check them by expanding the brackets to check your working.

2. $R^2 + 5R - 14$

3. $x^2 - 7x + 10$

Although this method is a bit 'trial and error', it is not always immediately obvious what the factors may be. Adding the following level of complication makes us wish for another method.

Expanding the following brackets results in a quadratic:

$$(2x + 4) (x - 3) = 2x^2 - 2x - 12$$

but the quadratic has a coefficient (which is more than 1) in front of the x^2 term.
Does this affect the way we would factorise?

Worked Example

Look closely at the following variations:

$$(2x + 2)(2x - 3) \rightarrow 4x^2 - 6x + 4x - 6 \rightarrow$$
$$4x^2 - 2x - 6$$

$$(4x + 2)(x - 3) \rightarrow 4x^2 - 12x + 2x - 6 \rightarrow$$
$$4x^2 - 10x - 6$$

$$(x + 2)(4x - 3) \rightarrow 4x^2 - 3x + 8x - 6 \rightarrow$$
$$4x^2 + 5x - 6$$

Factorisation is still 'trial and error' – but where do we start?

The first and last terms in all three quadratic equations are the same.

$4x^2$ could be derived from any of the pairs of factors shown, that being $4x$ multiplied by x, or $2x$ multiplied by $2x$.

As before, with reference to the pairs of brackets, the product of the two 'inner' terms plus the product of the two 'outer' terms is equal to the middle term in the quadratic.

Taking the first example: factorise the quadratic equation

$$4x^2 - 2x - 6$$

The first term is the product of the first term (the x term) in each bracket.
The third term is the product of the two second terms in each bracket.
The middle term is $-2x$, and this is the sum of the products of the two inner terms and the two outer terms.

The factors of '-6' are '6 and -1' or '-6 and 1' or '2 and -3' or '3 and -2'.

Trying each of these now becomes a tedious operation:

All combinations are shown below (although you would normally stop when you find one that fits).

Considering only the sum of the inner and outer products (trying to find -2x),

either

$$(2x - 1)(2x + 6) \rightarrow 12x - 2x = 10x$$

or

$$(4x - 1)(x + 6) \rightarrow 24x - x = 23x$$

either

$$(2x + 1)(2x - 6) \rightarrow -12x + 2x = -10x$$

or

$$(4x + 1)(x - 6) \rightarrow -24x + x = -23x$$

either

$$\mathbf{(2x + 2)(2x - 3) \rightarrow -6x + 4x = -2x} \rightarrow$$
the correct middle term

or

$$(4x + 2)(x - 3) \rightarrow -12x + 2x = -10x$$

either

$$(2x - 2)(2x + 3) \rightarrow 6x - 4x = 2x$$

or

$$(4x - 2)(x + 3) \rightarrow 12x - 2x = 10x$$

so the factors of $4x^2 - 2x - 6$ are $(2x + 2)(2x - 3)$.

One purpose of factorising is to obtain the 'roots' of the equation.

Key Term

The 'roots' of an equation – The values where the graph of that function passes through the horizontal axis.

The roots of $4x^2 - 2x - 6$ are easily obtained if we let $4x^2 - 2x - 6 = 0$.

Then

$$(2x + 2)(2x - 3) = 0$$

This means that at least one of the brackets must equal zero at any time, so let us consider them one at a time:

$$(2x + 2) = 0$$

(divide through by 2) to obtain $(x + 1)$

For $(x + 1)$ to equal zero, x must be -1 and this is one root of the equation – that is, one value of x where the graph would cross the horizontal axis.

Taking the other bracket:

$$(2x - 3) = 0$$

(divide through by 2) to get $(x - 1.5)$

For $(x - 1.5)$ to equal zero, x must be 1.5, and this is the other root of the quadratic equation.

Hence, for the function $4x^2 - 2x - 6 = 0$, the roots are 1.5 and -1.

If you have access to any graph-drawing packages – even Microsoft Excel®, or a graphical calculator – try it.

The working out of the factors was a little tedious; we needed a more straightforward method.

The quadratic formula

$$x = \frac{-b \pm \sqrt{b^2 - 4ac}}{2a}$$

where a, b and c are the values of the numbers (coefficients) taken from the general quadratic equation,

$$ax^2 + bx + c = 0$$

The equation is set to equal zero to allow the determination of the values of x where the graph of this function would cross the horizontal axis.

Go back to the potentially troublesome quadratic equation and give it a go.

Worked Example

Solve the following using the quadratic formula:

$$4x^2 - 2x - 6$$

From this equation, and comparing it to the general quadratic equation

$$ax^2 + bx + c = 0$$

$a = 4$

$b = -2$

$c = -6$

Now put the values into the quadratic formula:

$$\frac{-(-2) \pm \sqrt{(-2)^2 - 4(4)(-6)}}{2(4)}$$

$$\frac{2 \pm \sqrt{4 + 96}}{8}$$

$$\frac{2 \pm \sqrt{100}}{8}$$

$$\frac{2 \pm 10}{8}$$

$$\frac{12}{8} = 1.5$$

or

$$\frac{-8}{8} = -1$$

These are the same roots we obtained earlier, but with much less trial and error.

159

Exercise 4.30

Solve the following quadratic equations to find the roots in each case.

1. $x^2 - 2x - 15 = 0$

2. $y^2 - 11y + 28 = 0$

3. $z^2 + 0.5z - 3 = 0$

4. $2z^2 + 9z + 9 = 0$

5. A solid circular cylinder is to be made for a steel rolling mill. To fit into the existing mill, it will have a total length (cylindrical height) of 1.8 m and a total surface area of 4.5 m² to allow the heat to dissipate.

   Calculate the radius of the cylindrical roller.
   (Hint: the surface area of a cylinder comprises:
   - two circular ends, each with an area of $2\pi r^2$.
   - the cylindrical surface, with a surface area of $2\pi rh$.)

6. Create the equation for the total surface area, then rearrange to create a quadratic equation.

Make the Grade   M2

Grading criterion M2 requires that you solve one quadratic equation by factorisation and one by the formula method.

Other methods of factorisation

Where an expression contains a common element or a common term (or more than one), the common term(s) can be collected or grouped together to simplify the expression.

If the expression does not get any simpler (fewer elements), then it might as well be left in its original state.

Consider the expression $ab + ac + ad$.

The variable a is a common factor and can be 'removed' from the others.

Hence this can be factorised to $a(b + c + d)$.

The brackets need to be used to indicate that $(b + c + d)$ are all still multiplied by a.

Remove the brackets from the following expression

$$(a + 2b) - 2(b + 2c) + 3(c + 2d)$$

and write it in its simplest terms.

Removing the brackets (or multiplying them out), the expression becomes

$$a + 2b - 2b - 4c + 3c + 6d$$

which then simplifies further to

$$a - c + 6d$$

Exercise 4.31

Simplifying the following expressions uses the basic rules of arithmetic and algebra. Expand the brackets and simplify each expression into its simplest terms.

1. $a(3a - 2b) + 2a(3a - 3b)$

2. $4(2a + 3b) - 6(a + b) - a(3 - 2a)$

3. $4[p(p - 3q) - (p + q)(p - q)]$

Factorisation by extraction of a common factor

In the following expression, a is a factor which is common to both parts.

$$ax + ay$$

This will factorise to $a(x + y)$

Similarly,

$ax + 2a + bx + 2b$ can be factorised by removing common factors

$$a(x + 2) + b(x + 2)$$

where $(x + 2)$ is a common factor, thus

$$(a + b)(x + 2)$$

An alternative factorisation would be to take the x's and the 2's out initially

$$ax + 2a + bx + 2b$$

giving

$$x(a + b) + 2(a + b)$$

which still arrives at

$$(a + b)(x + 2)$$

Exercise 4.32

Investigate the following to identify any common factors.

1. $ax + 3x + a + 3$

2. $3pq - 2pr + 3sq - 2sr$

3. Spend some time considering all options for the following to see if any of the groupings can lead to common factors.

$2A + 3B + 4C + BC + AB + CD$

Factorisation by grouping

$$ax + bx - ay - by$$

becomes

$$x(a + b) - y(a - b)$$

being factorised by grouping, but there is not a common factor.

Also

$$2R + 3P + RP$$

factorises by grouping to become

$$R(2 + P) + 3P$$

or

$$P(3 + R) + 2R$$

but there is no common factor to factorise further.

Make the Grade — P3

Grading criterion P3 requires that you factorise by extraction and grouping of a common factor from expressions with two, three and four items respectively.

This section will cover the following grading criteria:

P4 P5 **P6** P7

In this section, you will be introduced to the radian as an alternative form of measurement for an angle. The radian forms an essential part of a circle, whereas a degree, or angles measured in degrees (such as 60° or 90°), only represents a division on a scale of 0° to 360°, into which a circle is artificially divided because 360° was decided upon many centuries ago – because it is a nice round number. The importance of this will become more apparent as you study the calculus, in the final section of this unit.

This whole section is concerned with measurements and ratios for objects, of all shapes and sizes, including triangles and trigonometric ratios, such as sines, cosines and tangents, all of which are extremely helpful for engineers, in whichever sector they are working. This will be illustrated by the numerous examples as you work through this section.

Circular measure

In this section you will develop a knowledge and understanding of radian and degree measure, angular rotation, problems involving areas and angles measured in radians, and how to find the length of an arc and area of a sector of a circle.

Circles easily split into six sectors as you doodle with a circle and a pair of compasses.

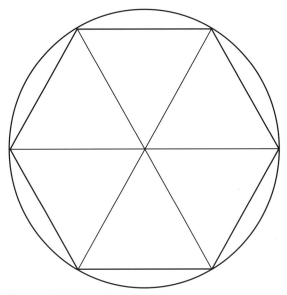

Figure 4.16 Drawing a circle

From this circle, hexagon and set of triangles grow all the ancient ideas about the number six being a magical number. The choice of 360° represents $6 \times 6 \times 10$, and each of the triangles within the circle has $3 \times 60°$ angles. A degree splits into 60

Activity 4.5
1. Set a pair of compasses to about 4 cm apart, and draw a circle in the middle of a sheet of paper.
2. Keeping the compasses at the same separation, place the pin at any point on the circle, and without letting the pencil touch the paper, revolve the compasses round until the pencil crosses the circle. Mark a small 'arc' where the pencil touches the circle.
3. Move the pin to the point you have just marked, repeat the operation, and continue doing this until you have worked round the circle and the pencil is finally on the original pinhole.
4. Draw a straight line between each of the six points, to create a hexagon. Now join each opposite pair of points, by drawing a line across the diameter three times.

You should end up with something like the diagram in Figure 4.16.

minutes, and a minute splits into 60 seconds; very precise measurements of angles can be quoted as, for example, 27° 15′38″ to represent 27 degrees, 15 minutes, 38 seconds.

If a second is a 60th of a minute and a minute is a 60th of a degree, what fraction of a circle is a second?

A 60th of a 60th of a 60th is a 21,600th.

That is a small angular measurement.

Small angular measurements, such as the second might be useful for measuring angles on a large plane or curved surface, like that of the Earth.

Next time you play with a satnav, check out the latitude and longitude settings, in degrees, minutes and decimal fractions of a minute.

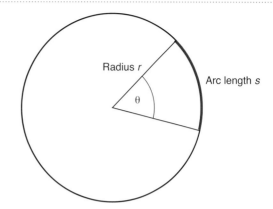

Figure 4.17 Radian measure

The relationship between the angle and the arc length is given as

$$\theta \text{ radians} = \frac{s}{r}$$

Rearranging this

$$s = r\theta$$

gives us a direct relationship between the angle we travel through as we move along an arc length s.

Activity 4.6

Search the internet for a degrees-minutes-seconds to decimal degrees calculator. Using this website, or any other source that you find, work out how many metres you would have to walk to travel through 1 second of arc, measuring the angle from the centre of the Earth.

Radian measure

After many years of working with degrees being a 360th of a circle, it may seem strange changing to radian measure.

The radian is determined by 'walking' round the circumference of a circle an exact same distance as the radius. As Figure 4.16 shows, if six radii can be fitted into a circle's circumference by using straight lines, and each line 'bi-passed' an arc of 60°, then walking the same distance around the arc will obviously take you through some angle which is less then 60°.

In Figure 4.17, the arc length s is the same length as the radius, r, of the circle.

The angle, θ, which this arc travels through, is one radian.

Exercise 4.33

1. If you continued drawing arcs of length r round the circle, how many would you manage to draw before you come back to the start point? (In other words, how many arc lengths of length r would fit into the circumference?)

There are 2π radians in a circle.

Travel around a semicircle, there are π radians. This is also 180° (180 degrees).

2. If there are 2π radians in a circle, how many degrees is 1 radian equivalent to?

With experience and practice, you will not need to convert from radians to degrees to understand what you are doing.

Exercise 4.34

Copy and complete the table below by converting the degrees to radians.

There are two ways to give your answer:

1. Divide the number of degrees by 57.3° and work to three or four significant figures (or more).

2. Work in multiples of π (for example, 360° is done for you in the table).

Degrees	Radians as a decimal fraction	Radians as multiples of π	Pronounced as
360	6.283185307..		2 pi
270		1.5π or $3\frac{\pi}{2}$	
180	3.141592654..		
90		$\frac{\pi}{2}$	pi by 2
60			
45		$\frac{\pi}{4}$	pi by 4
30	0.523598775..		
15			

Have a go at these before looking at the answers.

Which do you think is the easiest way to work – using decimal fractions of multiples of 57.3° or using fractions of π?

Length of arc of a circle ($s = r$)

The amount of drive belt that is in contact with a pulley or drive shaft affects the frictional forces and the power transmitted from one to the other.

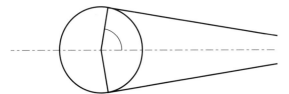

Figure 4.18 Length of arc of a circle

Consider the following system, where a drive belt sits on a pulley of 24 cm radius and is in physical contact with it, as shown in Figure 4.18.

Exercise 4.35

1. If angle A is 75°, what length of the pulley is in contact with the belt? (Hint, determine the angle of the pulley through which it is contact with the belt, then convert to radians, then use $s = r\theta$ to obtain the distance required.)

2. A pulley of diameter 32 cm drives a smaller pulley of diameter 12 cm by a drive belt. The two pulleys are situated 0.4 m apart, between centres. Draw the system to scale and determine the total length of drive belt that is in contact with the pulleys.

Mathematics for Engineering Technicians

Circular motion

Many objects and parts of machinery rotate or make angular movements. In electrical engineering, the generation of electricity depends on the rotation of conductors and magnetic fields. During the design processes, engineers will more than likely use calculus to carry out some of the calculations, and since the radian is a fundamental part of a circle and not just some arbitrary measurement by degree, the calculus works and gives answers that do not need to be constantly converted between degrees and radians, and vice versa.

Exercise 4.36

Convert the following angular speeds from r.p.m. to radians per second.

Remember that there are 2π radians in one revolution, and 60 seconds in one minute.

When a shaft or flywheel rotates, there are different ways to state the speed of rotation:

- revolutions per second
- revolutions per minute
- revolutions per hour
- degrees per second
- radians per second.

1. A flywheel rotates at $\frac{\pi}{2}$ radians per second. How many revolutions per minute is this?

2. A three-phase generator of electricity has three separate sets of windings, which are physically spaced 120° apart. The three lines are identified as L1, L2 and L3. If L1 is used as the reference voltage, the other two are lagging behind it by 120° and 240°, respectively. The angle is usually written as the reference plus the angle by which it lags behind the reference. What are 120° and 240° when written as multiples of π? (Hint: look back at the previous question to check for any similar values.)

Triangular measurement: sine, cosine and tangent functions

By the end of this section you will have developed a knowledge and understanding of functions; sine, cosine and tangent graphs; values of trigonometric ratios between 0° and 360°; periodic properties of trigonometric functions; the sine and cosine rules; and practical problems.

Have you ever wondered what the words 'sine', 'cosine' and 'tangent' actually mean? The best way to see these terms is to accept them as tools which help us to solve problems.

Sine, cosine and tangent (often abbreviated to sin, cos and tan) are ratios which were worked out for some sample triangles as mathematicians were busily looking for the secrets of the Universe and life itself.

You may know of SOHCAHTOA. This mnemonic may help us to remember the first letters of the trigonometric ratios for a right-angled triangle.

Assume that we are referring to an angle θ (lower-case Greek letter theta).

The ratios are:

$$\sin\theta = \frac{\text{opposite}}{\text{hypotenuse}}$$

$$\cos\theta = \frac{\text{adjacent}}{\text{hypotenuse}}$$

$$\tan\theta = \frac{\text{opposite}}{\text{adjacent}}$$

What exactly do we mean by opposite and adjacent? Opposite to what, and adjacent to what? Opposite to the angle θ and adjacent to the angle θ?

The right-angled triangle and use of θ

Figure 4.19 shows a right-angled triangle.

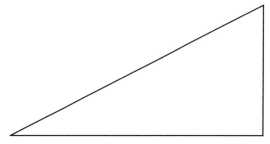

Figure 4.19 Right-angled triangle

Make the Grade

Grading criterion P4 requires that you solve circular and triangular measurement problems involving the use of radian, sine, cosine and tangent functions.

165

The right angle is in the bottom right-hand corner, for no reason at all, other than that is where I have drawn it. The longest side of a right-angled triangle is called the hypotenuse, and it is always opposite the right angle, but it is never referred to as the opposite side.

Which side is called the opposite and where is the adjacent one? We will come to the names of the sides very shortly, but first we need to consider one of the angles that is not a right angle.

Which angle is θ?

Key Points

The use of θ or any other symbol, to represent an angle is just to inform everyone that we are considering the angle that we have just decided to consider.

Which side is opposite to what, and where is the adjacent?

For example: consider the angle in the bottom left-hand corner, and let's call it θ.

Exercise 4.37

Using a pencil and rule, draw a right-angled triangle similar to the one drawn in Figure 4.19. The side lengths can be anything you want, as long as the bottom right-hand corner is at 90°.

In the bottom left-hand corner, using a pen or pencil, label this angle θ.

In relation to the angle θ that you have just labelled, write 'opposite' on the side that is opposite to the angle θ.

Ignoring the hypotenuse, which already has a name, write 'adjacent' on the side that is adjacent to the angle θ.

The third angle

The third and final angle has not yet been labelled. Let us call this α (lower-case Greek letter alpha).

Using a different colour pen, label the third angle α.

Look at the sides of the triangle and decide which side is opposite angle α. Write 'opposite' on that side (and you should now see why a different colour pen was needed).

The other side, which is adjacent to α can now be labelled 'adjacent'.

Confusion over names and labels

You now have a right-angled triangle with a hypotenuse, two angles labelled θ and α along with two sides, each labelled opposite and adjacent – hopefully in different colours.

How can a side be opposite *and* adjacent? Because the labels are not absolute – they are relative, relative to the angle to which they relate.

Obtaining the fundamental trig ratios

Exercise 4.38

Start with a new right-angled triangle, but this time draw it with some specific side lengths.

Make the base or horizontal side 4 cm and the upright or vertical side 3 cm.

You should find that the hypotenuse is 5 cm long. If not, check the other measurements.

When you have drawn the triangle, label the bottom left-hand angle with a letter or symbol such as α or A or *x* or whatever you like.

Now, make use of SOHCAHTOA, or any other mnemonic that you use to remember the ratios, and complete the following table.

Ratio	Values and calculation	Final value
sine	$\frac{\text{opp}}{\text{hyp}} =$	
cosine	$\frac{\text{adj}}{\text{hyp}} =$	
tangent	$\frac{\text{opp}}{\text{adj}} =$	

Production of values and look-up tables

Remember the previous section where the log tables were referred to? Someone had to calculate all the values by hand and compile the tables. The same thing was done with sin, cos and tan tables as well. These are also in your calculator.

How do we know they are in there for 'look up' and they are not calculated when we press a calculator button? With your calculator in degree mode, enter sin 30°. You should get 0.5, or you may get 0.50000.

For this calculation, how did the calculator know what the length of the opposite side and the hypotenuse were? Does it matter if the lengths are different? For different angles, you will get different sine values, but the sine (and the others) are just 'ratios' of two sides.

Exercise 4.39

1. Draw a horizontal line across a sheet of paper, using a pencil and rule. From the bottom left-hand corner, using a protractor, measure 30° and draw a line at 30° to the horizontal line. Experiment and investigate the ratios of opposite and hypotenuse, for any length, by drawing three or four vertical lines (at 90° to the horizontal line), to create three or four triangles. For each triangle in the set, measure the opposite side and the hypotenuse, and find the sine ratio. What do you observe?

2. You could use the same set of triangles to carry out the same ratio calculations for the cosine and tangent of 30°, or even for the 60° angle too.

They are just ratios and are not related to the size lengths, just the ratio of side lengths. Hence, if a calculator was calculating the sine, cosine and tangent values, what side lengths would it use?

What do trigonometric functions look like?

In this section, we will produce sketches of each of the three fundamental trigonometric functions.

Exercise 4.40

Draw a table similar to the following example and complete the trig ratio values (sin, cos and tan) in the appropriate column.

angle θ°	sinθ°
0	0.0000
30	0.5000
60	
90	
120	
150	
180	0.0000
210	−0.5000
240	
270	
300	

330	−0.5000
360	0.0000

angle θ°	cosθ°
0	1.0000
30	0.8660
60	
90	
120	
150	
180	−1.0000
210	−0.8660
240	
270	
300	
330	0.8660
360	1.0000

angle θ°	tanθ°
0	0.0000
30	0.5774
60	
90	
120	
150	
180	0.0000
210	
240	
270	
300	
330	−0.5774
360	0.0000

You may notice something strange about the values for the tangent, and it may help if you work in steps of 10° instead of 30°.

After you have completed the tables, transfer this data onto a graph, or a separate graph for each function if this makes it clearer for you and easier to refer to.

Make the Grade — P5

Grading criterion P5 requires that you sketch each of the three trigonometric functions over a complete cycle.

You may recognise the sine wave, and possibly the cosine wave, but the tangent usually causes some confusion, particularly if you try to use Microsoft Excel® to draw the graph.

The problem is that the tangent curve is **asymptotic** at 90° and 270°.

Key Term

Asymptote – Something that gets closer and closer to something, but never actually touches it, except at infinity.

As the angle approaches 90°, the tangent of the angle becomes extremely large. At 'almost' 90° it will be several million or more.

The tangent of 89.99° = 5,729.577

tan89.9999° = 572,957.795

At tan89.999999° calculators start to struggle and the value exceeds the number of digits it can show.

tan 90° = 'error' because this is infinity and in maths, and calculators, infinity is not defined, because it cannot exist.

As the angle then goes slightly beyond 90°, something even stranger happens.

Exercise 4.41

Copy and complete the following table, using your calculator to obtain the tangents.

Angle (°)	Tangent of the angle
89.99	
89.9999°	
89.999999°	
Ignore 90°	
90.000001°	
90.0001°	
90.01°	

Another way to obtain the tangent waveform

Think back to the right-angled triangle and the fundamental trig ratios for sin, cos and tan.

Also remember **SOHCAHTOA**.

Exercise 4.42

Remembering that $\sin\theta = \frac{opp}{hyp}$ and $\cos\theta = \frac{adj}{hyp}$ (and remembering some algebra, simplifying equations, and so on) rewrite $\frac{\sin\theta}{\cos\theta}$.

Substitute for $\sin\theta$ and $\cos\theta$ in terms of the triangle side names – that is, write $\frac{opp}{hyp}$ divided by $\frac{adj}{hyp}$. Then compare your result to the ratio for $\tan\theta$.

As the angle approaches 90°, the value of $\sin\theta$ approaches 1 and $\cos\theta$ approaches 0. Trying to divide anything by zero is not possible (it is not defined) – hence the inability of your calculator and Microsoft Excel® to work out tan90°.

Looking ahead

The combination and manipulation of trig functions is an essential skill, which will develop further as you progress your studies to a higher level, if your career intentions and aspirations take you that way. If your course includes further mathematics later on, you will see a few more of the trig functions, and although they may seem cumbersome, they all have a real use in engineering.

Periodic properties of the trigonometric functions

Now you have investigated some of the fundamental trig functions, look again at the graphs or waveforms that you produced for the sine, cosine and tangent ratios.

They all follow a pattern which repeats every 360° (or every 2π radians)

This property is called periodic and periodicity is the repeating of the rises, falls and polarity changes at a fixed rate or frequency.

For example, the electricity supply is alternating current and it follows a sinusoidal pattern.

Activity 4.7

What other examples of periodicity can you think of? Make a list to discuss with your classmates.

So where might these ratios be useful?

All branches of engineering use measurements of angles and rotation, whether these are angles for cutting pieces of material for joining together, shaping and bending, or for the angle of the crankshaft of a petrol engine, to ensure that the ignition ignites at the correct angle before top-dead-centre, or for the rotational speed of a generator, to determine the frequency of the electricity it is producing.

Practical problems include:

- calculation of the phasor sum of two alternating currents
- resolution of forces for a vector diagram.

Key Terms

Phasors and vectors – These are two aspects of the same concept:

- Vectors represent quantities that have magnitude and direction.
- Phasors are the vectors of electrical quantities.

Vectors can be used to combine forces, which you may have come across in the science unit, or in your previous studies.

Scale drawing

Worked Example

Consider the two tugboats pulling a large ocean-going vessel.

If both tugboats are applying equal forces to move the large vessel, their combined or resultant force will not simply be the sum of the two.

They are combined using vector addition.

To see this, consider that both tugboats are applying a force of 10 kN in the direction of their respective arrows.

What is the total (or resultant) force being applied to the large vessel, and in what direction?

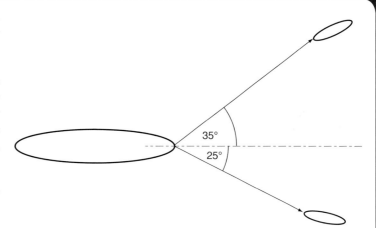

Figure 4.20 Tugboats pulling an ocean-going vessel

To solve this, there are two methods:

- by scale drawing
- by calculation.

169

Assume a scale of 1 cm = 1 kN.

From the ends of each of the respective vectors, draw a line parallel to the other vector, to create what is known as the parallelogram of forces.

Draw the resultant from the origin to the opposite point where the two other lines meet.

The length of the resultant indicates the combined force, using the same scale of 1 cm = 1 kN.

The angle between the resultant and the centre line indicates the sense or direction of this force.

Draw this to scale and record the value for the resultant force, and the angle between it and the horizontal centre line.

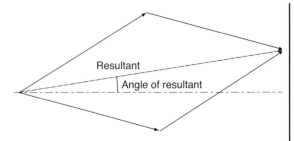

Figure 4.21 Scale drawing for tugboats

In Figure 4.20, the forces are drawn in the same directions as the tugs are indicating, and their lengths are drawn to a suitable scale, and in this case, with reference to the centre line shown.

Calculation

The two (or more) original coplanar forces need to be resolved into their vertical and horizontal components.

If a force is pulling at 90° to the horizontal, it is reasonable to expect that the object being pulled will move vertically. Similarly, if a vertical force is applied to a lift, then we would not expect it to move sideways.

Where forces are applied at some angle to the horizontal or vertical other than 90°, we consider the force as being a mixture or combination of a vertical component and a horizontal component.

Consider the force being applied by the tugboat at 35° to the straightforward direction from our example. This is shown in Figure 4.22, with its components indicated in the horizontal and vertical planes.

The vertical and horizontal components of the first force are shown in Figure 4.22.

The horizontal component of the force is given by $F_h = F\cos\theta$.
The vertical component of the force is given by $F_v = F\sin\theta$.

When these components are in the opposite directions (down or to the left) they are worked out in a similar fashion, but have negative values).

For the diagonal forces at $+35°$ and $-25°$ in the example, the components equate to the following;

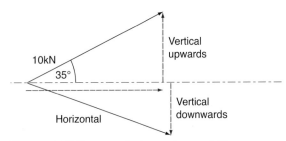

Figure 4.22 Force applied to tugboat at 35°

Force and angle, (with direction)	Horizontal components ($\cos\theta$)	Vertical components ($\sin\theta$)
10 kN, 35°	8192 N (to the right)	5736 N (upwards)
10 kN, −25°	9063 N (to the right)	−4226 N (downwards)
Totals	17 255 N (to the right)	1510 N (upwards)
Resultant	17 321 N, angle 5°	

Let us have a look in more detail at how the resultant was obtained.

Worked Example

For the resultant vector or force, use Pythagoras' theorem to combine two lines or vectors at 90° to each other.

Tangent $= \dfrac{\text{opposite}}{\text{adjacent}}$.

17 255 N is the horizontal $\dfrac{\text{distance}}{\text{length}}$ (adjacent to the angle)

1510 N is the vertical length, at a right angle (opposite to the angle)

Hence, $\tan\theta = 1510/17\,255 = 0.0875$

The angle whose tangent is $0.0875 = \tan^{-1} 0.0875 = 5°$

Check that these values are the same as you obtained when you drew to scale and measured them.

If your calculator has an 'R-P' button, you could use this. R-P means 'rectangular' to 'polar' conversions and, effectively, does both the above calculations for you in one go, but stores one answer in a second memory. Check your calculator instructions to find this second memory. This could be labelled X→Y, which means 'swap the contents of the two memories, X and Y, over'.

Here is an example for a three, four, five right-angled triangle.

Use a calculator.

Enter the following four steps: 3, R-P, 4 =

The value 5 should be on your screen, being the hypotenuse length.

Now swap memories, which could be X→Y, and you are shown 53.13°.

Now spend some time getting to know your calculator and the use of R-P conversions and/or Pythagoras and the tangent function.

Activity 4.8

Working in pairs or on your own, create a range of problems involving the combination of vectors that relate to the branch of engineering that you are studying.

These could be forces being applied at different angles, with sketches or diagrams to illustrate the force magnitude and direction, or they could be in the form of a table, like the ones in Exercise 4.43.

Swap the problems you have created with someone else, who will then solve them while you solve someone else's.

Create your own explanation of the processes involved in what is also known as vector addition, using no more than a single page of A4.

Your teacher will select some of the group to stand up and explain the summary page to the rest of the group.

Exercise 4.43

Find the length of the hypotenuse for each of the following right-angled triangles, and the angle between the hypotenuse and the base in each case.

Base length (cm)	Opposite side (cm)	Hypotenuse (cm)	Angle (°)
5	9		
2	9		
3	12		
5	35		
576	852		

Electrical vectors are called phasors

Electricity is generated and distributed in three phases, each 120° apart. Each single phase taken from that supply consists of a phase (or line) and neutral. Depending on the type of circuit that is connected to the supply, the current behaves differently. For example, if the load is a heating element, the load is resistive. As the voltage increases, so does the current, and as the voltage decreases (within the sinusoidal waveform), so does the current. They are said to be 'in phase'.

When a capacitor or inductor (coil/transformer/choke) is connected, the current either leads or lags behind the voltage, causing a phase difference.

For a series capacitor/resistor circuit, the phasor diagram in Figure 4.24 illustrates the combination of capacitance and resistance by considering the phasor (or electrical vector) addition of two voltages.

Where do the two voltages come from?

Consider the series circuit in Figure 4.23:

In any series circuit, the sum (phasor sum) of the voltage drops around any closed loop (and here there is only one such loop) that is equal to the supply voltage.

Hence, from the circuit shown,

$$V_s = V_R + V_C$$

(but these are phasors and this is phasor addition, so, more correctly)

$$\overline{V}_s = \overline{V}_R + \overline{V}_C$$

(where the 'bar' on the top indicates that vector or phasor quantities are being represented).

Some textbooks also use a circumflex accent ˆ instead of the bar to indicate vector quantities.

To complete the phasor diagram, you need to know that the voltage across a capacitor lags behind the current in the circuit by 90°.

Hence, a phasor diagram for the CR circuit would look like Figure 4.24.

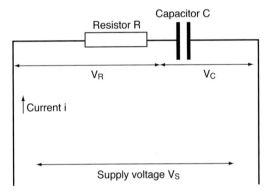

Figure 4.23 Series circuit

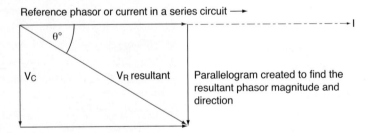

Figure 4.24 Phasor diagram

Exercise 4.44

1. If the value of V_C is 184 V and the value of V_R is 169 V, calculate the supply voltage V_S.

2. Using CAH from SOHCAHTOA or $\cos\theta° = \dfrac{\text{adjacent}}{\text{hypotenuse}}$, determine the angle between the supply voltage V_S and the supply current I.

Remember that the value for V_C will be negative, due to pointing 'downwards', which is equivalent to negative upwards.

3. In pairs, or on your own if you prefer, solve the following problem. An electrical coil possesses two electrical properties, resistance (R) and inductance (L), from which the inductive reactance, X_L, can be determined. These two properties allow us to treat the coil as a 'lump' of resistance in series with a 'lump' of inductive reactance.

A coil has a resistance R of 50 Ω and an unknown value of inductive reactance, X_L. When connected to a supply of 100 V d.c., a current of 2 A flows, which confirms the resistance value, using Ohm's law, $I = \dfrac{V}{R}$.

When connected to 260 V a.c., a current of 2 A then flows through the coil. The a.c. current is opposed by the resistance and the inductive reactance in combination, where

$$\overline{Z} = \overline{R} + \overline{X}_L \text{ ohms}$$

given that $V_s = 260$ V a.c.

and that current $I = 2$ A

also given that $R = 50$ Ω

and the value of Z (the circuit's impedance) $= V_s/I$

then find:

- the value of the inductive reactance X_L.

What about non-right-angled triangles?

So far, we have looked at triangles where one angle is 90° and have investigated how they can be made use of in engineering, along with the algebra and trig ratios that have been developed to help us solve problems.

In this section, we will be considering some examples where none of the sides is 90°; and not having a right angle, there is no hypotenuse, only three sides. These sides are generally labelled a, b and c, and the angle opposite each labelled side is labelled with the same letter, but as a capital, A, B or C (see Figure 4.25).

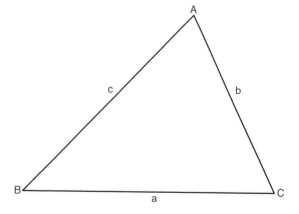

Figure 4.25 Non-right-angled triangle

The sine rule

For any triangle, the following ratios apply:

$$\frac{a}{sinA} = \frac{b}{sinB} = \frac{c}{sinC}$$

Try this with a right-angled triangle that you have worked on before, and see if it really does work for ALL triangles. You should find that it does.

Using the sine rule is like using any other equation. You decide which two of the three side/angle ratios you wish to work with (although this is usually decided by the information with which you are presented).

Then you rearrange and substitute the known values in order to obtain the unknown.

Limitations for use of the sine rule are that you must have:

- one side length and any two angles, or
- two sides and an angle which is opposite to one of these sides.

The second option can lead to an ambiguous situation where you end up with two answers.

Remember the sine values and sine wave diagrams you drew?

Look back at the waveform you plotted and select the angle that has a sine of 0.5.

Alternatively, using your calculator, find the sine of 30° and 150°.

Two angles have the same value of sine, hence the ambiguity with the second option.

All questions refer to the labelling on the previous triangle, where A is opposite a, B opposite b, and C opposite c.

$$\frac{a}{sinA} = \frac{b}{sinB} = \frac{c}{sinC}$$

Hint: always sketch and label the triangle or other shape for reference, to allow you to see the question and how you are working towards the answers.

Exercise 4.45

1. If $a = 12\,cm$, and given that $A = 50°$ and $C = 38°$, find the other two lengths and the unknown angle.

2. Triangle ABC has unknown base (side b) length, and $a = 6\,cm$, $c = 4\,cm$. If $B = 40°$, find the two possible values for angle A.

When might the sine rule not be useful?

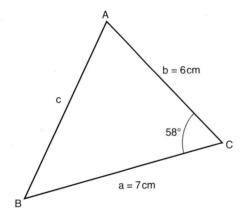

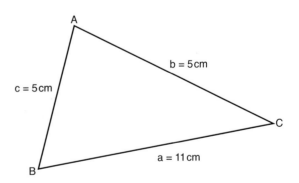

Figure 4.26 Examples where the sine rule is not useful

The cosine rule

This is the great fallback rule, which is used whenever the sine rule cannot be used.

In Figure 4.26, there are two triangles. One of them has two sides and the included angle (such as angle $C = 58°$). The other one has all sides, but no angles. These require the cosine rule to solve them.

The cosine rule is a general expression which links all three sides and one angle.

Hence, if we have two sides and the included angle, we can calculate the side opposite the angle.

Also, if we have three sides, we can rearrange the expression to obtain the required angle.

The cosine rule expressions:

$$a^2 = b^2 + c^2 - 2bc\,cosA$$
$$b^2 = a^2 + c^2 - 2ac\,cosB$$
$$c^2 = a^2 + b^2 - 2ab\,cosC$$

Worked Example

In a triangle, side $a = 7\,cm$, side $b = 5\,cm$ and the included angle $(C) = 75°$.

Use the cosine rule to find the other side, c, then solve to find the other two angles, using the sine rule (because it is easier to rearrange).

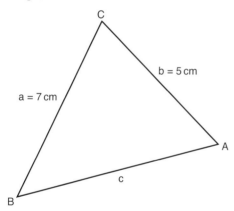

Figure 4.27 Cosine rule

Start by sketching the triangle and labelling what you already know.

Using

$$c^2 = a^2 + b^2 - 2ab\,cosC$$

$$c^2 = 49 + 25 - (2 \times 7 \times 5)\,cos75°$$

$$c = \sqrt{(74 - 18.117)} = 7.48\,cm$$

Now, using the sine rule, another angle can be found, say A,

using

$$\frac{a}{sinA} = \frac{b}{sinB} = \frac{c}{sinC}$$

$$\frac{7}{sinA} = \frac{7.48}{sin75°}$$

$$sinA = 7\,sin75°/7.48$$

$$sinA = 0.9039$$

$$A = sin^{-1}\,0.9039 = 64.68°$$

$$B = 180° - 75° - 64.68° = 40.32°$$

Exercise 4.46

Revisit the two triangles that were given as examples which cannot be solved using the sine rule (see Figure 4.26).

Solve these to find all unknowns.

Make the Grade
P6

Grading criterion P6 requires that you produce answers to two practical engineering problems including the sine and cosine rule.

Mensuration

By the end of this section you will have developed a knowledge and understanding of how to use standard formulae to solve surface areas and volumes of regular solids.

Key Term

Mensuration – This is the use of mathematical tools to help with physical measurements.

In engineering, there are many measurements that are made regularly as part of everyday work. These measurements are generally applications of units and quantities which have been derived from the fundamental SI units.

For example, the measurement of length or distance is a fundamental quantity and uses the fundamental SI unit of length, the metre.

Area is length $\times$ length, so the derived units are metres $\times$ metres or m^2.

It follows, then, that volume being area $\times$ length, $= m^3$

and so on.

However, in engineering, things are not that straightforward. There are squares and cubes, rectangles, trapezoidal shapes, cones, cylinders, and so on.

Common shapes and regular solids

For the objects and shapes we make regular use of, there are standard formulae to work out surface areas (to determine the amount of material required) and volume (to allow us to work out the weight of an object if we know it is density).

Activity 4.9

Either working on your own or in small groups, complete the following table with ideas about where you might need to know this detail in real life and in engineering. At the bottom are two blank rows for you to choose another shape to include in the list.

Some initial ideas have been included to help you get started. Modify these as you think fit, to make more interesting challenges.

Once complete, exchange your table with someone else so that they can work out some values to explain your suggested application problems.

Quantity	Formula	Application at home	Application in engineering
Volume of a cylinder	$\pi r^2 h$		Pneumatic or hydraulic cylinder or car engine piston displacement. Why is a 3.4 litre V6 engine called 3.4 litre?
Total surface area of a cylinder	$2\pi rh + 2\pi r^2$	Amount of insulation to cover a hot-water cylinder (with only one end – the top) – the formula becomes $rh + r^2$ if the cylinder is, for example, 1.4 m high and 0.45 m diameter.	
Volume of a sphere	$\frac{4}{3}\pi r^3$		
Surface area of a sphere	$4\pi r^2$	Someone makes spherical candles of 150 mm diameter and wants to coat them with glitter. The adhesive for the glitter is supplied in tubes of 100 ml, and each has a surface coverage of 0.33 m². How many tubes would be needed to totally cover 24 candles?	
Volume of a cone	$\frac{1}{3}\pi r^2 h$		

Quantity	Formula	Application at home	Application in engineering
Curved surface area of a cone	$\pi r \times$ slant height		An exhaust pipe is made of sheet metal formed into a cone, with a small cone cut off the end. Determine the amount of sheet metal required for a cone-shaped exhaust 0.45 m, long with diameters of 2 cm at one end and 8 cm at the other.
Volume of another shape of your choice			
Surface area of your chosen shape			

Recap and summary

Standard formulae used to solve surface areas and volumes of regular solids:

- volume of a cylinder $= \pi r^2 h$
- total surface area of a cylinder $= 2\pi rh + 2\pi r^2$

- volume of sphere $= \frac{4}{3}\pi r^3$
- surface area of a sphere $= 4\pi r^2$
- volume of a cone $= \frac{1}{3}\pi r^2 h$
- curved surface area of cone $= \pi r \times$ slant height.

Exercise 4.47

1. Assume that ice cream has the same volume when frozen as it does after melting. Given a cone, what diameter sphere of ice cream can be put on the top to ensure that all the cone will be full, with no overflow, after it melts?

2. A company making metal fishing weights buys metal stock on rolls of 0.25 cm diameter wire to melt into the moulds for fishing weights. The weights are in three regular shapes: spheres, cylinders and cones. Determine the relationships between the length of stock metal required for each shape, and test this mathematically for three different sizes of each. The sizes and cone slope angle, and so on, can be any reasonable value of your choice.

Figure 4.28 Ice cream cone

3. A hydraulic machine consists of four cylinders, with inside dimensions of 15 cm diameter and operation length of 3.6 m. The total length of pipes used from the tank to the cylinders is 42 m and the inside bore of the pipe is 22 mm. Work out the amount of oil that would be required to completely fill the pipes and extended cylinders. Then double this to obtain a recommended volume of oil to be poured into the system to allow for leakage and cooling.

Make the Grade
P7

Grading criterion P7 requires that you use standard formulae to find surface areas and volumes of regular solids for three different examples respectively.

4.3 Using statistical methods to display data

This section will cover the following grading criteria:

Statistics is a branch of maths where values and data are manipulated to produce meaningful reports of what has happened, and possibly to forecast what is likely to happen in the future, assuming that things continue to progress in more or less the same fashion.

Working with and manipulating data (not changing it) allows us to represent them using a range of methods to suit our needs. In this section you will see how statistical diagrams are used to represent data, including:

- bar charts
- pie charts
- frequency distributions
- class boundaries and class width
- frequency tables.

You will also see how the following are used to help us to make sense of data and numbers:

- discrete variables
- continuous variables
- histograms
- cumulative frequency curves.

Within all these, you will carry out a range of calculations and observations, involving:

- mean
- mode
- median
- discrete data
- grouped data.

Statistical measurement

By the end of this section you should be familiar with meanings and uses of arithmetic mean, median, mode, discrete and grouped data.

Make the Grade

Grading criterion P9 requires you to determine the mean, median and mode for two statistical problems and explain the relevance of each average as a measure of central tendency.

As the introduction to this section has shown, there is much mathematical or statistical jargon to cover.

Activity 4.10

1. Do this on your own. To introduce you to some jargon, spend about half an hour searching for the definitions and explanations of the following terms. You might know some of them already.
   - Term
   - Variable
   - Mean
   - Mode
   - Median
   - Discrete data
   - Continuous data
   - Grouped data
   - Bar chart
   - Pie chart
   - Histogram
   - Frequency distribution
   - Class boundary

- Class width
- Frequency table
- Variant
- Cumulative frequency
- Cumulative frequency curve

You will have found that many of these terms have a range of explanations; it might help you to learn a little more if you organise a discussion about your findings and come to a class or group definition.

2. Share your findings. Using the information that you have written down, share your findings with others in your class using a snowball technique. Snowballing means that, in pairs, you should share your definitions, meanings and explanations to arrive at one agreed combined meaning or definition. You should then share these with another pair, combining them into one agreed meaning. Next, share in groups of 8, then 16, and so on, until you have a shared group definition or meaning for each of the terms. If you cannot agree, leave both definitions in.

3. Write them out, pin them on a wall, type them up, put them on a virtual learning environment (VLE) or print them out and put them at the start of this section in your notes. Use them in any way that will help you to learn them and provide an accessible reference whenever you need to remind yourself of the meanings.

Data – used and abused in many instances

The need to attract customers can sometimes lead to news headlines using statistics and data that are designed to shock.

Imagine what you would think if you saw the following headline (created for illustrative purposes):

50 per cent less likely to die before the age of 18 if you eat raw vegetables.

One's immediate reaction is that 50 per cent means half – so, many of us think that this is telling us that half of us will live longer if we eat raw vegetables. Further investigation might reveal that a sample of 10 000 teenagers, from 14 to 18 years' old, were monitored over four years. In this sample, it was reported that around half of them ate raw vegetable, (about 5000 teenagers). The others ate cooked vegetables.

The data also showed that three of the 10 000 teenagers had died, and one of these ate raw vegetables. No details of the causes of the deaths could be found, and there was no apparent definite link between eating cooked or raw vegetables.

The headline could have also said 'twice as likely to die if you eat cooked vegetables', which makes cooked vegetables look like killers.

Figure 4.29

Exercise 4.48

For the example given on the previous page, how could the data have been represented more realistically and made less shocking? What is your overall opinion of the use of such data and what the details actually represent?

Next time you pick up a newspaper or listen to some news, try to find examples of how statistics and data have been used to attract customers. You might be surprised.

What do numbers tell us?

Everyone has some understanding of numbers. We can all look at a series of numbers and interpret some meaning from it.

Exercise 4.49

Look at the following set of numbers and write down a few comments to describe the information that the numbers portray to you.

3, 7, 5

So what do the numbers actually mean? Until we know what we are interpreting, the numbers mean absolutely nothing, other than the things you imagined or pieced together from your previous knowledge, and the human instinct of trying to put things into recognisable patterns.

Using a larger set of numbers

When the set of numbers is larger, there is obviously more data, but we still need some tools to interpret what it is telling us and we need to know what we are looking for. Look at the following set of numbers:

3, 5, 2, 7, 4, 4, 1, 9, 3, 5, 2, 6, 8, 4, 7, 4, 2, 5, 4, 3

What meaningful understanding do they offer? Again, none, really, or not until we are asked to interpret them, or told to look for some kind of pattern or relationship that the numbers are conveying.

Here, then, is a context: assume the set of numbers represents the number of cars that people have owned so far in their lives. Their ages are irrelevant, particularly for the task at hand.

Exercise 4.50

To establish what you already know, from GCSE or other studies, and using the given data, answer the following questions.

1. What is the highest number of cars that someone has owned?

2. What is the lowest number of cars that someone has owned?

3. What is the most common number of cars that the people have owned?

4. What is the least common number of cars that people have owned?

5. What is the median number of cars that people have owned?

6. What is the average number of cars that people have owned?

How did you get on with the six questions?

Questions 1 and 2 could be answered by scanning across the row of numbers, looking for the highest or lowest number.

Questions 3 and 4 need some thought. They could be answered by making a tally of frequencies for each number – for example, how many 3s, how many 4s, and so on, then identifying the highest and lowest frequencies of occurrence.

Question 5 requires us to put the numbers in numerical order to identify the value of the number which is physically (not numerically) in the middle.

Question 6 requires some calculations, but before we start with these, we need to understand exactly what is meant by the term 'average'. The question should really have asked you to find the arithmetic mean, or the mean value of the number of cars.

'Average' is a term which is used to refer loosely to a distribution of values, and there are a few types of average, as you have seen already, from your work on finding the definitions of terms in Activity 4.10.

Go back to question 3 in Exercise 4.50.

Did you use a tally chart looking something like the one opposite?

Number of cars owned	Tally of occurrences	Frequency of this number occurring
I	I	I
2	I I I	3
3	I I I	3
4	I I I I I	5
5	I I I	3
6	I	I
7	I I	2
8	I	I
9	I	I

Now consider question 5. First of all, we usually work better with numbers when they are in numerical order, because that is what we are used to.

Original sequence

3, 5, 2, 7, 4, 4, 1, 9, 3, 5, 2, 6, 8, 4, 7, 4, 2, 5, 4, 3

Sequence in numerical order

1, 2, 2, 2, 3, 3, 3, 4, 4, **4, 4**, 4, 5, 5, 5, 6, 7, 7, 8, 9

As there are 20 numbers, the median is halfway between the 10th and 11th number (these are shown in bold).

In this case, the 10th and 11th numbers are both 4, so the median is 4.

Finally, we come to question 6. The average, or the arithmetic mean (to use the proper term), is calculated by taking the sum of all the numbers = 88, and dividing this by 20 (the number of numbers); we see that the average or arithmetic mean is 4.4.

A quick recap: this means that from the data provided, the average (arithmetic mean) number of cars that the sample of 20 people have owned in their life so far is 4.4 cars.

To summarise what the data have told us so far: 4.4 is the average number of cars across the sample taken (this is called the mean). When all the numbers of cars that people in the sample have owned are placed in numerical order, the number of cars which is the mid value is four cars (this is called the median). From our sample of 20 people, four is the most common (which is called the mode) number of cars that people have owned in their life, so far.

All these figures tell us something about the spread of the data and the central tendency of the set of numbers. These are sometimes referred to as measures of central tendency, measures of location, measures of central value or simply the average values.

Key Terms

Median – The number that is in the middle when all the numbers are placed in numerical order.

Mode – The number that occurs most often. The mode is used to represent a typical item or person – for example, Mr Average – and the type of television he would buy to watch sport, the type of car he buys, and so on.

Mean – The arithmetic mean is the sum of all numbers divided by the number of numbers. This is what most people generally refer to when they talk about the average.

Unfortunately for the average person, all three terms are commonly referred to as average values.

To define loosely what each kind of average represents, consider the following:

- Mean – the arithmetic mean – the value obtained when all the numbers are added together and the sum divided by the number of samples.
- Mode – most common or most often, the 'fashion' – with all the values being considered, this is the value that appears most often, providing some kind of 'averageness' to the value.
- Median – the mid value when all the values are put into numerical order. There are as many values higher than the median as there are lower than it, so it is a representation of 'middleness' or 'average'.

In statistics, or engineering (and life in general), the use of the correct 'average' or 'measure of central tendency' is important.

Worked Example

The lengths of ten bolts are measured and six of them are 51 mm long, two of them are 55 mm long, one of them is 49 mm long and the final one is 42 mm in length.

How could one or more of the three types of 'average' provide us with information that would not be considered useful for the use of a bolt, which must be no shorter than 48 mm (to allow enough threads to provide sufficient grip) and no longer than 52 mm (because of the depth of the hole or the space available)?

Figure 4.30

Mean = 50.7 mm
Median = the value in the middle when lined up in numerical order = 51 mm
Mode = most common = 51 mm

The mean value tells us that, on average, the bolts are 50.7 mm long. Common sense tells us that *none* of the bolts is 50.7 mm long.

This does not mean that the mean is a useless value, but it does mean that the mode or the median have more significance than the mean in this instance.

Exercise 4.51

What if the numbers of cars was as follows?

1, 1, 2, 2, 2, 2, 3, 3, 3, 3, 4, 4, 5, 5, 5, 6, 7, 7, 8, 9

1. What are the values of the median and the mode for the new range of values?

2. 1–15 resistors are bought to construct part of a circuit. Their values are indicated in the following table. What are the mean, the mode and the median values?

Resistor	Resistance (Ω)
1	33.2
2	33.1
3	32.1
4	33.5
5	33.0
6	33.2
7	33.7
8	33.0
9	29.9
10	31.4
11	33.1
12	33.4
13	29.7
14	29.1
15	33.1

Again, you should have found that the mean value does not represent a real value of any of the components.

3. Assume that the purchaser wanted resistors with a value of 33 Ω exactly.

Only two would be acceptable, with 13 rejected.

(a) If the acceptable value was 33 ± 10 per cent, how many would be suitable?

(b) If the acceptable value was 33 ± 5 per cent, how many would be suitable?

The arithmetic mean is a value that is determined by arithmetic processes, whereas the mode and the median seem to represent actual sizes that occur. But this is not always the case.

Finding the mean value of a frequency distribution

Worked Example

One task that has to be carried out by a maintenance engineer is maintaining the lighting, either in a factory, an office or any workspace. Assume that strip lights are in use in a factory, and the level of illumination has to be maintained at a certain level for the sake of the quality of the product being manufactured and for safety reasons.

The following data have been collected over a few years and the planning engineer has to determine when to have the tubes changed.

Lifetime (days)	150–159	160–169	170–179	180–189	190–199	200–209	210–219
Frequency	2	16	53	80	61	32	12

Use these grouped data to determine the mean lifetime of the strip lights.

Then explain how the data and the mean lifetime can be used to determine a rota for changing/replacing the tubes before they fail.

This problem requires a slightly different technique.

The data are telling us that two lights failed within 150–159 days of use.

You could take the pessimistic approach and assume that they both 'popped' at close to 150 days, or an optimist might accept that they lasted 159, or thereabouts.

Generally, the limit that you would use would depend on the requirements of the problem, but we will work with the middle of the ranges, or the approximate average of each range.

Hence, the table becomes:

Lifetime (days)	155	165	175	185	195	205	215
Frequency	2	16	53	80	61	32	12

Now multiply each value by its frequency (to get xf), then divide by the sum of the frequencies (the total number of lamps) to get f.

Exercise 4.52

1. Work through this problem to determine the average (mean) life of a lamp.

2. What values would the pessimist and the optimist obtain if they used the lower class boundaries and the upper class boundaries, respectively (that is, 150, 160, and so on, or 159, 169, and so on?)

3. In small groups, or on your own if you prefer, discuss what these calculations tell the maintenance engineer about the frequency of replacement (before they blow) of the lamps.

Data handling

By the end of this section you should be able to interpret and produce data represented by statistical diagrams – for example, bar charts, pie charts, frequency distributions, class boundaries and class width, frequency tables, variables (discrete and continuous), histograms (continuous and discrete variants), cumulative frequency curves.

Make the Grade   P8

Grading criterion P8 requires you to collect data and produce statistical diagrams, histograms and frequency curves.

Collecting data or large sets of numbers and putting them on a page does not help with their interpretation. What are the numbers trying to tell us?

For example, consider a pizza place which has five motor scooters for use by the delivery team.

Over a period of a month, the daily mileage of each of the motor scooters is monitored.

	Motor scooters' mileage				
Days	a	b	c	d	e
1	234	201	35	77	128
2	52	211	197	194	152
3	176	179	145	140	134
4	38	134	179	201	166
5	145	88	130	117	193
6	132	149	82	172	15
7	201	160	170	126	69
8	98	52	201	159	129
9	108	93	77	72	163
10	211	106	29	35	231
11	78	168	107	9	149
12	156	202	170	125	192
13	143	183	140	172	169
14	187	55	162	169	56
15	123	238	94	110	170
16	99	174	157	218	147
17	75	196	129	57	184
18	43	121	185	179	156
19	199	153	63	183	43
20	165	104	221	147	147
21	135	218	184	125	204
22	123	93	149	165	77
23	109	42	182	121	149
24	62	12	62	135	129
25	26	160	129	53	174
26	8	23	55	178	162
27	193	182	158	93	79
28	49	111	21	124	92
29	108	77	126	201	111
30	45	93	53	42	44

Exercise 4.53

For each motor scooter (a–e), work out the mode, median and mean mileage covered in the month. Also, identify any that are bimodal or trimodal.

Creating manageable groups of data

From the data given for the pizza delivery motor scooters, the initial mileage chart was a grid containing 150 numbers, which were difficult to visualise or make use of. Putting them in numerical order helped to identify the median and modal values. This also gives us some indication of the 'spread' of numbers or mileages.

Because 150 items of data is a large amount, we could group them into manageable 'sets' of 'classes' – for example, in 'class widths' of 25. This would then put all the data into ten groups.

Now let us count the frequency of occurrences of each mileage value, but this time, indicating where it occurs within a group, set, or class.

Statistical diagrams

In statistics, there is a wide range of types of diagrams that can be used to create a picture of the data, making it easier to see a trend or to interpret the information contained within the data.

Activity 4.11
Look through a range of newspapers and/or magazines to find any examples of pie charts, bar charts or histograms being used to represent data.

Make a note of what you found, identifying the source of the data (which magazine or newspaper, the date, edition, page, etc.), to allow others to locate them.

When you have found the information, put your findings onto one or more posters, which would make good revision aids.

On the poster, include the example(s) that you found and a brief explanation that you have created from previous knowledge or from your research.

Exercise 4.54

1. Copy and complete the table below by counting up the number of occurrences of each mileage range of group.

Grouped mileage data	Number of occurrences for all riders taken together
1–25	
26–50	
51–75	
76–100	
101–125	
126–150	
151–175	
176–200	
201–225	
226–250	

2. Add another column to the right of the number of occurrences, which should add up to 150 at the 226–250 group.

3. Plot a graph to show the values of groups against the cumulative frequencies.

The pie chart

Earlier in this section, you were asked to identify and explain any obvious relevance in the numbers 3, 7 and 5, and there was no obvious significance, because they needed putting into some context. The worked example below puts these numbers in context.

Worked Example

Fifteen resistors (or drive belts) were purchased from a supplier and they did not all meet the required tolerances. Of the 15, 3 were too small or too low a value and 5 were too high or too long; 7 were perfect.

Show this detail on a pie chart.

Drive belts

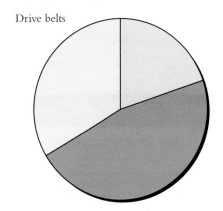

Figure 4.31 Pie chart

Resistors

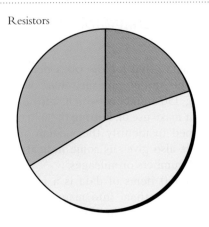

Figure 4.32 Pie chart

The advantage of using a pie chart (or any diagram) is that it saves the observer having to read words and/or numbers (which some people are scared of), the diagram immediately shows the information which the data carries.

One thing is immediately obvious from both pie charts and that is the fact that just less than half were perfect, or slightly more than half were rejected. This is easier to see on a pie chart, even with simple numbers like 3, 7 and 5.

More complex pie charts can be used, and some can become too complicated to make sense of. When using any kind of diagram to illustrate some data, make sure you do not try to include too much information.

Exercise 4.55

A car is made of many materials. Use the data provided to create a pie chart to illustrate the proportions of each material, by volume.

Ferrous metal	Non-ferrous metal	Plastics	Glass	Fluids	Rubber	Hardboard	Adhesives
27%	41%	18%	2%	4%	6%	1%	1%

You may find it helpful to use a spreadsheet such as Microsoft Excel® to produce this, and to save you having to do several calculations from percentages to degrees.

The pie chart is simple to draw and interpret, but it soon becomes overcrowded when there are more than five or six items of data being represented on it.

A better way of demonstrating such a large range of data items is to use a bar chart.

Bar charts

If you do these on a spreadsheet such as Microsoft Excel®, you may find different names. A bar chart is described and illustrated as a series of horizontal bars growing out of the vertical left-hand axis.

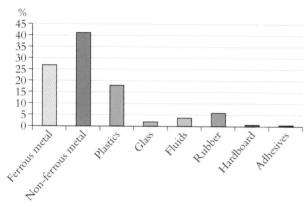

Figure 4.33 Bar chart

A bar chart that we recognise as such is called a column chart, because it more correctly describes the series of columns which grow out of the horizontal axis.

A bar (or column) chart of the data given for Exercise 4.55 would look like the diagram in Figure 4.33.

The histogram

A histogram is similar to the bar chart, but is fundamentally different in the following respects. With the bar chart (or column chart), each bar is a separate entity and generally represents discrete lumps or sets of data, such as types of material which make up a car, names of the months in a year when illustrating the rainfall over a year, or comparing two years' rainfall on a monthly basis, and so on. A histogram, on the other hand, is generally used to represent continuous data, and the nature of the continuity of the data is represented with a series of bars or blocks that are joined together, and the continuous data are fitted into the slots made by the bars.

To explain this, an example would be easier than words. Considering the example about pizza deliveries:

Grouped mileage data	Number of occurrences for all riders taken together
1–25	6
26–50	12
51–75	14
76–100	16
101–125	17
126–150	25
151–175	25
176–200	20
201–225	12
226–250	3

A histogram representing this data is shown in Figure 4.34.

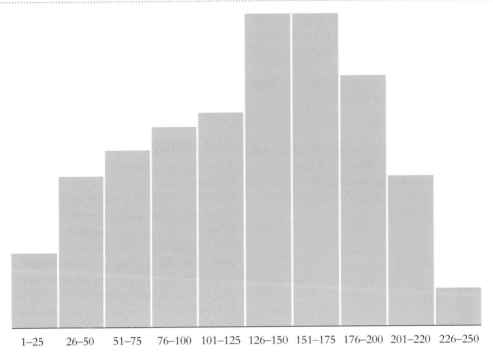

| 1–25 | 26–50 | 51–75 | 76–100 | 101–125 | 126–150 | 151–175 | 176–200 | 201–220 | 226–250 |

Figure 4.34 Histogram

Exercise 4.56

Select a topic that is related to your place or work or your subject area, and carry out a practical investigation to obtain some meaningful data. This could be the number of employees you work with – how many are in particular categories of jobs, such as clerical, part-time, full-time, maintenance, installation, design, management, labouring, production and so on, assuming the company is big enough. If not, you could look at the types of vehicle they drive – cars, type, age, motorbikes, and so on.

Alternatively, you could select an item of equipment that you work with regularly and collect your own data sets to work with. These could be electrical lights, fittings, components, relays, electronic components, mechanical components, such as screws, nuts and bolts, drive belts, bearings, wire ropes, moulds, pipework, and so on.

When you have decided what to do, you need to make sure that you can collect data that will allow you to compile a frequency table, histogram, frequency curve and as many of the other topics covered in this section as possible.

To help you with this, and as a final recap of the material covered in this section, have another go at completing the following table.

Term	Definition, meaning or explanation
Variable	
Mean	
Mode	

Term	Definition, meaning or explanation
Median	
Discrete data	
Continuous data	
Grouped data	
Bar chart	
Pie chart	
Histogram	
Frequency distribution	
Class boundary	
Class width	
Frequency table	
Variant	
Cumulative frequency	
Cumulative frequency curve	

4.4 Using elementary calculus techniques

This section will cover the following grading criteria:

Graphs and gradients

By the end of this section you will have developed a knowledge and understanding of how to work out the gradients of graphs.

You will be familiar with working out gradients of graphs from earlier studies, but to recap this important process, which becomes the basis of the calculus. Work through this at a pace to match your confidence with working with straight-line graphs.

Exercise 4.57

Using graph paper, to get useable results, draw a pair of vertical and horizontal axes, enter some clearly readable scales and draw four or five lines, with different slopes, which intercept the vertical axis at different points, some positive, one through the origin and at least one which intercepts the vertical axis in the negative region.

On the lines that you have drawn, create some accurate 'rise over run' triangles and calculate the gradient of each line (see page 148 for further details on rise over run).

Now write out the equation for each line.

Refer back to learning outcome 1 if you get stuck or you wish to check your working.

The equation tells us the relationship between the x and y axes, which depends on the gradient (m) and the intercept (c).

For example, given that $y = mx + c$

with values added:

$$y = 3x + 2$$

This tells us that for any value of x, the corresponding value of y will always be three times the value of x, plus 2, the value 2 being the starting point of the graph or the value of y when x = zero (also known as the intercept).

Exercise 4.58

Now select one of your lines, or draw another, as in Figure 4.35, and calculate the gradient or slope at two or three points along its length. You may already know the answer, but what level of accuracy can you state for the gradient?

You should find that the gradient is the same rise divided by run, no matter what values of y and x you use to do the calculation, as long as they come from the graph. Because the line rises at the same rate, the slope is the same all the time, and the relationship between y and x is a constant, called the gradient.

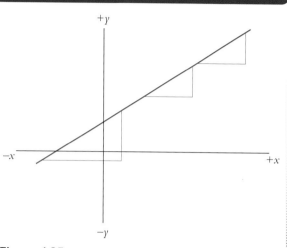

Figure 4.35

Key Points

We can confidently say that the graph of $y = mx + c$ is a straight line, which is why it is often referred to as the straight-line equation.

The reason why the calculus was invented

Not all lines are straight. However, we still want to be able to calculate the relationship between x and y (or any other variables/coordinates) with a high level of accuracy.

The next exercise will take you through the initial steps which the inventors of the calculus might have gone through.

Exercise 4.59

Again, using graph paper with a suitable scale, plot the graph of $y = x^2$ from $x = -3$ to $x = +3$, using increments of whole numbers.

Always start with a table of values to show your working (which allows you to refer back to it in case you forget where the values came from, AND it allows someone to see that you have worked it out correctly.

Your immediate observation of this curve should be to notice that it is symmetrical about the vertical axis. The y axis acts as a mirror and reflects the positive values and shape in the negative section.

Now look more closely at the problem of evaluating the gradient of the curve of $y = x^2$.

If you place a straight edge, such as a rule, on the horizontal axis, it should be just touching the curve at $x = 0$.

At $x = 0$, the slope $= 0$.

Now place the straight edge against the curve, at, or as close as you can get to, the value of $x = 1$, where y also $= 1$, so that it is just touching it. Draw a line along the straight edge which touches the curve at the point $x = 1$.

You will have observed that as you rotated the straight edge from horizontal to the point where $x = 1$, the slope of the straight edge became steeper. This indicates that the gradient at $x = 1$ is different to the gradient at $x = 0$.

This line you've drawn is called a tangent to the curve. It is tangential, which means that the line just touches the curve and doesn't cross it, and it allows us to work out the approximate gradient at that point. Use the rise over run method and work to a couple of decimal places if you can.

Notice the use of the word 'tangential'. This is not used by accident. The tangential line is at 45°, and the slope $= 1$.

Exercise 4.60

1. What is the gradient at $x = 1$?

2. What is the tangent of 45°?

3. Repeat the process to obtain the rise over run values and work out the gradient of the curve at the points where $x = 2$ and $x = 3$.

191

When you draw the lines, tangential to the curve, extend them until they touch the horizontal axis. Compare the tangent of the angle they make between the x axis, and the gradient.

Remember, $\tan\theta = \frac{\text{rise}}{\text{run}}$. You should also remember that for a right-angled triangle, the opposite is the rise, and the adjacent is the run.

Did you get approximately 4 at $y = x^2$?
And did you get approximately 6 at $y = x^3$?

By using calculus we can confirm the values are exactly 4 and 6.

The calculus

The calculus is a mathematical technique that comprises two distinctly different, but related, operations:

- differentiation
- integration.

Make the Grade — P10

Grading criterion P10 requires that you apply the basic rules of calculus arithmetic to solve three different types of function by differentiation and two different types of function by integration.

They are technically opposite activities, in much the same way as multiplication is opposite to division, or addition is opposite to subtraction, squaring to square-rooting, and so on.

Differentiation

By the end of this section you will have developed a knowledge and understanding of the differential coefficient; the gradient of a curve $y = f(x)$; rate of change; Leibniz notation; differentiation of simple polynomical functions, exponential functions and sinusoidal functions; and problems involving evaluation.

If you pick up a textbook or search the internet for details on the calculus, you will find the names of two eminent scientists – one of whom you may not have heard, and one of whom you *will* have heard of. Their names are prominent when you investigate the development of the techniques involved in differentiation.

The two mathematicians (and probably many more) worked on the problem of solving the gradient at any point along a non-straight line graph (a curve), and they are reputed to have made their discoveries within a few years of each other, independently. They were Gottfried Willhelm Leibniz and Isaac Newton.

To convey recognition to both people, their names are used to identify the two different methods of notation when using the calculus, $\frac{dy}{dx}$ (Leibniz notation) and $f(x)$ (Newton notation), which are spoken as: dy by dx and f of x.

So what is differentiation?

Here is an introduction which starts at basic principles and develops from the straight-line graph and graph of $y = x^2$ that you have just plotted and worked with.

Consider again the graph of $y = x^2$ and the following attempt to determine the gradient of the curve at $x = 2$, as accurately as possible.

Start by drawing a straight line from $x = 2$ through the point where $x = 3$.

Notice that this line is inside the curve of $y = x^2$ and not outside, touching the curve tangentially. This will reduce the uncertainty in the estimation of drawing an accurate tangent to the curve.

We could try to estimate the gradient of the curve using rise over run measurements, but is this the gradient of the curve at $x = 2$ or at $x = 3$?

Actually, it is neither. The line represents the gradient of the curve at some point between $x = 2$ and $x = 3$, which could be estimated by drawing a line parallel to the original line, but tangential to the curve.

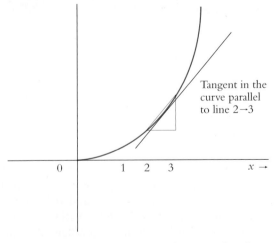

Tangent in the curve parallel to line 2→3

0   1   2   3 $x \rightarrow$

Figure 4.36 Tangent of the curve from $x = 2$ to 3

Draw the lines, measure them and calculate the rise over run value.

However, we want to know the tangent at $x = 2$, so draw a shorter line joining 2 and 2.5.

Again, draw this line and measure the rise and run, then calculate the gradient (rise over run). It should be a bit less than the original value for the $2 \rightarrow 3$ line.

This is still not giving us the gradient of the curve at $x = 2$ exactly and with a high level of accuracy.

We could draw the graph 1,000 times the size and reduce the incremental step from 2, to say $x = 2.002$.

Try a few more gradients, each one starting at $x = 2$ and finishing at $x = 2.25$, then 2.1, and so on, to try to see a pattern evolving.

When the small line is very very small, it becomes almost impossible to draw and measure the rise and run, because you are almost working 'on top' of the point where $x = 2$, but not exactly. Therefore, although the gradient given by that straight line must be getting very close to the gradient at the point where $x = 2$, we want to know the gradient at the point where $x = 2$ *exactly*, and approximately is no longer an option.

By looking at this problem and the outcomes of each incremental step closer to 2 for the second point (from 3, 2.5, 2.25, 2.1, 2.01, and so on), your results should be tending to suggest that the value for the gradient (by rise over run calculations) is approaching the value 4.

The gradient of $y = x^2$ for any value of x along the curve is equal to $2x$. At $x = 2$, the gradient $= 2 \times 2 = 4$; at $x = 3$, the gradient $= 2 \times 3 = 6$, and so on.

Exercise 4.62

Now repeat this for the graph of $y = x^3$

Using a range of values for x, plot x against $y = x^3$ using a suitable scale, and try to spot any pattern which seems to be appearing as you work out the reducing gradients, and compare each one to its corresponding value of x.

To save you from spending for ever working out gradients in this way, for such an expression, where y is given in terms of x to some power, the following is true:

Where

$$y = x^n \text{ (where } n \text{ is any number)}$$

the gradient at any point is given by

$$nxn^{-1}$$

Exercise 4.61

Now try the same method to try to determine accurately the value of the gradient of this curve at $x = 3$.

If you are doing this in a class with several others, you could try using a large range of values for a much larger curve, drawn from, say, $x = 0$ to $x = 10$. From this, you would be able to share a very wide set of results.

From the largest set of results that you can obtain, complete the following table.

x value	Gradient at x value
0.5	
1	
2	
4	
7	
10	

Your skills of observation and interpretation are important for the next bit, where you will be looking for patterns in numbers.

Can you see any pattern evolving or a fundamental relationship emerging from your results?

(Hint: compare the values of x and the apparent gradient of the curve at each corresponding value of x.)

Does this agree with the earlier findings and all the graph drawings?

For $y = x^2$ the gradient at any point is given by $2x$.

For $y = x^3$ the gradient at any point is given by $3x^{3-1} = 3x^2$.

Hence, the gradient is given by multiplying x by the power or index and reducing the index by 1.

Exercise 4.63

Try this for the following equations, and check some of them by drawing, as before.

1. $y = x^{1/2}$ (which is a way of saying $y = \sqrt{x}$ by using indices)

2. $y = x^4$

3. $y = x$

4. $y = 3x^4$

You may have thought that it was a bit clumsy to keep writing the gradient at any point =, so we have some shorthand which simplifies this.

Key Points

Newton said: The gradient at any point on a curve is a function of x and he wrote this as $f(x)$, which is pronounced as f of x.

Leibniz said:
The gradient is a change or difference in x with respect to the corresponding difference in y, and is written as dy/dx

or

$$\frac{dy}{dx}$$

The gradient at any point can also be called the tangent to the curve – so we now have three ways to say what we mean.

So where does this have a place in engineering? Any aspect of life which involves movement or change can be solved using some branch of the calculus.

Velocity is the rate of change of position, so $\frac{dp}{dt}$ which is a difference in position with respect to the corresponding difference or change in time.

Acceleration is the rate of change of velocity, so $\frac{dv}{dt}$ which is the difference or change in velocity with respect to the corresponding difference in time.

Control system theory uses multiple integrals and double differentiation.

Key Term

$\frac{dp}{dx}$ and $\frac{dv}{dt}$ – A change in p or v with respect to ... *not* divided by.

An obvious question might be: if $\frac{dy}{dx}$ did mean dy divided by dx, what algebraic error could be easily made?

A less obvious answer would be: in algebra, if there is a common term on the top and the bottom, they cancel out, so $\frac{dy}{dx}$ would become $\frac{y}{x}$ which is single point value of y divided by a single point value of x, which has a completely different meaning.

In calculus, $\frac{dy}{dx}$ means much more than division. It means the gradient of a tangent to the curve at any point $[x, y]$ when the lengths of the rise and the run are infinitesimally small. (In mathematical language, this is described as the limit where the incremental change in y and x are approaching zero, but they are not in fact zero.)

In other words, the increment in x (run) could be 0.0000004; the increment in y (rise) could be 0.000007. Rise over run would still give us a gradient.

For an electrical example: the e.m.f. induced in a coil of wire or a conductor is given by $-L\frac{di}{dt}$ where L is the inductance of the coil or loop of wire and $\frac{di}{dt}$ is the rate of change of current. The negative sign is there because the induced e.m.f. is always such that it tries to cause a current that will oppose the current which was changing in the first place, or (put simply) because a negative change in e.m.f. is caused by a positive change in current.

To add a few more complexities, you may be faced with a problem or equation which looks like the following:

$$y = x^2 + 3x + 4$$

and you are asked to differentiate this equation.

First of all – if you have access to a graphical calculator, Microsoft Excel® or any other graphs package, or some graph paper and a pencil – plot the curve of this equation to see what it looks like.

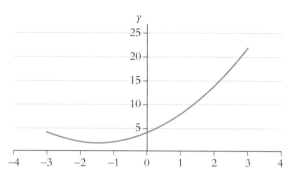

Figure 4.37 Graph of $y = x^2 + 3x + 4$

The problem is no more difficult than the others we have seen, with each item needing to be treated as an individual expression when we use the calculus.

Differentiation of simple polynomial functions

Key Terms

Prefix – Used to represent or describe the number of numbers or expressions contained within an expression.

Polynomials – Expressions which contain more than one item, where 'poly' means 'some' or many.

This prefix 'poly' is commonly used in engineering, particularly in materials science, where modern materials such as polyethylene, polystyrene, polyvinyl chloride (PVC), polycarbonate, and so on, are in regular use.

Consider the equation

$$y = x^2 + 3x + 4$$

as meaning $y = x^2$, and added to this we have $3x$, and finally we add 4.

These are differentiated individually, then added together.

Hence:

$$\frac{dy}{dx} = 2x + 3$$

What happened to the 4?

Exercise 4.64

Spend a few minutes trying to explain why the number 4 disappears when it is differentiated.

Exercise 4.65

1. Plot the graph of $2x + 3$ on the same axes (or same graph paper) as $y = x^2 + 3x + 4$, using the manual gradient estimation method, by using your rule or other straight edge and sketching the rise over run as you did for the graph of $y = x^2$.

Does your estimated gradient plot look anything like the gradient given by the calculus?

2. Compare the graphs.

This is a comparison between the calculus and graphical differentiation, which compares mathematical techniques with a real-life sketch that is visible and allows the picture to be seen.

Another term that you will encounter is the 'differential coefficient'. This simply means the expression that is obtained after differentiation.

For $y = ax^n$

Then $\frac{dy}{dx} = nax^{n-1}$

The expression nax^{n-1} is known as the differential coefficient of ax^n.

Key Term

The differential coefficient – The equation you get when you differentiate an equation.

195

Exercise 4.66

Make some notes to help you explain the relationship between

$$y = ax^n$$

and

$$dy/dx = nax^{n-1}$$

(Hint: n just means any number, a is any number that is a multiplier of the expression.)

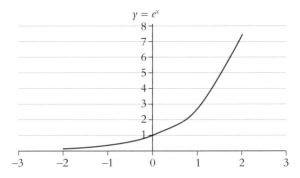

Figure 4.38 $y = e^x$

Exercise 4.67

Differentiate the following polynomials, with respect to x.

1. $y = x^3 + 3x^2$

2. $y = 3x^3 + 3x^2 + 4x - 6$

3. $y = x^3 - 3x^2 - 19$

4. $y = 3x^4 - 0.5x^{-3}$

Exercise 4.68

Using a full page of graph paper, prove this for yourself.

Plot the graph of $y = ex$ for a range of, say, $x = 12$ to $x = +2$, and sketch the curve as accurately as you can.

Access to software may allow this to be done quicker, if you can use it, and the software may allow you to read off the gradient at any point.

Either way, and particularly if you are drawing this manually, at $x = $ zero, $y = e^x = 1$.

Try to determine the gradient at the point $y = 1$ using the rise over run method.

Then try to obtain the value of the gradient at $x = 1$.

Differentiation of exponential functions

You may remember from the section about logarithms and the natural number e, that an activity was introduced where a graph of ex was plotted and the gradient of the curve $y = ex$ was plotted. You were then asked to use rise over run at several points on that curve to determine the gradient.

The interesting thing about the value e is that if you plot a graph of x against values of e^x, you obtain a curve, and at any point on that curve, the gradient is equal to the corresponding value of e^x.

It follows from this statement, and observation, that if the gradient is equal to the e^x value, then the differential coefficient of x (where $y = e^x$) will also be e^x.

This is better written as an algebraic statement with its differential coefficient.

If

$$y = ex$$

then

$$\frac{dy}{dx} = e^x$$

You may be given a list of differential coefficients, but it is always a good idea to make your own as you come across new ones.

Once the expression starts becoming a little more complicated than $\frac{dy}{dx}$, and as an alternative way to write the expression, we tend to use a slightly different way to write it down.

For example, if

$$y = e^x$$

then

$$\frac{dy}{dx} = e^x$$

is fine.

But it may be written as

$$\frac{d}{dx}(e^x)$$

which reads as the differential with respect to x of e^x without mentioning the variable y at all.

Also, as with

$$y = ax^n$$

or

$$\frac{d}{dx}(ax^n)$$

having a differential coefficient of

$$\frac{dy}{dx} = nax^{n-1}$$

$$\frac{d}{dx}(4e^x)$$

has a differential coefficient of

$$4e^x$$

Finding the differential coefficients is one achievement, but normally we want to find the value of this for some value of x.

If $y = 5e^x$, find the differential coefficient and evaluate this when $x = 2$.

The answer would be as follows:

$$\frac{d}{dx}(5e^x) = 5e^x$$

at $x = 2$, this evaluates to $5e^2$

$$36.945$$

In order to solve this, you need to find the button(s) on your calculator which allow you to work out powers. Hence, if you find the method to enter ex into your calculator, and enter e^2, then multiply this by 5.

Exercise 4.69

Solve the following.

1. $y = 4e^x$

Find d/dx and evaluate this to three decimal places when $x = 2.5$.

2. $y = 6.8e^x$

Differentiate this with respect to (w.r.t.) x. Evaluate the gradient when $x = 0.5$

Note: when the exponent is of the following form,

$$y = 5e^{2x}$$

the differential is not as straightforward, and you will see more of this later, particularly in further maths.

What we have to do is involve the index within the differentiation by multiplying the overall expression by the differential of the index.

Differentiate $2x$, and we get 2.

Multiply this by $5e^{2x}$

which gives

$$\frac{dy}{dx} = 10e^{2x}$$

(This is called the function of a function rule – more on this later.)

Differentiation of sinusoidal functions

Sinusoidal functions – for example, sines and cosines – can be considered as a waveform which shows the variation of a voltage, displacement or current, and so on, with time (or angle of rotation).

Plot a sine wave over one complete cycle, with peak values of 1 unit to -1 unit.

It should look like this.

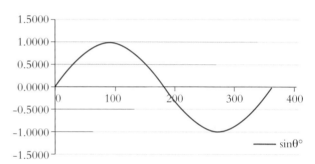

Figure 4.39 The sine curve

Now for some more graphical differentiation. On the same diagram, using the same peak values of 1 and -1, mark the following points and plot a graph which shows how the gradient changes depending on the angle.

Points to plot:

- At $0°$, the gradient is at its steepest, and is rising in a positive direction, so plot a value of 1, above $0°$.
- At $90°$, the gradient is zero, because it is a flat line, so plot a value of zero at $90°$ on the horizontal axis.

Key Term

Turning point – Where a graph stops going in one direction, then goes in the other direction. The point is referred to as a '.

You came across this earlier with the graph of $x^2 + 3x + 4$.

At 180°, the gradient is at its steepest, but reducing, or falling, in a negative direction. At 180°, plot the value of −1 to indicate that the gradient is at a maximum value, but negative. This is a turning point for the gradient plot.

Continue with this process through to 270°, where the sine curve is at a maximum negative value (−1) and the gradient is zero because it is another turning point.

Finally, at 360°, plot a point to represent the gradient or slope of the sine curve, which you should see as increasing in a positive value at its maximum rate (steepest), so plot the value of 1 above 360°.

Now join up the points, bearing in mind that if you plotted more than these five points, they would show that the plot of the gradient's behaviour follows a sinusoidal shape, but is shifted by 90° to the left-hand side.

You have seen this curve before.

The cosine curve is shown in Figure 4.40.

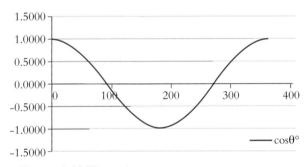

Figure 4.40 The cosine curve

So, what has this to do with the calculus? Using the graph of a sine wave and plotting the gradient of this curve, across a range of points or values, you have graphically determined that the gradient of a sine wave is equal to a cosine wave at any point on its curve. (That is graphical differentiation, again.)

You have found, by observation, an expression which allows us to calculate the gradient of a sine wave at any value.

Exercise 4.70

1. Repeat the process for a cosine wave. What is the curve that represents the gradient of a cosine wave at any point?

2. Use the same process to graphically differentiate the inverse sine wave. Then carry out the same process again on the result of that one. What do you observe?

Differential coefficients of sine and cosine

Key Points

When referring to angles in the calculus, remember that they must be in radians, not degrees.

Value of y	Value of $\frac{dy}{d\theta}$
$\sin\theta$	$\cos\theta$
$\cos\theta$	$-\sin\theta$
$\sin\theta$	$-\cos\theta$
$\cos\theta$	$\sin\theta$

Further to this, then, the angles are multiples of θ, for such as 4θ or 2θ, the differentials become:

Value of y	Value of $\frac{dy}{d\theta}$
$\sin 2\theta$	$2\cos 2\theta$
$\cos 2\theta$	$-2\sin 2\theta$
$\sin 2\theta$	$-2\cos 2\theta$
$\cos 2\theta$	$2\sin 2\theta$

Or in general terms:

Value of y	Value of $\frac{dy}{d\theta}$
$\sin a\theta$	$a\cos a\theta$
$\cos a\theta$	$-a\sin a\theta$
$-\sin a\theta$	$-a\cos a\theta$
$-\cos a\theta$	$a\sin a\theta$

Because $a\theta$ and 2θ are functions of θ:

- $a\theta$ is a multiplied by θ
- 2θ is 2 multiplied by θ.

You have seen how we differentiate a function, so differentiating, say, $a\theta$ w.r.t. θ we would get a, in

Exercise 4.71

1. Referring to the right-angled triangle and SOHCAHTOA, you could determine the values of sine, cosine and tangents with some accuracy using the trig ratios from SOHCAHTOA. Plot these on a graph and accurately determine the lengths of the sides for a right-angled triangle at $10°$ through to $90°$ or $180°$, and plot the gradients to obtain the graphs of sin, cos and tan, along with the differentials of these.

2. Graphically differentiating $\tan\theta$ will allow you to see the shape of the secant curve. If you have access to mathematical graph-plotting software, plot the graph of $\tan\theta$, then $\sec\theta$ and finally $\sec^2\theta$, to compare with your graphical sketches. Note: $\sec^2\theta$ is how we say it, but calculators, say it as $(\sec\theta)^2$.

3. If $v = \sin\theta$, differentiate to obtain $\frac{d}{d\theta}(\sin\theta)$.

4. The instantaneous voltage of a waveform is $v = 24\sin\theta + 12\cos\theta$; use differentiation to obtain the rate of change of v w.r.t θ, or $\frac{dv}{d\theta}$.

exactly the same way as we arrive at 2 when we differentiate $2x$ w.r.t. x.

The problem becomes complicated because $\cos 2\theta$ is a function (cosine) of the function 2θ and solving this requires the function of a function rule.

The function of a function rule is beyond this textbook and is covered in further maths.

What about differentiating the tangent?

It has already been mentioned a couple of times that when using the calculus, angles must be given in radians and *not* degrees.

This is because the degree is a 360th of a complete revolution. It is a fraction of a circle using a value that was chosen arbitrarily and not by using any fundamental section or value which directly relates to the circle. A radian, on the other hand, is a fundamental part of a circle. When the calculus becomes more complex in further studies, this becomes more obvious.

Activity 4.12

From your notes, what fundamental part of a circle is used to determine an angle on one radian?

In the same way that a degree is arrived at by dividing a circle into 360 increments, the tangent is derived from $\frac{\sin}{\cos}$, which is a division.

Differentiating terms that involve division (or quotients) requires a deeper understanding of the complexities of the calculus and will be covered in further maths or AS/A2 maths (applied or calculus), as well as any other higher-level study which has a mathematical element, such as engineering.

If $y = \tan\theta$

$$\frac{dy}{d\theta} = \sec^2\theta$$

(where sec is an abbreviation for secant).

Integration

By the end of this section you will have developed a knowledge and understanding of integration as a reverse of differentiating; indefinite integrals; constant of integration, definite integrals; limits; evaluation of simple polynomial functions; and area under a curve.

Generally described as the reverse or inverse of differentiation, integration follows similar rules, and is, again, quite a common feature in any engineering qualification from the BTEC National upwards.

Effectively, integration can be seen as being presented with the differential coefficient of a function, and we are asked to find the original function.

Looking at it as a step-by-step process, consider the following function, which was covered earlier in this section:

When given a function such as $y = x^2$ and asked to differentiate it, the process is

- multiply by the power
- reduce the power by 1.

Hence $\frac{dy}{dx}$ became $2x$.

Exercise 4.72

Treating integration as the reverse process to differentiation, write down the two steps we should carry out on $2x$ to get back to the original function $y = x^2$.

Worked Example

When given $y = 2x^3$, we found the differential coefficient to be

$$\frac{dy}{dx} = 6x^2$$

The differentiating process being:

- multiply by the power
- reduce the power by 1.

So working through the reverse processes gives us the original function:

- add 1 on to the power
- divide by the new power.

We get $2x^{3+1}$ divided by $(3 + 1)$

or

$$0.5x^4$$

Did you see where the 0.5 came from? Think about it. What is 2 divided by $(3 + 1)$?

When adding numbers together, we call it addition. The reverse process of addition has its own name, subtraction.
In calculus we call the reverse process of differentiation by its own name, integration, and we also use other symbols to represent it when we are writing out expressions and functions.

Again, using the example of $y = 2x^3$, we found the differential coefficient to be

$$\frac{dy}{dx} = 6x^2$$

We use a slight rearrangement of this, and another symbol to indicate that we are integrating the expression w.r.t. x

$$\frac{dy}{dx} = 6x^2$$

which reads as the change in y, w.r.t. $x = 6x^2$

When integrating, this becomes

$$dy = 6x^2\, dx$$

which almost reads as change in y is

$$6x^2 \text{ w.r.t. } x$$

And to complete the integral function, using a symbol which is derived from an elongated, stretched S for summation, we get

$$y = \int 6x^2.dx$$

which reads as y equals the integral of $6x^2$ w.r.t. x.

This is where some confusion can arise, because the dx is only telling us that we are integrating with respect to x, and it should not be seen as multiplied by dx, as in algebra.

Exercise 4.73

Integrate the following, using the notion that integration is the reverse process to differentiation.

1. $\int 3x^3\, dx$
2. $\int 5x^4\, dx$
3. $\int x^2\, dx$
4. $\int 3\, dx$
5. $\int \cos x\, dx$
6. $\int \cos x - \sin x\, dx$
7. $\int \sin x - \cos x\, dx$

(For the trig functions, you do exactly the opposite of the differentiating process.)

There is a problem that you need to watch out for. In fact, there are always problems, and the techniques which are developed do not always work in every case.

Remember that so far we have said, the basic rule for integrating is to add 1 to the power, then divide by the new power.

Exercise 4.74

Try this, and write down why it does not work.

Find the integral of $\frac{1}{x}$

(Hint: convert the reciprocal to a power of, and $\frac{1}{x} = x^{-1}$ and try following the standard rule.)

A mathematician would disagree. To a mathematician, divide by zero is not defined, so it cannot be done because it is not allowed.

In any table that provides a list of standard integrals, the following would be indicated:

$$\int x^n \, dx$$
$$= \frac{x^{n+1}}{n+1} \quad x \neq -1 \text{ (except when } x = -1)$$

Integrating such a term (and that is not the only one) requires different techniques to be used.

Integration of exponential functions

To start with a recap of the differentiation of exponentiation:

- remember when we had to differentiate $\frac{d}{dx}(4e^x)$
- the differential coefficient is $4e^x$
- differentiating e^x gives us e^x

It must also follow that when we integrate e^x we will get e^x.

Standard integrals summary

$$\int x^n \, dx \quad = \frac{x^{n+1}}{n+1} \, x \neq -1 \text{ (except when } x = -1)$$
$$\int e^x \, dx \quad = e^x$$
$$\int sinx \, dx = -cosx$$
$$\int cosx \, dx = sinx$$

Exercise 4.75

This part requires you to think back to straight-line graphs and working out the gradients.

1. Look back at Exercise 4.57 and the piece of graph paper you used when you drew this out yourself. If you cannot find it, start again, and using a sheet of graph paper, draw two axes, x and y. Draw one line through the origin, one above it, cutting through the y axis in the positive region, and one below the x axis, cutting through the y axis in the negative region, as shown below in Figure 4.41.

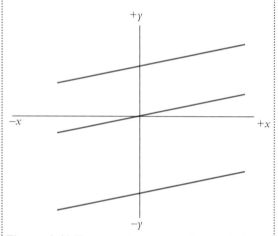

Figure 4.41 Three graphs with identical gradient

2. Work out the gradient of the three lines that you have drawn.

3. Write out the equation for each of the lines, in the $y = mx + c$ format.

4. Differentiate each line and write down your answers.

You should have observed that all the constants (c values) disappear, leaving us with the gradient, which is what differentiation does. Differentiation tells us the rate of change of the variables, with no information about where the intercept was.

If you are faced with the results of your differentiation and asked to integrate them, what information do you have about the intercept?

Exercise 4.76

Using the results you obtained by differentiating the line equations, now integrate them. You should not need to do many to observe that they all give the same integrals. What has gone missing?

The value of c or the intercept, does not exist. It can be found if we are given a couple of values of the integral, so all is not lost, but in the meantime, we need to assume that a value of c *could* exist, and until we can evaluate it, we simply put $+ c$ to indicate that there may be some constant of integration, which may not have a value of zero.

The integration you have done so far is called indefinite integration, because we only convert the gradient back into the original equation, and do not know the intercept.

Key Term

Indefinite – The result of integration, because the equation cannot be fully defined because there is no intercept.

Worked Example

We are told that the gradient of a curve has been determined and the gradient is x^2.

We are also told that one point on this curve at $x = 2$, $y = 3$.

Find the equation of this curve.

(The point $x = 2$, $y = 3$ is normally written as (2, 3).

The x value (the point along the horizontal axis) always appears first.

We need to integrate x^2 and put the values (2, 3) into it to obtain a specific value for c. The specific value of c tells us the value of the intercept.

$$y = \int x^2 \, dx$$

which gives us

$$y = \frac{x^3}{3} + c$$

Now, returning to the algebra done at the start of this unit, substitute the values $x = 2$, $y = 3$ to solve for a specific value of c.

$$3 = \frac{2^3}{3} + c$$

$$c = 3 - \frac{8}{3}$$

or

$$c = \frac{1}{3}$$

Now that we have a real value for c, we can write down the full equation,

which gives us

$$y = \frac{x^3}{3} + \frac{1}{3}$$

which could factorise to

$$y = \frac{1}{3}(x^3 + 1)$$

Exercise 4.77

1. To arrive at the full equation in the worked example opposite, we had to make use of the known values from the curve, typically $x = 2$ and $y = 3$. What if the values had been $x = 3$, $y = 2$?

2. Experiment with different values of x and y.

3. Can you identify any relationship between the equations after c has been given a specific value?

Evaluation of the constant of integration c

So far, we have seen how the constant of integration is used to represent a value which may have disappeared during the process of differentiation, and we need to allow for its reappearance when we integrate.

That said, we do need a method that will allow us to evaluate the constant of integration. To do this, we need to be given some values of the variables following integration; then we can substitute these values into the integral equation, and evaluate c.

Evaluation of polynomial functions

For polynomials (more than one number), each 'nomial' is integrated as a standalone item, and the mathematical operators ($+$ and $-$) are left unchanged, unless the integral value becomes negative (as with some trig identities), in which case the normal techniques for combining $+$ve and $-$ve values apply.

Note, when multiples (products) and divisions (quotients) are involved, the use of calculus becomes a bit more complicated.

For example, if you are required to integrate the following:

$$y = 2x(x - 5)$$

we immediately see that a product is involved, but it can be removed if we expand (multiply out) the brackets; we covered this at the start of this unit.

Hence

$$y = 2x^2 - 10x$$

which can be handled much more easily.

The integral of y with respect to x can now be carried out.

Exercise 4.78

Integrate the equation

$$\int 2x^2 - 10x.dx$$

(Note: the . before dx is used here to make it clear that dx is not part of the equation.)

The definite integral

So far, the integrals you have seen are what are referred to as indefinite integrals. Although we can evaluate the constant of integration c when we are given the coordinates of one point on the curve, we need to look a bit closer at how we would use integration. Then the concept of a definite integral will become clear.

The integration symbol was described as an elongated S for summation for a good reason. We use integration to find the sum of infinitely narrow strips of a graph, under the curve, to calculate the area accurately.

Calculating the area under a curve is difficult because of the shape of the curve. If it were a rectangle or triangle, it would be easy, but usually it is not. As with finding the gradients of curves, we use the calculus to find areas under curves.

Consider the following graph, which represents some expression for y, which is a function of x.

We introduced this notation (Newtonian) at the start of this section, so $y = f(x)$.

An example of the types of problem that we are able to solve using definite integrals is to find the area under a specific section of the curve, by considering the problem, as shown in Figure 4.42.

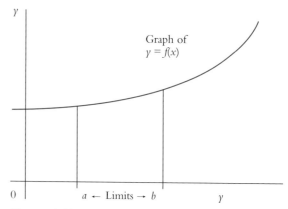

Figure 4.42 Diagram of the curve, indicating the area between $x = a$ and b

203

To find the area, we integrate the equation of the curve and evaluate it between the limits a and b.

$$\int_a^b f(x).dx$$

The two limits a and b are shown at the bottom (a) and top (b) of the integral symbol.

Assuming that the equation is $y = x^2 + 2$, find the area under the curve between $x = 0.5$ and $x = 1.5$.

$$\int_{0.5}^{1.5} x^2 + 2.dx$$

$$= \left[\frac{1}{3}x^3 + 2x\right]_{0.5}^{1.5}$$

The integrated expression has been placed in square brackets, and the upper and lower limits are indicated outside the end bracket. These are then substituted into the equation for x, one at a time.

The value $x = 1.5$ is substituted first to get one part of the final calculation.

The value $x = 0.5$ is then substituted, and this value will be subtracted from the first.

The following demonstrates this:

$$[\tfrac{1}{3}(1.5)^3 + 2(1.5)] - [\tfrac{1}{3}(0.5)^3 + 2(0.5)]$$

$$= [4.125] - [1.04167]$$

$$= 3.0833 \text{ square units}$$

Note: if you included c in your integrals, it would not be wrong, and as we worked through the process of subtracting the lower limit from the higher limit, we would have $c - c$, which would disappear, so it is not wrong to leave it out when working on definite integrals.

Examples which you may come across in your specific branch of engineering could involve electrical or mechanical applications of the material you have encountered in this unit.

Areas under a curve, which require definite integration to solve between known (given) limits, always look complicated because of the symbols in use, but they all involve the integration of an equation for the variables under consideration, and obtaining a definite solution between the two given limits.

Exercise 4.80 is an example of such an application of the work from this section, with reference back to the overall unit.

Make the Grade   D2

Grading criteria D2 requires that you apply the rules of definite integration to two engineering problems that involve summation.

Exercise 4.79

1. Using graph paper, with cm squares, plot and draw the curve of $y = x^2 + 2$ from $x = 0$ to $x = 2$.

2. Mark a vertical line from the x axis to touch the curve at $x = 0.5$, and another one at $x = 1.5$.

3. Estimate the area of the section under the curve, between the limits of $x = 0.5$ and $x = 1.5$. Does it look like just over 3 m^2?

4. How accurately could you determine the area under the curve, without using integration?

   If you drew it to ten times the size, or more, and worked on a much smaller scale, you would get closer and closer to the value given by using definite integration. That is why it was invented.

5. Now find the area under each of the following curves:

   (a) $y = x(x - 3)$ between $x = 0.5$ and $x = 2.5$

   (b) $y = x^2 + 3x - 5$ between $x = 2$ and $x = 4$

   (c) $y = \sqrt{x} - 2x + 3$ between $x = 1$ and $x = 2$

   (In each case, sketch or draw the curve to 'see' what you are doing.)

Plotting exponential functions and checking by differentiation

Near the start of this unit, you were introduced to exponential functions and the use of *e* as a natural number.

You also saw that this is used to help interpret capacitor and resistor circuits and other applications.

It may help if you read that section again, or any notes you made as you read it, to help you prepare for the next section.

Exercise 4.80

Consider a circuit containing capacitors being connected to a 24 V d.c. supply. Before being connected, the capacitors are discharged and the voltage across them is zero. When first connected, the charge rushes into the circuit, charging the capacitors, and the rate of charge reduces as the capacitors become fully charged. The voltage across the capacitors also increases as the charge increases.

The voltage across the capacitor, v_c, and the time, t, are related by the following equation:

$$v_c = 24(1 - e^{-4t})$$

1. Produce a table to show the voltage, v_c, at a range of times, from $t = 0$ s to $t = 0.5$ s, in intervals of 0.1 s.

2. Plot an accurate curve of this relationship to show the increasing voltage, and, from the graph, find the time taken to reach 15 V. Check this using the equation for v_c.

3. Use differentiation to find another expression for the rate of change of the voltage, dv_c/dt.

4. Use this last equation to obtain the initial rate of change of voltage when first connected to the supply (that is, evaluate it at $t = 0$ s). Check this by measuring rise over run from the curve.

Note that for (4), the differentiation requires the use of the function of a function rule, which is not difficult, and was explained a little earlier in this section. Remember, differentiate the index, then multiply the overall expression by that differential, to get the final result.

Many of the concepts covered throughout this unit will be enhanced by further study, particularly of the further mathematics BTEC unit, which is generally covered in the second year of a two-year National Certificate or National Diploma qualification.

Beyond that, HNC/D or degree study uses mathematics which builds on what has been covered in this unit, and an explanation of any specific problems that you may have can usually be solved by spending a short while discussing the problem with an engineer or engineering tutor from your chosen or intended sector of engineering.

To achieve a pass grade you will have:	To achieve a merit grade you will also have:	To achieve a distinction grade you will also have:
P1 manipulated and simplified three algebraic expressions using the laws of indices and two using the laws of logarithms	**M1** solved a pair of simultaneous linear equations in two unknowns	**D1** applied graphical methods to the solution of two engineering problems involving exponential growth and decay, analysing the solutions using calculus
P2 solved a linear equation by plotting a straight-line graph using experimental data and used it to deduce the gradient, intercept and equation of the line	**M2** solved one quadratic equation by factorisation and one by the formula method.	**D2** applied the rules for definite integration to two engineering problems that involve summation.
P3 factorised by extraction and grouping of a common factor from expressions with two, three and four terms, respectively		
P4 solved circular and triangular measurement problems involving the use of radian, sine, cosine and tangent functions.		
P5 sketched each of the three trigonometric functions over a complete cycle		
P6 produced answers to two practical engineering problems involving the sine and cosine rule		
P7 used standard formulae to find surface areas and volumes of regular solids for three different examples respectively		
P8 collected data and produced statistical diagrams, histograms and frequency curves		
P9 determined the mean, median and mode for two statistical problems and explained the relevance of each average as a measure of central tendency		
P10 applied the basic rules of calculus arithmetic to solve three different types of function by differentiation and two different types of function by integration.		

By the end of this unit you should be able to:

- determine the effects of loading in static engineering systems
- determine work, power and energy transfer in dynamic engineering systems
- determine the parameters of fluid systems
- determine the effects of energy transfer in thermodynamic systems.

In order to pass this unit, the evidence you present for assessment needs to demonstrate that you can meet all of the above learning outcomes for this unit. The criteria below show the levels of achievement required to pass this unit.

To achieve a pass grade you need to:	To achieve a merit grade you also need to:	To achieve a distinction grade you also need to:
P1 calculate the magnitude, direction and position of the line of action of the resultant and equilibrant of a non-concurrent coplanar force system containing a minimum of four forces acting in different directions	**M1** calculate the factor of safety in operation for a component subjected to combined direct and shear loading against given failure criteria	**D1** compare and contrast the use of d'Alembert's principle with the principle of conservation of energy to solve an engineering problem
P2 calculate the support reactions of a simply supported beam carrying at least two concentrated loads and a uniformly distributed load	**M2** determine the retarding force on a freely falling body when it impacts upon a stationary object and is brought to rest without rebound, in a given distance	**D2** evaluate the methods that might be used to determine the density of a solid material and the density of a liquid.
P3 calculate the induced direct stress, strain and dimensional change in a component subjected to direct uniaxial loading and the shear stress and strain in a component subjected to shear loading	**M3** determine the thermal efficiency of a heat transfer process from given values of flow rate, temperature change and input power	
P4 solve three or more problems that require the application of kinetic and dynamic principles to determine unknown system parameters	**M4** determine the force induced in a rigidly held component that undergoes a change in temperature.	
P5 calculate the resultant thrust and overturning moment on a vertical rectangular retaining surface with one edge in the free surface of a liquid		
P6 determine the up-thrust on an immersed body		
P7 use the continuity of volume and mass flow for an incompressible fluid to determine the design characteristics of a gradually tapering pipe		
P8 calculate dimensional change when a solid material undergoes a change in temperature and the heat transfer that accompanies a change of temperature and phase		
P9 solve two or more problems that require application of thermodynamic process equations for a perfect gas to determine unknown parameters of the problems.		

Introduction

Mechanical principles and how they are applied form the basis of engineering systems. Science and scientific principles are the key elements which allow us to understand how engineering systems work. These principles are used in the design, manufacture and maintenance of mechanical systems. In this unit you will learn about static engineering systems and the principles of equilibrium. You will learn about Newton's laws and how they are applied to dynamic systems and energy transfer. The principles of energy transfer are further explored in the section on thermodynamics and fluids.

Key Points

In this unit you will be using formulae to solve engineering problems and present data. The use of standard form is widely used to represent large or small numbers. The most commonly used values are shown in Table 5.1.

Key Terms

Greek letters are often used to represent engineering terms:

σ — direct stress

ϵ — linear strain

α — coefficient of linear expansion

ρ — density

θ — angle

τ — shear stress

γ — shear strain

μ — coefficient of friction

π — the ratio of the circumference of a circle to its diameter ($\varnothing$)

Number	Standard form	Prefix	Symbol
1 000 000 000 000	$\times 10^{12}$	tera	T
1 000 000 000	$\times 10^{9}$	giga	G
1 000 000	$\times 10^{6}$	mega	M
1 000	$\times 10^{3}$	kilo	k
0.001	$\times 10^{-3}$	milli	m
0.000 001	$\times 10^{-6}$	micro	μ
0.000 000 001	$\times 10^{-9}$	nano	n
0.000 000 000 001	$\times 10^{-12}$	pico	p

Table 5.1

This section of the book is organised as follows; each of the subsections can be readily linked to the learning outcome (LO), pass (P), merit (M) and distinction (D) criteria.

Section/content	LO	P	M	D
5.1 Determining the effects of loading in static engineering systems	1	1, 2, 3	1	
5.2 Determining work, power and energy transfer in dynamic engineering systems	2	4	2	1
5.3 Determining the parameters of fluid systems	3	5, 6, 7		2
5.4 Determining the effects of energy transfer in thermodynamic systems	4	8, 9	3, 4	

5.1 Determining the effects of loading in static engineering systems

This section will cover the following grading criteria:

Static engineering systems are those systems, structures and components that are subject to loading. There is no internal displacement and the structures are subject to a variety of loading conditions, including tension, compression and shear. It is important to consider the application of loads in the design of structures; to determine these loads and their effects is a key technique for engineers, particularly when considering appropriate safety factors.

Non-concurrent coplanar force systems

By the end of this section you should have developed a knowledge and understanding of what a non-concurrent coplanar force system is. You will also be able to determine the resultant and equilibrant forces, both in magnitude and direction, for these systems.

A force system refers to any situation where forces act upon an object, either to move the object or to keep it in balance. When an object is in balance owing to the action of forces, it is said to be in equilibrium. The most straightforward examples of this apply when two forces push against each other. For example, when you sit on a chair, your weight acts downwards and the seat pushes upwards with a force that is equal and opposite to your weight. Because both forces act through the same point, this system is said to be concurrent.

However, if you consider a see-saw or a balance scale, the system is in balance. This is because the balancing effect (moments) about the pivot point is the same. In this section we are going to investigate how forces and moments are balanced and how we can determine the resulting forces, and the forces that are required to bring a system into balance.

Key Term

Concurrent – Something that is happening at the same time. Concurrent forces are forces that meet at a point.

Make the Grade   P1

Grading criterion P1 requires you to calculate the magnitude, direction and position of the line of action of the resultant and equilibrant of a non-concurrent coplanar force system containing a minimum of four forces acting in different directions. For a given problem, you will need to resolve each force into its horizontal and vertical components and consider the rules of static equilibrium, along with Pythagoras' theorem and simple trigonometry, to determine the required values and directions.

Graphical representation

Key Term

F = force or load and is measured in newtons (N)

When we consider force systems, we need to understand what a force is. It is useful to consider that load and weight are both forces, and we shall consider weight to illustrate how forces are often in balance.

When you stand on a scale you will see your 'weight' in kilogrammes (kg). This is a common misunderstanding; you are actually seeing your mass. This is because weight is a kind of force and is measured in newtons.

It was Sir Isaac Newton who first developed the principles of force, mass and acceleration, when he observed an apple falling from a tree. He realised that it was gravity acting on the apple which caused it to fall to the ground.

From this principle we know that:

weight = mass × acceleration due to gravity

Key Term

a = acceleration and is measured in $\dfrac{\text{metres}}{\text{second}^2}$ (m/s²)

In Figure 5.1, we can see that a person who has a mass of 70 kg is standing on a spring. The spring will deflect due to the weight of the person; we say that the spring is in compression.

The acceleration due to gravity varies very slightly, depending on whereabouts on earth you happen to be. However, to two decimal places, it can be taken to be 9.81 m/s².

weight of the person = mass × acceleration due to gravity = 70 × 9.81 = 686.7 N

Because the person and the spring are in balance, the compressive force in the spring acting upwards is equal to the weight of the person acting downwards.

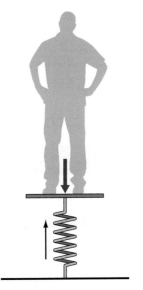

Figure 5.1 Forces in balance vertically

To simplify the diagram, we can represent the upward and downward forces as arrows and these are said to be scalar quantities.

Key Terms

Scalar quantity – Indicates direction and is normally represented by an arrow.

Vector quantity – Indicates magnitude and direction, and is normally represented by an arrow and drawn to scale.

Because the forces are balanced, we can say that they are in equilibrium. Similarly, because there is no movement, we describe this situation as a system that is in a state of *static equilibrium.*

Key Term

Static equilibrium – An object is in static equilibrium if the forces on the object are balanced and the object is at rest.

Similarly, Figure 5.2 shows a person pulling a spring horizontally with a steady force of 20 N. The spring will begin to stretch until it reaches a point where the tension in the spring is equal to the pulling force.

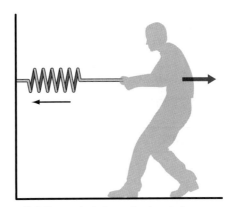

Figure 5.2 Forces in balance horizontally

Once again, the forces are in balance; so the system is in static equilibrium.

In Figure 5.1,

upward force = downward force.

In Figure 5.2,

leftwards force = rightwards force.

Using the rules of static equilibrium, we can determine unknown forces for a coplanar system.

Key Term

Coplanar – Forces that all act on the same plane.

Triangle of forces

Worked Example

The force system shown in Figure 5.3 is in static equilibrium. The diagram in Figure 5.3(b) is known as a free-body diagram (sometimes called a space diagram); this shows the forces and directions as **scalar** quantities. The forces are drawn from the concurrent point, showing their direction. This can be followed by constructing a **vector** diagram; in Figure 5.3(c), a vector diagram shows the forces drawn 'nose to tail'.

Because we know the mass being suspended, we can determine the force (weight).

$$F_1 = 50 \times 9.81 = 490.5\,\text{N}$$

This allows us to draw a vector to represent F_1. This is drawn to scale.

From the 'nose' of this vector we can plot another line from right to left, at 90° to the vector F_1. Of course, we do not know the size of F_2, only the direction. Finally, we can draw a guideline at 50° from the 'tail' of the first arrow, representing F_3. By measuring the length of the arrows, and using the scale we have chosen, we can determine the sizes of the unknown forces.

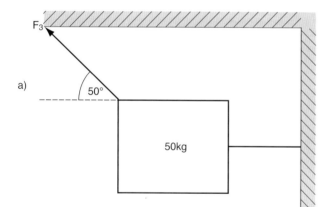

a)

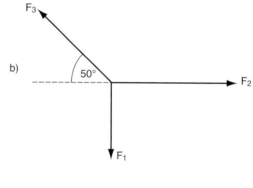

b)

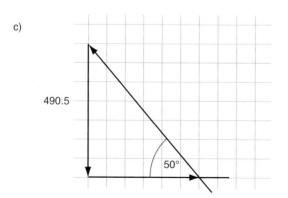

c)

490.5

50°

Figure 5.3 Load suspended from two ropes

The technique for drawing vector diagrams involving three forces, and hence calculating unknown forces, is known as the 'triangle of forces' method.

Example

Figure 5.4 shows three forces in equilibrium and the triangle of forces constructed to allow the unknown forces to be calculated.

By drawing a vector diagram, using a suitable scale, the vectors for the tension in the cable T and load P can be constructed. This allows the vector for the reaction force at point A (R_A) to be constructed, and the length of the arrows and the angles to be measured.

Polygon of forces

For a system of more than three forces, a similar technique is used. This arrangement is known as the polygon of forces. The process for solving this type of problem is very similar:

- choose an appropriate scale for drawing the force vectors
- draw the first force vector to scale
- draw the second force vector to scale from the 'nose' of the previous force vector

- continue in this manner until all force vectors have been constructed
- if the resulting polygon is 'closed', then all forces are in equilibrium.

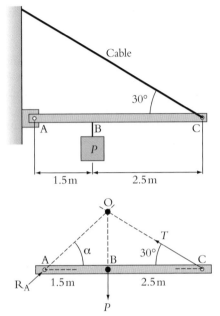

Free-body diagram

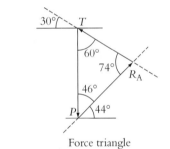

Force triangle

Figure 5.4 Force system

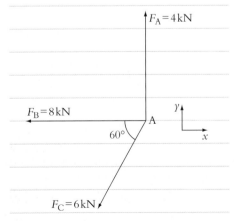

Figure 5.5 Force system

Example

The force system shown in Figure 5.5 is not in equilibrium. However, by using the polygon of forces technique, it is possible to construct a vector diagram which allows us to determine the resulting force and the force required to bring the system into a state of equilibrium.

The resulting vector F_R is known as the resultant; this represents how the three given forces could be represented as one force. If the force were to act in the opposite direction, it would be known as the equilibrant.

Key terms

Resultant – When several forces are combined to form one force.

Equilibrant – A force which, when applied, brings a system into a state of equilibrium.

Activity 5.1

For the force systems shown in Figures 5.6 to 5.10, construct a vector diagram in order to determine the resultant and equilibrant forces.

6kN

7kN

Figure 5.6

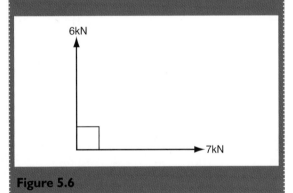

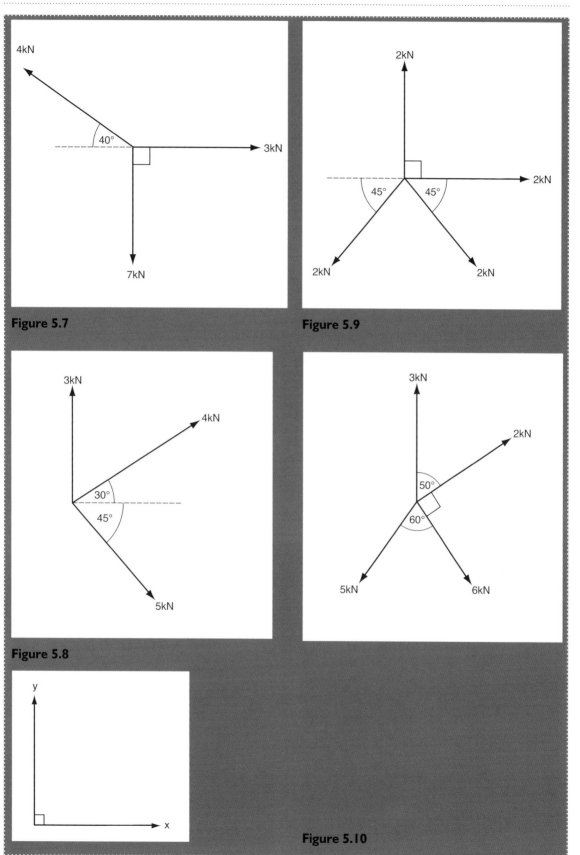

Figure 5.7

Figure 5.9

Figure 5.8

Figure 5.10

213

Conditions for static equilibrium

Having studied how we can determine unknown forces by graphical means, let us now consider an analytical approach.

We know that:

- leftwards forces = rightwards forces ($\Sigma F_x = 0$)
- upwards forces = downwards forces ($\Sigma F_y = 0$).

There is a third condition for static equilibrium; this relates to turning moments.

A moment is a turning or twisting effect caused when a force is applied. It is related to the perpendicular distance between where the force is applied and the point where turning occurs.

Figure 5.11 shows a force (F) applied to a spanner; this will cause a clockwise moment. However, it is not just the force applied that will affect the size of the turning moment, but the distance as well. The greater the distance (d), which is the perpendicular distance, the greater the turning moment.

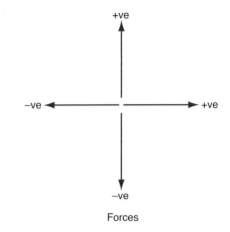

Forces

Key Term

Moment – The product of force multiplied by distance.

$$M = FD \ (\text{Nm})$$

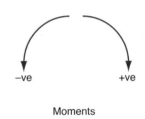

Moments

Figure 5.12 Sign convention

If we consider clockwise (CW) and anticlockwise (ACW) moments, if these are in balance, then a system will be in equilibrium.

In other words, CW moments = ACW moments ($\Sigma M = 0$).

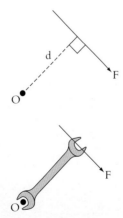

Figure 5.11 Turning moment

Worked Example

In Figure 5.13:

$$CW = ACW$$

$$8 \times 5 = 10 \times 4$$

Therefore, the system is in equilibrium; we consider CW moments to be positive and ACW moments to be negative.

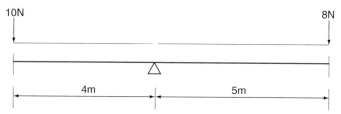

Figure 5.13 Balanced moments

Resolution of forces in perpendicular directions

Forces can be split into horizontal and vertical components and this helps us to determine unknown forces using an analytical approach.

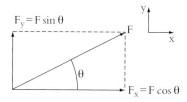

Figure 5.14 Resolution of forces

In Figure 5.14, the force F is split into a horizontal component $F\cos\theta$ and a vertical component $F\sin\theta$.

Worked Example

$F_R = 11.064\,\text{kN}$

Figure 5.15 Resolving forces

In Figure 5.15, the downward force of 1.196 kN and the leftwards force of 11 kN can be resolved into one force using trigonometry and Pythagoras' theorem.

$$\sqrt{(1.196^2 + 11^2)} = 11.064\,\text{kN}$$

$$\tan\theta = \frac{\text{opposite}}{\text{adjacent}} = \frac{1.196}{11} = 0.11$$

$$\theta = 6.2°$$

Key Term

Resultant – When several forces are combined to form one force.

Similarly, if you have two perpendicular forces, they can be resolved into one force.

Example

Consider the force system in Figure 5.16. In this example it is useful to draw a table. The sign convention previously used is adopted here.

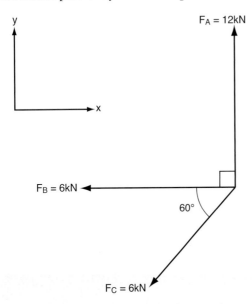

Figure 5.16 Force system

Force	Horizontal force (F_x)	Vertical force (F_y)
12 kN (F_A)	0 kN	12 kN
6 kN (F_B)	−6 KN	0 kN
6 kN (F_C)	−6 cos 60 = −3 kN	−6 sin 60 = −5.12 kN
Total	**−9 kN**	**−6.88 kN**

This can be represented as a simple vector diagram, as shown in Figure 5.17. The directions of the horizontal and vertical components are determined by noting whether the total figures are positive or negative.

To determine the resultant force, use Pythagoras' theorem:

$$F_R = \sqrt{(9^2 + 6.88^2)} = 11.33 \text{ kN}$$

The angle, measured from the horizontal, can be determined from trigonometry:

$$\tan\theta = \frac{6.88}{9} = 0.76$$

$$\theta = 37.23°$$

This allows you to determine the resultant force. If you are trying to find the force that will restore equilibrium to the system (the equilibrant), this will be exactly the same force; however, it will act in the opposite direction (see Figure 5.18).

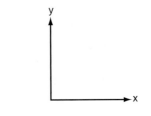

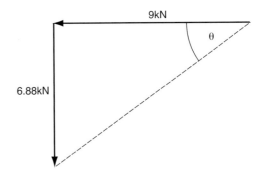

Figure 5.17 Resultant horizontal and vertical elements

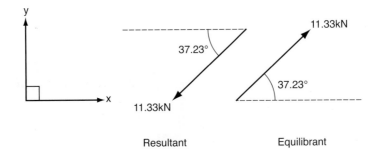

Figure 5.18 Resultant and equilibrant

Activity 5.2

For the force systems shown in Figures 5.19 to 5.22, determine the resultant force by analytical methods.

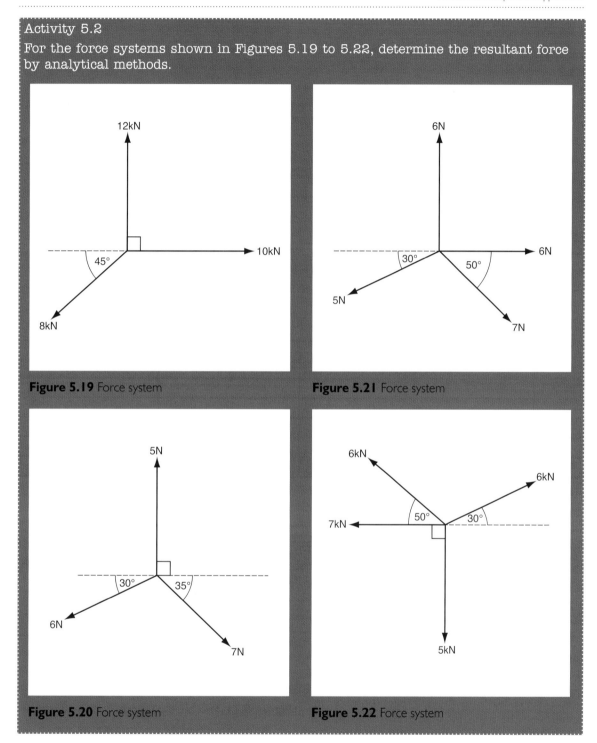

Figure 5.19 Force system

Figure 5.20 Force system

Figure 5.21 Force system

Figure 5.22 Force system

Vector addition of forces

Up to this point we have considered coplanar concurrent force systems. However, things become somewhat more complicated if we consider a non-concurrent force system.

In Figure 5.23, the four forces do not pass through the same point; therefore, they are non-concurrent.

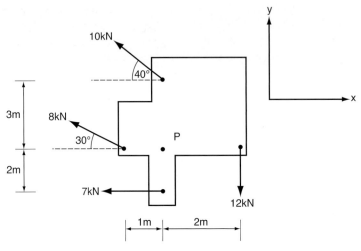

Figure 5.23 Non-concurrent force system

Force	Horizontal force (F_x)	Vertical force (F_y)	Moment of horizontal force (F_x) about point P	Moment of vertical force ($F_y y$) about point P
10 kN	$-10\cos40 = -7.66$ kN	$10\sin40 = 6.43$ kN	$-7.66 \times 3 = -22.98$ kNm	$6.43 \times 0 = 0$ kNm
12 kN	0 kN	-12 kN	0 kNm	$12 \times 2 = 24$ kNm
7 kN	-7 kN	0 kN	$7 \times 2 = 14$ kNm	0 kNm
8 kN	$-8\cos30 = -6.93$ kN	$8\sin30 = 4$ kN	$-6.93 \times 0 = 0$ kNm	$4 \times 1 = 4$ kNm
Total	**-21.59 kN**	**-1.57 kN**		**-8.98 kNm $+$ 28 kNm $=$ 19.02 kNm**

Table 5.2

To analyse this system will require us to consider all three of the conditions for static equilibrium.

$\Sigma M = 0$ where CW moments are positive and ACW moments are negative

$\Sigma F_x = 0$ where upwards forces are positive and downwards forces are negative

$\Sigma F_y = 0$ where rightwards forces are positive and leftwards forces are negative

Again, a tabular approach is used and a reference point is decided upon, in this case, point P. This process is the same as for concurrent forces; however, we now use the reference point P and determine the moment of each force about that point, using the perpendicular distance to the line of action.

As before, we can draw a vector diagram (Figure 5.24) and determine the size and direction of the resultant force.

To determine the resultant force, use Pythagoras' theorem:

$$F_R = \sqrt{(21.59^2 + 1.57^2)} = 21.65 \text{ kN}$$

The angle, measured from the horizontal, can be determined from trigonometry:

$$\tan\theta = \frac{1.57}{21.59} = 0.073$$

$$\theta = 4.16°$$

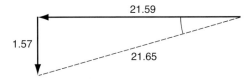

Figure 5.24 Resultant showing horizontal and vertical elements

Because the line of action of the resultant force does not pass through point P, remember that the force system is non-concurrent; the final step is to determine the perpendicular distance to the line of action of the force.

To determine where the force vector is positioned, we need to consider whether the sum of the moments is CW or ACW.

The sum of the moments gives us a total of 19.02 kNm, and as this number is positive, the moment is clockwise and must lie below point P, as shown in Figure 5.25.

Resultant moment about point P = sum of all moments about point P

$$21.65 \times Z = 19.02$$
$$Z = 0.88\,\text{m}$$

If you wanted to determine the equilibrant force, this would be exactly the same size and distance from P; however, the force would act in the opposite direction.

Simply supported beams

By the end of this section you should have developed a knowledge and understanding of the conditions that apply for a simply supported beam to be in static equilibrium. This will allow you to determine the support reactions for the beam.

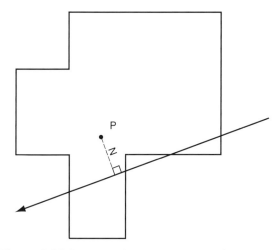

Figure 5.25 Perpendicular distance to line of action

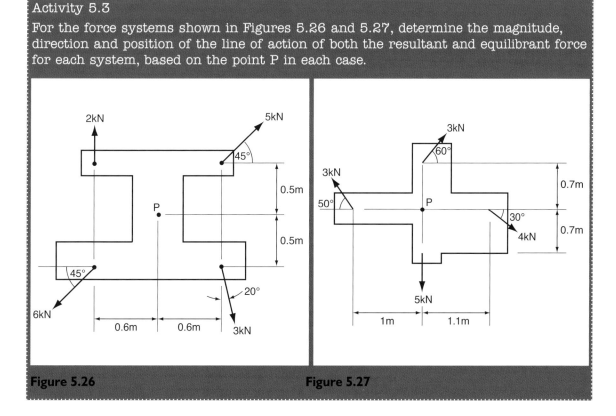

Activity 5.3

For the force systems shown in Figures 5.26 and 5.27, determine the magnitude, direction and position of the line of action of both the resultant and equilibrant force for each system, based on the point P in each case.

Figure 5.26

Figure 5.27

A simply supported beam is a beam that is supported on a point or roller. Consequently, it is free to move horizontally. Simply supported beams are subject to point or uniformly distributed loads.

> ### Key Points
>
> When we analyse beams, we make use of the same principles of static equilibrium that we used in the previous section. However, in this case, we refer to the forces acting downwards on a beam as *active*. The balancing forces supporting the beam and maintaining equilibrium are known as *reactive* and are called reactions.

Key Term

Uniformly distributed load (UDL) – A load that is evenly distributed along a beam; this is often the self-weight of the beam and is measured in $\frac{N}{m}$.

Make the Grade   P2

Grading criterion P2 requires you to calculate the support reactions of a simply supported beam carrying at least two concentrated loads and a uniformly distributed load. You will need to use the rules of static equilibrium to determine these values.

Conditions for static equilibrium

As previously discussed, the analysis of simply supported beams requires us to follow the rules for static equilibrium:

$\Sigma M = 0$ where CW moments are positive and ACW moments are negative

$\Sigma F_x = 0$ where upwards forces are positive and downwards forces are negative

$\Sigma F_y = 0$ where rightwards forces are positive and leftwards forces are negative

In our study of beams, we will not need to consider horizontal forces, however, which makes the analysis somewhat easier.

Loading

Beams are generally subject to point loading or uniformly distributed loading.

Point loading

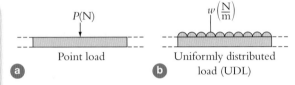

Figure 5.28 Types of beam loading

A beam can be subject to point loads, UDLs or a combination of both. Point loads are represented by downwards arrows, indicated by the load P (N) in Figure 5.28(a) whereas the UDL is denoted by the load w indicated in Figure 5.28(b).

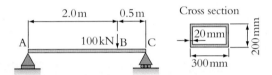

Figure 5.29 Loaded beam

The forces acting on the beam supports are known as reactions and can be determined using the rules of static equilibrium. In Figure 5.29 there is a point load at point B applied to the beam. We can use the principle of moment to determine the unknown reactions.

It is customary to use the left-hand support as a point to take moments from. Taking moments about point A, the 100 kN force will cause a CW moment and the reaction force at C will oppose this with an ACW moment.

Remember that CW moments = ACW moments:

$$(100 \times 2) = (R_C \times 2.5)$$

where R_C is the reaction force at point C:

$$\therefore \frac{200}{2.5} = R_C$$

$$R_C = 80\,kN$$

To determine the reaction force at point A, you could take moments about point C. However, there is an alternative method, as, for the beam to be in balance, the upwards forces must equate to the downwards forces.

Upwards forces = downwards forces:

$$R_C + R_A = 100$$
$$80 + R_A = 100$$
$$R_A = 20\,kN$$

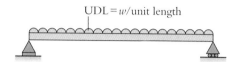

UDL = w/unit length

Figure 5.30 Beam with UDL

Determining the support reaction for a beam with a single point load is relatively straightforward. However, when a UDL is introduced, things become a little more complicated.

The beam in Figure 5.30 has a UDL with a load of w expressed in Newton per metre. Because the load is uniform, we can determine the overall value for the UDL and, for the purpose of analysing the beam, 'replace' it with a point load acting at the centre.

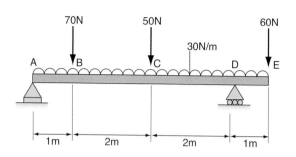

Figure 5.31 Beam with UDL and three point loads

This approach can be taken to solve problems involving a combination of point loads and UDL loads on a simply supported beam. Consider the beam shown in Figure 5.31.

Referring to Figure 5.31, and taking moments about A (CW moment = ACW moments)

$$(70 \times 1) + (50 \times 3) + (60 \times 6) + (30 \times 6 \times 3)$$
$$= (R_D \times 5)$$
$$R_D = \frac{(70 + 150 + 360 + 540)}{5}$$
$$R_D = 224\,N$$

Upwards forces = downwards forces:

$$R_A + R_D = 70 + 50 + 60 + (30 \times 6)$$
$$R_A + 224 = 360$$
$$R_A = 136\,N$$

Worked Example

The total weight of the UDL will be $5 \times 6 = 30\,N$

Taking moments about A (CW moment = ACW moments)

$$(30 \times 3) = (R_B \times 6)$$
$$R_B = 15\,N$$

Upwards forces = downwards forces:

$$R_A + R_B = 30$$
$$R_A + 15 = 30$$
$$R_A = 15\,N$$

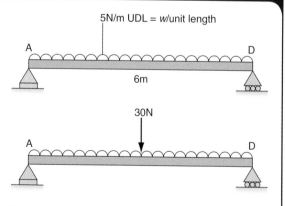

5N/m UDL = w/unit length

6m

30N

Figure 5.32

221

Activity 5.4

For each of the simply supported beams shown in Figures 5.33 to Figure 5.36, determine the unknown reactions.

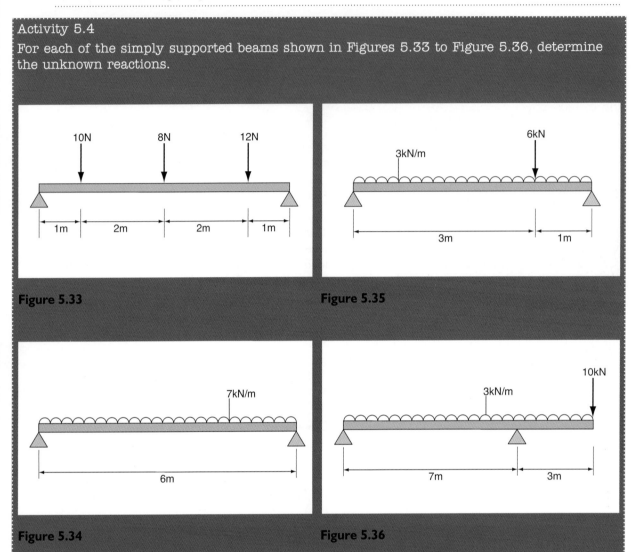

Figure 5.33

Figure 5.35

Figure 5.34

Figure 5.36

Loaded components

By the end of this section you should have developed a knowledge and understanding of elastic constants used in the analysis of components subject to uniaxial and shear loading. You will use this knowledge to determine dimensional change due to loading; direct stress and strain; shear stress and strain and factor of safety.

Loaded components display different behaviour depending upon their elastic properties and the type of loading. Engineers are concerned with the amount of stress or strain to which components can be subjected and how they react to given loading conditions. In order to design components that can withstand a variety of different loading conditions, it is important to understand how the elastic properties of the material react to loading. Engineers use this information to determine safe working loads and the factor of safety that can be applied.

Key Points

Units of stress are N/m², however:

$$10^6 \, mm^2 = 1 \, m^2$$

and

$$10^6 \, N^2 = 1 \, MN^2$$

So

$$\frac{1 \, MN}{m^2} = 1 \, N/mm^2$$

Make the Grade   P3

Grading criterion P3 requires you to calculate the induced direct stress, strain and dimensional change in a component subjected to direct uniaxial loading and the shear stress and strain in a component subjected to shear loading. You will be able to determine these values for a given problem, or series of problems, by applying the appropriate formulae carefully and methodically.

Loading

Components are often subjected to a variety of loading; however, for our purposes, we can subdivide these into uniaxial loading and shear loading. Uniaxial loading refers to loading along an axis.

Figure 5.37(a) shows an object being pulled along its axis; this type of loading will cause the object to stretch and is known as **tension**.

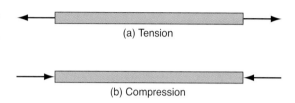

(a) Tension

(b) Compression

Figure 5.37 Uniaxial loading

Figure 5.37(b) shows an object being squeezed along its axis; this type of loading will cause the object to compress and is known as **compression**.

When an object is loaded and the lines of action of the force are not uniaxial, as shown in Figure 5.38, the object will tend to deform, as one part of the material slides over another, rather like a deck of cards. This type of loading is known as shear force.

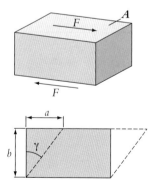

Figure 5.38 Diagram of shear loading

223

Effects

Consider first the effects of uniaxial loading, whether in tension or compression.

Direct Stress

Key Terms

σ – The symbol for direct stress.

$\dfrac{N}{m^2}$ – The units for direct stress.

$$\sigma = \frac{F}{A}$$

When a force is applied to a component, corresponding forces act within the material to resist this external force. Stress is the term we use to indicate the magnitude of the loading.

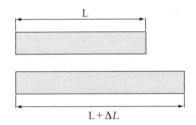

Figure 5.40 Loaded component

$$\text{direct stress} = \frac{\text{load}}{\text{cross-sectional area}}$$

Worked example

The component in Figure 5.39 is subject to a tensile force, as shown. Determine the direct stress induced in the component.

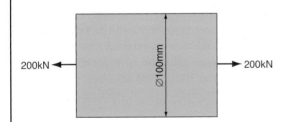

Figure 5.39 Loaded component

Cross-sectional area $= \dfrac{\pi d^2}{4}$

$$A = \pi \times \frac{100^2}{4}$$

$$A = 7.85 \times 10^3\,mm^2$$

$$\sigma = \frac{F}{A} = \frac{200 \times 10^3}{7.85 \times 10^3}$$

$$\sigma = 25.48\,N/mm^2$$

Direct strain

Strain is a measure of the deformation of a material. If a component is subject to a compressive or tensile load along its length, the length will change.

Direct strain is a ratio. Strain is $\dfrac{\text{change in length}}{\text{original length}}$.

Key Term

ϵ – The symbol for direct strain.

There are no units for direct strain as it is a ratio.

$$\epsilon = \frac{\Delta L}{L}$$

Worked Example

The component in Figure 5.41 is subject to a tensile force, as shown. Determine the direct strain induced in the component.

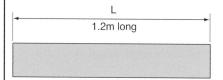

L
1.2m long

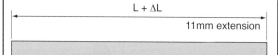

L + ΔL
11mm extension

Figure 5.41 Component under strain

$$\epsilon = \frac{\Delta L}{L}$$

$$\Delta L = 11\,\text{mm}$$

$$L = 1200\,\text{mm}$$

$$\epsilon = \frac{11}{1200} = 9.17 \times 10^{-3}$$

Shear stress

Shear stress is determined in a similar way to direct stress; however, the area in shear is different, as shown in Figure 5.42.

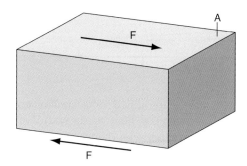

Figure 5.42 Component under shear

Key Terms

τ – The symbol for shear stress.

$\dfrac{\text{N}}{\text{m}^2}$ – The unit for shear stress.

$$\tau = \frac{F}{A}$$

Worked Example

The component in Figure 5.43 is subject to a shear force of 25 kN, as shown. Determine the shear stress induced in the component.

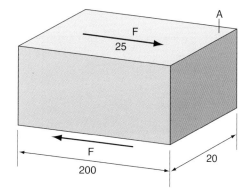

Figure 5.43 Component in shear

225

Area in shear $= 200 \times 20$

$$A = 4 \times 10^3 \, mm^2$$

$$\tau = \frac{25 \times 10^3}{4 \times 10^3}$$

$$\tau = 6.25 \, N/mm^2$$

Shear strain

Shear strain is determined in a similar way to direct strain. The key here is the ratio of the relative movement between the two shear surfaces and the distance between these surfaces. Referring to Figure 5.38, the ratio is $\frac{a}{b}$.

Key Term

γ – The symbol for shear strain.

There are no units for shear strain as it is a ratio.

$$\gamma = \frac{a}{b}$$

Worked Example

The component in Figure 5.44 is subject to a shear force, as shown. Determine the shear strain induced in the component.

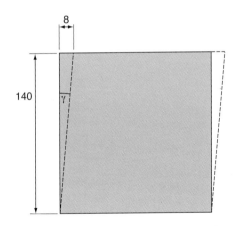

Figure 5.44 Loaded components showing shear area

$$\gamma = \frac{8}{140}$$

$$\gamma = 57.14 \times 10^{-3}$$

Elastic constants

Elastic constant are values that have been determined by experimenting and testing samples of a given material. Elasticity is the property that allows a material to return to its original shape after being deformed under load. If a material is loaded excessively, to the point where it becomes deformed, it is said to have gone beyond its elastic limit.

Modulus of elasticity

The modulus of elasticity is a material constant and is determined from tensile testing, where a test piece is loaded uniaxially. Test pieces are stretched under controlled conditions and a reading of load and extension can be recorded. The modulus can be demonstrated by plotting stress against strain for an elastic material. Before the material reaches its elastic limit, the graph is a straight line, and the slope of this graph is known as the modulus of elasticity, as shown in Figure 5.45.

Key Terms

E – The symbol for modulus of elasticity.

N/m^2 – The units for modulus of elasticity.

$$E = \frac{\sigma}{\epsilon}$$

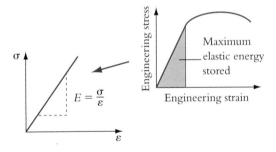

Figure 5.45 Stress/strain graph

Modulus of rigidity

The modulus of rigidity is often called the shear modulus, as it refers to the resistance to shear exhibited by a material when loaded. It is the ratio between shear stress and shear strain.

Key Terms

G – The symbol for shear modulus.

N/m^2 – The units for shear modulus.

$$G = \frac{\tau}{\gamma}$$

Worked Example

A test sample is found to extend 0.2 mm under the loading condition shown in Figure 5.46. Determine the modulus of elasticity for the material.

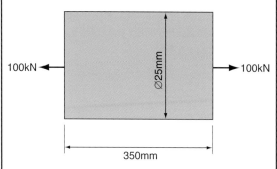

Figure 5.46 Round component shows length, diameter and load

Cross-sectional area $= \dfrac{\pi d^2}{4}$

$$A = \pi \times \frac{25^2}{4}$$

$$A = 490.87\,\text{mm}^2$$

$$\sigma = \frac{F}{A} = \frac{100 \times 10^3}{490.87}$$

$$\sigma = 203.72\,\text{N/mm}^2$$

$$\sigma = 203.72\,\text{MN/m}^2$$

$$\epsilon = \frac{\Delta L}{L}$$

$$\Delta L = 0.2\,\text{mm}$$

$$L = 350\,\text{mm}$$

$$\epsilon = \frac{0.2}{350} = 571.43 \times 10^{-6}$$

$$E = \frac{\sigma}{\epsilon}$$

$$E = \frac{203.72 \times 10^6}{571.43 \times 10^{-6}}$$

$$E = 356.51 \times 10^9\,\text{N/m}^2$$

$$E = 356.51\,\text{GN/m}^2$$

Worked Example

A test sample is found to deflect under the loading condition, as shown in Figure 5.47. Determine the modulus of elasticity for the material.

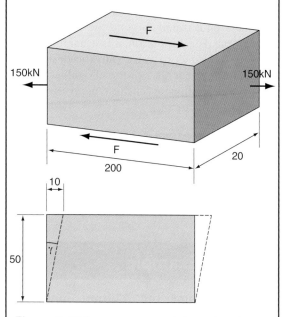

Figure 5.47 Square component shows length, deflection and load

Shear area $= 200 \times 20$

$$A = 4 \times 10^3\,\text{mm}^2$$

$$\tau = \frac{F}{A} = \frac{150 \times 10^3}{4 \times 10^3}$$

$$\tau = 37.5\,\text{N/mm}^2$$

$$\tau = 37.5\,\text{MN/m}^2$$

$$\gamma = \frac{a}{b}$$

$$\gamma = 10\,\text{mm}$$

$$b = 50\,\text{mm}$$

$$\gamma = \frac{10}{50} = 0.2$$

$$G = \frac{\tau}{\gamma}$$

$$G = \frac{37.5 \times 10^6}{0.2}$$

$$G = 187.5 \times 10^6\,\text{N/m}^2$$

$$G = 187.5\,\text{MN/m}^2$$

Factor of safety

When engineers design components, it is not always clear what the conditions of service for that component might be. It may be subject to a variety of steady loading or the loading may fluctuate. Changes in dimensions may occur or imperfections in the component material may cause weaknesses. In addition, the ultimate stress for a material is usually beyond the elastic limit of the material. For this reason, it is usual for a factor of safety to be incorporated into the design. The factor of safety is a ratio of the allowable working stress and the ultimate tensile/compressive/shear stress for the material.

Make the Grade — M1

Grading criterion M1 requires you to calculate the factor of safety in operation for a component subjected to combined direct and shear loading against given failure criteria. This requires you to use the methods and techniques previously studied, but you will have to apply resolution of forces techniques to identify direct and shear loading, as shown in Figure 5.49.

Key Term

Factor of safety (FOS) – Ultimate stress/working stress.

Worked Example

An aluminium rod 125 mm long and 17 mm diameter has an ultimate tensile stress of 200 MN/m^2. Using a factor of safety of 5, find the maximum load and change in dimensions for the rod. Take $E_A = 70$ GN/m^2.

Working stress $= \dfrac{\text{ultimate stress}}{\text{factor of safety}}$

Working stress $= \dfrac{200}{5} = 40$ MN/m^2

$$A = \frac{\pi \times 17^2}{4}$$

$$A = 226.98 \text{ mm}^2$$

$$A = 226.98 \times 10^{-6} \text{ m}^2$$

$$\sigma = \frac{F}{A}$$

$$F = 40 \times 10^6 \times 226.98 \times 10^{-6}$$

$$F = 9.079 \times 10^3 \text{ N}$$

$$F = 9.079 \text{ kN}$$

$$E = \frac{\sigma}{\epsilon}$$

$$\epsilon = \frac{40 \times 10^6}{70 \times 10^9}$$

$$\epsilon = 571.43 \times 10^6$$

$$\epsilon = \frac{\Delta L}{L}$$

$$\Delta L = 571.43 \times 10^{-6} \times 125$$

$$\Delta L = 71.43 \times 10^{-3} \text{ mm}$$

Worked Example

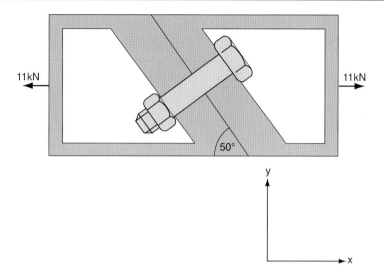

Figure 5.48

The 12 mm diameter bolt shown in Figure 5.48 holds two components together. For the given loading, determine the factor of safety in operation. Assume that ultimate tensile stress is 550 MN/m² and ultimate shear stress is 350 MN/m².

$$A = \frac{\pi \times 12^2}{4}$$

$$A = 113.10 \, \text{mm}^2$$

$$A = 113.10 \times 10^{-6} \, \text{m}^2$$

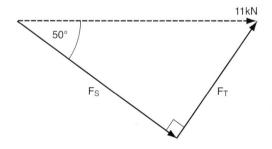

Figure 5.49 Force triangle

The tensile force acts uniaxially, at an angle of 50° to the horizontal (F_T).

The shear force acts perpendicular to the tensile force (F_S).

$$F_T = F\sin\theta = 11 \times 10^3 \times \sin 50$$

$$F_T = 8.43 \times 10^3 \, \text{N}$$

$$F_S = F\cos\theta = 11 \times 10^3 \times \cos 50$$

$$F_S = 7.07 \times 10^3 \, \text{N}$$

Tensile stress

$$\sigma = \frac{F}{A}$$

$$\sigma = \frac{8.43 \times 10^3}{113.10 \times 10^{-6}}$$

$$\sigma = 74.54 \times 10^6 \, \text{N/m}^2$$

$$\sigma = 74.54 \, \text{MN/m}^2$$

$$\text{FOS} = \frac{\text{ultimate tensile stress}}{\text{working tensile stress}}$$

$$\text{FOS} = \frac{550 \times 10^6}{74.54 \times 10^6}$$

$$\text{FOS} = 7.38$$

Shear stress

$$\sigma = \frac{F}{A}$$

$$\sigma = \frac{7.07 \times 10^3}{113.10 \times 10^{-6}}$$

$$\sigma = 62.51 \times 10^6 \, \text{N/m}^2$$

$$\sigma = 62.51 \, \text{MN/m}^2$$

$$FOS = \frac{\text{ultimate shear stress}}{\text{working shear stress}}$$

$$FOS = \frac{350 \times 10^6}{62.51 \times 10^6}$$

$$FOS = 5.60$$

The factor of safety in operation is the smaller of two numbers calculated

Therefore

$$FOS = 5.6$$

Activity 5.5

1. A 10 mm diameter steel rod is subject to a compressive load applied uniaxially. If the force applied is 25 kN, determine the compressive stress induced in the material.

2. A 750 mm long tie bar is subject to a uniaxial tensile load. If the bar extends 0.7 mm, determine the strain induced.

3. An 8 mm square metal bar of 1.5 m length is loaded uniaxially in tension. If the load applied is 125 kN and the modulus of elasticity for the material is 200 GN/m², determine the extension in mm.

4. The block in Figure 5.50 has a modulus of rigidity of 15 GN/m². Find the horizontal deflection indicated.

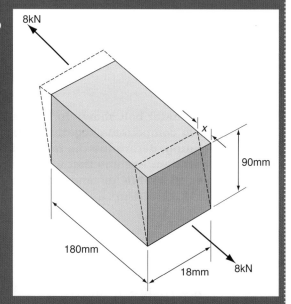

Figure 5.50

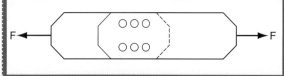

Figure 5.51

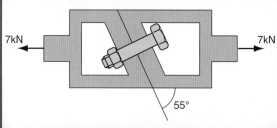

Figure 5.52

5. Two plates are held together by rivets, as shown in Figure 5.51. The rivets are 10 mm diameter and the material has an ultimate shear stress of 300 MN/m². Determine the maximum force to be applied if a factor of safety of 5 is used.

6. The 10 mm diameter bolt shown in Figure 5.48 holds two components together. For the given loading, determine the factor of safety in operation. Assume that the ultimate tensile stress is 275 MN/m² and the ultimate shear stress is 475 MN/m².

5.2 Determining work, power and energy transfer in dynamic engineering systems

This section will cover the following grading criteria:

Dynamic engineering systems refer to those systems where movement is present. Engineering often involves the transfer of energy or the conversion of energy to power and work. Newton's laws are fundamental to an understanding of this process.

Kinetic parameters

By the end of this section you should be able to state the key kinetic parameters used by engineers.

When we consider how fast we are going, we usually refer to our speed; similarly, if we want to explain how far we are going, we refer to the distance travelled. However, in engineering, we need to be more precise in our terminology.

Key Points

Distance travelled refers to how far we have gone. If you are at home and you walk to work or college and back again, you might have travelled several miles. However, your displacement is zero, as you start and finish at the same place.

Make the Grade   P4

Grading criterion P4 requires you to solve three or more problems that require the application of kinetic and dynamic principles to determine unknown system parameters. To do this, you will need to be able to use the appropriate symbols and terminology in your solutions.

Displacement

Displacement refers to the distance from where an object starts to where it finishes. This is not the same as the distance travelled, as indicated in Figure 5.53.

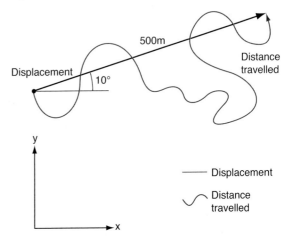

Figure 5.53 The difference between distance and displacement

The symbol for displacement is *s* and the units are metres.

Initial velocity

Initial velocity is the starting velocity of an object in a given direction; consequently, it is a vector quantity.

The symbol for initial velocity is *u* and the units are metres per second.

Final velocity

Final velocity is the ending velocity of an object in a given direction; consequently, it is a vector quantity.

The symbol for final velocity is *v* and the units are metres/second.

Uniform linear acceleration

Uniform linear acceleration is the change in rate of velocity of an object in a given time.

The symbol for uniform linear acceleration is *a* and the units are metres per second squared.

Key Points

Uniform linear acceleration can be a negative value; this represents a deceleration and shows the change in rate of velocity is decreasing.

Key Terms

u = initial velocity (m/s)

v = final velocity (m/s)

a = acceleration (m/s^2)

t = time (s)

s = displacement (m)

Kinetic principles

By the end of this section you should have developed a knowledge and understanding of how to determine kinetic parameters from given information using appropriate formulae.

Using standard formulae allows us to determine specific parameters from limited information. Newton's laws first allowed the analysis of objects in motion, and it is from these principles that the laws of linear motion can be derived.

Key Points

The equations in this section only apply for uniform acceleration and where objects are travelling in a straight line.

Make the Grade   P4

Grading criterion P4 requires you to solve three or more problems that require the application of kinetic and dynamic principles to determine unknown system parameters. To do this, you will need to be able to use the appropriate formulae to determine the required unknown values.

Equations for linear motion with uniform acceleration

Using the symbols for kinetic parameters:

$$v = u + at$$
$$s = ut + 0.5\,at^2$$
$$v^2 = u^2 + 2\,as$$
$$s = 0.5\,(u + v)t$$

Key Points

Vehicles travelling in km/h have to have their velocity converted to m/s. Remember, there are 1,000 metres in a kilometre and 3,600 seconds in an hour.

Worked Example

A vehicle moving in a straight line decelerates from 50 km/h to 30 km/h in 35 s. Determine the rate of deceleration and displacement of the vehicle.

$$u = \frac{50 \times 1,000}{3,600}$$

$$u = 13.89\,\text{m/s}$$

$$v = \frac{30 \times 1,000}{3,600}$$

$$v = 8.33\,\text{m/s}$$

$$v = u + at$$

$$a = \frac{(8.33 - 13.89)}{35}$$

$$a = -0.16\,\text{m/s}^2$$

Remember, the $-ve$ sign indicates a deceleration.

$$s = 0.5\,(u + v)t$$

$$s = 0.5(13.89 + 8.33)35$$

$$s = 388.89\,\text{m}$$

233

Activity 5.6

1. A vehicle is accelerated at $1.2\,\text{m}/\text{s}^2$ for 12 s; if the initial velocity is $40\,\text{m/s}$, find:
   - the final velocity
   - the displacement.
2. A train travelling at $40\,\text{km/h}$ is brought to rest in 2 min, with uniform deceleration. Find:
   - the rate of deceleration
   - the displacement.
3. A machine component moves in a straight line with a velocity of $12\,\text{m/s}$. If the component accelerates to $24\,\text{m/s}$ over a distance of 3 m, find:
   - the time taken
   - the rate of acceleration.

Dynamic parameters

By the end of this section you should have developed a knowledge and understanding of the parameters used in determining the relationship between work, energy and power.

In order to analyse dynamic systems and the relationship between work, energy and power, it is important to understand the parameters and terminology used. This section introduces you to the key concepts required to fully analyse dynamic systems.

Key Points

Energy is measured in joules (J) and the rate that energy is used or consumed (J/s) is known as power in watts (W).

Tractive effort

Tractive effort can be defined as the force required to move an object and overcome resistance to movement such as friction, gravity, and so on. It is measured in newtons (N).

Braking force

Braking force can be defined as the force required

to stop an object moving. It can be considered to be the opposite of tractive effort and is measured in newtons (N).

Inertia

The inertia of an object refers to the resistance to motion it exhibits. The greater the mass of an object, the more inertia that is evident and, consequently, a greater amount of force is required to accelerate it and give it momentum.

Frictional resistance

Frictional resistance is not related to inertia; although it is the resistance to movement, it is determined by the surfaces in contact. An ice skater will demonstrate low friction, while a brake being applied to a wheel will demonstrate high friction. This is because the coefficient of friction between ice and metal skate is low, while the coefficient of friction between brake friction material and metal wheel is high.

Frictional force (the force required to overcome friction) can be calculated from the weight and the coefficient of friction.

Key Points

$$F = \mu N \ (\text{N})$$

where F is the friction force, μ is the coefficient of friction and N is the normal reaction (which is equal and opposite to the weight).

Gravitational force

The weight of an object is another way of expressing gravitational force. As previously discussed, it is the product of mass and acceleration due to gravity.

Momentum

Momentum is the product of mass and velocity. If an object displays a large inertia, it will inevitably have a large amount of momentum when moving.

Key Term

Momentum $= mv$ (kg m/s)

Mechanical work

Mechanical work is related to energy and is a measure of the amount of energy being transferred. It is the product of force and displacement and is measured in joules.

Key Term

Mechanical work $W = Fs$ (J)

Power dissipation

Power dissipated is a measure of the amount of energy being used or transferred by an object. Instantaneous power is the product of force and velocity, while average power is the amount of work transferred in unit time. Power is measured in watts.

Key Terms

Instantaneous power $P = Fv$ (W)

Average power $P = \dfrac{W}{t}$ (W)

Gravitational potential energy

Potential energy (PE) refers to the amount of energy available due to an object being raised into the air. It is stored energy and is the product of weight and height.

Kinetic energy

Kinetic energy (KE) is the energy due to movement and is related to the mass of an object and its velocity.

Key Terms

$PE = mgh$ (J)

$KE = 0.5\,mv^2$ (J)

Worked Example

A 50 kg boulder is pushed off a cliff into the sea. The cliff is 75 m above sea level. Determine the potential energy possessed by the boulder at the top and the velocity with which it will hit the sea, assuming no losses.

$$PE = mgh = 50 \times 9.81 \times 75$$

$$PE = 36.79 \times 10^3\,\text{J}$$

$$PE = 36.79\,\text{kJ}$$

At the moment the boulder hits the water, the PE will be zero and all of the energy will be kinetic. Assuming no losses:

$$KE = 36.79 \times 10^3\,\text{J}$$

$$0.5\,mv^2 = 36.79 \times 10^3$$

$$v^2 = \frac{36.79 \times 10^3}{(50 \times 0.5)}$$

$$v^2 = 1471.5\,\text{m/s}$$

$$v = 38.4\,\text{m/s}$$

Worked Example

A car of mass 1850 kg has velocity of 40 km/h; a braking force of 3 kN is applied. Determine the distance travelled in stopping.

$$u = \frac{40 \times 1000}{3600} = 11.11\,\text{m/s}$$

$$KE = 0.5\,mv^2 = 0.5 \times 1850 \times 11.11^2$$

$$KE = 114.17 \times 10^3\,\text{J}$$

$$KE = 114.17\,\text{kJ}$$

The energy dissipated in stopping is equivalent to the mechanical work.

Mechanical work $= Fs$

$$KE = Fs$$

$$114.17 \times 10^3 = 3 \times 10^3 \times s$$

$$s = 38.06\,\text{m}$$

235

Dynamics principles

By the end of this section you should have developed a knowledge and understanding of how to apply Newton's laws of motion and d'Alembert's principle to the solution of engineering problems. In addition, you will demonstrate an understanding of the conservation of energy principle and the conservation of momentum principle.

Some of the key concepts of engineering have been developed from Newton's laws of motion. It is these laws and their application that we will be considering in this section.

> **Key Points**
>
> You will need to use the parameters from the previous section and understand which engineering units are appropriate in order to solve equations involving dynamic engineering systems.

Newton's laws of motion

Newton's first law of motion

An object remains in a state of rest, or continues in a state of uniform motion in a straight line, unless it is acted upon by an externally applied force.

It is this principle which allows us to consider the inertia of an object.

Newton's second law of motion

The acceleration of an object acted upon by an external force is proportional to the force and is in the same direction as the force.

It is this principle that gives us the formula:
$F = ma$ (N)

Newton's third law of motion

For every force, there is an equal and opposite reacting force.

It is this principle that gives us the conditions for static equilibrium.

D'Alembert's principle

D'Alembert's principle states that if the internal reaction to the acceleration or deceleration (retardation) of a body is imagined to be an external force, then the body can be treated as though it were in static equilibrium, under the action of a system of external forces. A free body diagram can then be drawn to aid the solution of a dynamic problem. Essentially, the moving system is frozen in time and the forces acting are applications of Newton's laws.

Worked Example

Consider the block shown in Figure 5.54. The forces are shown and you will notice that we have the equivalent of a free body diagram. The gravity force or weight (W) is shown acting downwards and this is opposed by the normal reaction (N) acting upwards. The force being applied in order to move the block (F_A) is being opposed by the frictional force F.

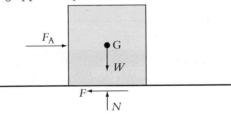

Figure 5.54

If the mass is 200 kg and the coefficient of friction between the block and the surface is 0.3, we can determine the force to overcome friction.

$$W = mg = 200 \times 9.81$$

$$W = 1.962 \times 10^3\,\text{N}$$

$W = 1.962\,\text{kN}$ – this will be the same as the normal reaction (N)

$$F = \mu N = 0.3 \times 1.962 \times 10^3$$

$$F = 588.6\,\text{N}$$

This is relatively straightforward; now let us consider an example of d'Alembert's principle with a moving component.

Worked Example

The 300 kg block in Figure 5.55 is being raised up a slope using a cable and pulley system, as shown. The slope is 40 m long and the block is accelerating up the slope at 2 m/s². It takes two minutes to winch the block, and the coefficient of friction between the slope and the block is 0.25. Determine:

- the tractive effort
- the work done in winching the block up the slope
- the power required by the winch.

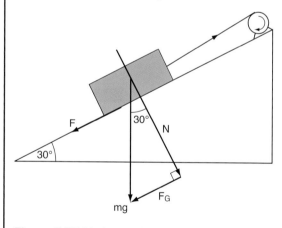

Figure 5.55 Block on a slope

Weight of 300 kg mass = 300 × 9.81

$$W = 2.94 \times 10^3\,\text{N}$$

$$W = 2.94\,\text{kN}$$

Because the normal reaction acts perpendicular to the slope, we refer to the force diagram in Figure 5.55.

$$N = 2.94 \times 10^3 \cos 30$$

$$N = 2.55 \times 10^3\,\text{N}$$

$$N = 2.55\,\text{kN}$$

Frictional force $F_\text{F} = \mu N = 0.25 \times 2.55 \times 10^3 = 637.18\,\text{N}$

The component of the gravity force acting parallel to the slope = F_G

$$F_\text{G} = W \sin 30 = 2.94 \times 10^3 \sin 30$$

$$F_\text{G} = 1.47 \times 10^3\,\text{N}$$

$$F_\text{G} = 1.47\,\text{kN}$$

Total resisting force = $(1.47 \times 10^3) + 637.18 = 2.11 \times 10^3\,\text{N}$

Resisting force = 2.11 kN

From Newton's second law, $F = ma = 300 \times 2 = 600\,\text{N}$

Therefore, tractive effort = $600 + 2.11 \times 10^3 = 2.71 \times 10^3\,\text{N}$

Tractive effort = 2.71 kN

Work done = $Fs = 2.71 \times 10^3 \times 40 = 108.4 \times 10^3\,\text{J}$

Work done = 108.4 kJ

$$P = \frac{w}{t} = \frac{108.4 \times 10^3}{120}$$

$$P = 903.3\,\text{W}$$

237

Principle of conservation of momentum

Conservation of momentum is a principle developed from Newton's laws. The principle states that the total momentum of a system, in a given direction, remains constant, providing that the masses remain unaltered and no external forces act on the system.

Principle of conservation of energy

The principle of the conservation of energy states that energy cannot be created or destroyed; it simply changes its form. So, for example, when a vehicle brakes, the energy is converted into heat, noise, sound, and so on.

Key Points

When vehicles collide, the total momentum before impact is equivalent to the total momentum after impact.

Worked Example

A spacecraft with a mass of 6000 kg and travelling at 50 m/s connects with another spacecraft of 5000 kg travelling at 40 m/s. If both spacecraft are travelling in the same direction, determine the velocity of the combined craft.

Momentum $= mv$, so using the principles of the conservation of momentum:

$$m_1 v_1 + m_2 v_2 = m_f v_f \text{ where } m_f \text{ and } v_f \text{ are the final masses and velocity after impact.}$$

$$(6 \times 10^3 \times 50) + (5 \times 10^3 \times 40) = [(6 \times 10^3) + (5 \times 10^3)] \, v_f$$

$$v_f = \frac{500 \times 10^3}{11 \times 10^3}$$

$$v_f = 45.45 \, \text{m/s}$$

Worked Example

If we consider the spacecraft in the previous example, we can determine the loss of energy due to the collision.

Loss of KE at impact = total KE before impact − total KE after impact

KE after impact $= 0.5 \times (6 \times 10^3 + 5 \times 10^3) \, 45.45^2 = 11.36 \times 10^6 \, \text{J}$

KE before impact $= [0.5 \times (6 \times 10^3) \times 50^2] + [0.5 \times (5 \times 10^3) \times 40^2] = 11.5 \times 10^6 \, \text{J}$

Loss of KE at impact $= (11.5 \times 10^6) − (11.36 \times 10^6)$

Loss of KE at impact $= 140 \, \text{J}$

D'Alembert's principle and the principle of conservation of energy can be compared when solving problems that involve momentum and loss of energy.

Make the Grade M1

Grading criterion M2 requires you to determine the retarding force on a freely falling body when it impacts upon a stationary object and is brought to rest without rebound, in a given distance. To do this you will need to consider the way potential energy is converted into kinetic energy and the theory of momentum.

Make the Grade D1

Grading criterion D1 requires you to compare and contrast the use of d'Alembert's principle with the principle of conservation of energy to solve an engineering problem. Using given data, you will need to solve the same problem twice, using the different methods. You can then consider which method you found most straightforward and draw conclusions about the techniques used.

Worked Example

A sledgehammer is used to drive a stake into the ground. The sledgehammer has a mass of 10 kg and the stake has a mass of 15 kg. The sledgehammer falls from a distance of 0.8 m and drives the stake 250 mm into the ground. Determine the resisting force in the ground, assuming that the sledgehammer remains in contact with the stake and does not bounce.

D'Alembert's principle
Assuming the initial velocity $= 0$ m/s and the accelerating force on the sledgehammer is equivalent to gravity:

$$v^2 = u^2 + 2as$$

$$v = \sqrt{(0^2 + 2 \times 9.81 \times 0.8)}$$

$$v = 3.96 \text{ m/s}$$

This is the velocity as the sledgehammer connects with the stake.

Using the conservation of momentum principle where:

$v_1 =$ velocity of hammer $= 3.96$ m/s
$v_2 =$ velocity of stake $= 0$ m/s

$$m_1v_1 + m_2v_2 = m_fv_f$$

$$(10 \times 3.96) + (15 \times 0) = (10 + 15)\, v_f$$

$$v_f = \frac{39.6}{25} = 1.58 \text{ m/s}$$

The velocity of the sledgehammer and stake combination $= 1.58$ m/s as it starts to penetrate the ground.

The final velocity $= 0$ m/s, so we can determine the rate of deceleration.

$$v^2 = u^2 + 2as$$

$$a = \frac{(0^2 - 1.58^2)}{(2 \times 0.25)}$$

$$a = -5.0 \text{ m/s}^2$$

(Remember, the $-$ sign indicates deceleration.)

Applying d'Alembert's principle

resisting force $= mg + ma = (10 + 15)9.81 + (10 + 15)5$

resisting force $= 370$ N

Conservation of energy principle

Loss of KE $=$ gain of KE

For the sledgehammer:

$$mgh = 0.5\, mv^2$$

$$v^2 = \frac{(10 \times 9.81 \times 0.8)}{(0.5 \times 10)}$$

$$v^2 = 15.70$$

$$v = 3.96 \text{ m/s}$$

Using the conservation of momentum principle where:

$v_1 =$ velocity of hammer $= 3.96$ m/s
$v_2 =$ velocity of stake $= 0$ m/s

239

$$m_1v_1 + m_2v_2 = m_fv_f$$

$$(10 \times 3.96)+(15 \times 0) = (10 + 15) \, v_f$$

$$v_f = \frac{39.6}{25} = 1.58 \, \text{m/s.}$$

For the combination:

work done = energy used (potential + kinetic)

$$Fs = 0.5mv^2 + mgh$$

$$Fs = [0.5 \times (10 + 15) \times 1.58^2] + [(10 + 15) \times 9.81 \times 0.25]$$

$$F = \frac{257.33}{0.25}$$

$$F = 370 \, \text{N}$$

The answers should be the same, allowing for rounding errors in calculations.

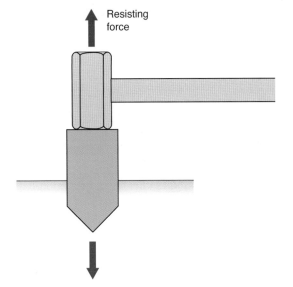

Figure 5.56 Hammer and stake

Resisting force

Activity 5.8

1. A railway wagon of mass 700 kg travelling at 30 m/s catches up with and connects to a railway wagon of mass 900 kg travelling at 20 m/s. Determine the velocity of the combination and the loss of KE due to coupling.

2. A winch is used to pull a small lifeboat up a slipway. The lifeboat has a mass of 15 tonnes and is accelerated to the top of the slope, from a stationary start, to a velocity of 1.2 m/s in 45 s. The slope is 22° to the horizontal and the coefficient of friction between the slipway and the lifeboat can be taken as 0.3. Determine:
   - the tractive effort of the winch
   - the work done in climbing the slipway
   - the average power required of the winch.

3. A 160 kg drop hammer used in a forging operation falls from a height of 1.8 m to strike a forging of mass 75 kg. The impact causes a 65 mm deformation of the forging and the drop hammer remains in contact with the forging without rebounding. Determine the resistance of the forging.

5.3 Determining the parameters of fluid systems

This section will cover the following grading criteria:

The behaviour of liquids at rest and in motion is of key importance to engineers when designing fluid systems and components. Fluids are defined as substances that will flow and, consequently, have no physical form. They take up the shape of the container they are within. Although fluids can be considered as gases or liquids, it is the analysis of liquids that interests us in this section. The forces applied by fluids on tanks, lock gates, walls is considerable and the up-thrust on boat hulls and other immersed objects is also of interest to engineers and designers, as are the characteristics of steady flow.

Thrust on a submerged surface

By the end of this section you should have developed a knowledge and understanding of how to calculate thrust and overturning moments on a vertical rectangular retaining surface with one edge in the free surface of a liquid.

The pressure and force exerted by fluids on vertical surfaces is considerable. The weight of water on a sea wall, lock gate or dam wall is a potentially destructive force. and the design of these structures requires an understanding of pressure and density of fluids.

Key Points

Density ρ is the relationship between the mass of a substance and its volume.

$$\rho = \frac{m}{v} \ (kg/m^3)$$

Hydrostatic pressure

Hydrostatic pressure is a measure of the weight of a liquid for unit area. If you consider a tank of liquid, as shown in Figure 5.57, the pressure of water acting on the base of the tank is a result of the density (ρ), the gravitational effect (g) and the amount of water in the tank, determined by its height (h).

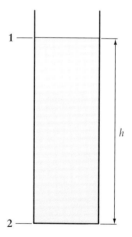

Figure 5.57 Tank of water

Key Term

Hydrostatic pressure $p = \rho g h \ (N/m^2)$

241

Key Points

Pressure is very similar to stress in that it is a measure of the internal forces in a liquid or gas. We can use the same formula as $P = F/A$ (N/m²).

To help avoid confusion, we refer to pressure in pascals (Pa). However, Pa and N/m² are exactly the same unit.

Worked Example

A tank of oil is 3 m deep. If the tank is half full and the density of oil is 875 kg/m³, determine the pressure on the base of the tank.

Height of oil = 1.5 m

$$p = \rho g h = 875 \times 9.81 \times 1.5$$

$$p = 12.88 \times 10^3 \, \text{Pa}$$

$$p = 12.88 \, \text{kPa}$$

Activity 5.9

1. A water tank is full of water and exerts a pressure of 15 kPa on the base. If the water has a density of 1000 kg/m³, determine the height of the tank.

2. A column of oil 3 m high exerts a pressure of 13 kPa on the base of a container. Determine the density of the oil.

3. A container is 6 m high; the container is one-third full of seawater. Determine the pressure on the base if the density of seawater is taken as 1013 kg/m³.

Hydrostatic thrust on an immersed plane surface

A typical example of the thrust on an immersed plane surface would be the force acting on a hatch or gate in a river or reservoir, as shown in Figure 5.58. The distance from the free surface to the centroid of the area is $\bar{y}$ and the distance to the centre of pressure from the base is $\frac{h}{3}$. The wetted area (A) is $B \times h$ (where B is the breadth).

The thrust, which is a force, is given by:

$$F = \rho g A \bar{y}$$

Key Points

For a surface of unit width (where $B = 1$), the area will be equivalent to the height, so:

$$F = \rho g h \bar{y}$$

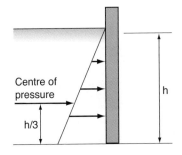
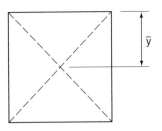

Figure 5.58 Pressure and force acting on a submerged surface

Centre of pressure of a rectangular retaining surface with one edge in the free surface of a liquid

In Figure 5.58, it is evident that the centre of pressure will not be at the centroid of the submerged surface. In fact, for rectangular shapes, it will be one-third of the distance from the base to the surface of the fluid.

Therefore, $\bar{y} = \frac{h}{2}$, and the centre of pressure, from the base, is $\frac{h}{3}$.

Therefore, the thrust for unit width will be:

$$F = \rho g h \times \frac{h}{2} = \frac{\rho g h^2}{2}$$

Moment about the base = force × distance

$$M = \frac{\rho g h^2}{2} \times \frac{h}{3}$$

$$M = \frac{\rho g h^3}{6}$$

Immersed bodies

By the end of this section you should have developed a knowledge and understanding of Archimedes' principle and the use of a gravity bottle to determine density.

Archimedes of Syracuse is widely considered to be one of the most influential mathematicians and scientists of Greek legend. He is remembered for many great scientific inventions; however, the principle of Archimedes is probably the most well known of his discoveries.

> **Key Points**
>
> Although we consider fluids to be liquids, it is useful to remember that gas is also a fluid.

Worked Example

The vertical retaining wall of a dock contains seawater to a depth of 15 m. Calculate the thrust, for unit width, and the turning moment about the base. Assume that the density of seawater = 1030 kg/m³.

For unit width, the area will be equal to the height:

$F = \rho g h \bar{y}$

$F = 1030 \times 9.81 \times 15 \times 7.5 = 1.137 \times 10^6 \, \text{N}$

$F = 1.137 \, \text{MN}$

$M = \frac{\rho g h^3}{6}$

$M = \frac{1030 \times 9.81 \times 15^3}{6} = 5.68 \times 10^6 \, \text{Nm}$

$M = 5.68 \, \text{MNm}$

Activity 5.10

1. A dam wall is 12 m high and 3 m wide. Determine the hydrostatic thrust on the dam wall if the water level is 2 m below the top of the wall. Assume that the density of water is 1000 kg/m³.

2. A dam wall is 15 m high and the surface of the water is 3 m below the top of the wall. Determine the turning moment about the base of the wall. Assume that the density of water is 1000 kg/m³.

Make the Grade — P6

Grading criterion P6 requires you to determine the up-thrust on an immersed body. To do this, you will need to use Archimedes' principle and given data.

Archimedes' principle

Archimedes' principle can be considered to be a statics problem. If you refer to Figure 5.59, for the mass to be floating, the weight (acting downwards) must be equal to the force (reaction) acting upwards (the up-thrust).

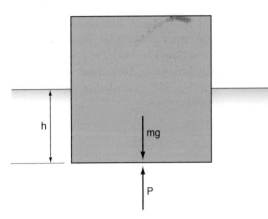

Figure 5.59 Archimedes' principle

Immersion of a body

$p = \rho gh$ and $p = \dfrac{F}{A}$, so it follows that $F = pA = \rho ghA$. Archimedes' principle states that a floating body displaces its own weight of liquid.

Partly immersed

Fully immersed

Fluid

As previously stated, fluids can be liquids or gases. When considering liquids, one of the most important features is the density; there are applications of Archimedes' principle that allow us to determine the density of a given liquid.

Key Term

Relative density – The density of a liquid, gas or solid substance that is *relative* to water.

If brass has a density of $8400\,\text{kg/m}^3$, it will have a relative density of 8.4.

Determination of density using flotation and specific gravity bottle

By using the principle of relative density, it is possible to determine the density of unknown liquids and components.

Immersion method

This technique requires careful use of measuring instruments and, ideally, should be undertaken under laboratory conditions.

The following conventions are followed:

mass of sample in air $= m_1$ (kg)
mass of sample immersed in water $= m_2$ (kg)
mass of sample immersed in liquid of unknown density $= m_3$ (kg)

The method typically requires a sample to be suspended on a spring balance in air. This allows the mass to be recorded; the mass is then submerged in water and the mass is recorded; finally, the sample is suspended in liquid of unknown density and the mass is recorded once more. This process is illustrated in Figure 5.60.

Relative density of unknown sample $= \dfrac{m_1}{(m_1 - m_2)}$

Relative density of unknown liquid $= \dfrac{(m_1 - m_2)}{(m_1 - m_3)}$

Specific gravity bottle method

This technique requires the use of a gravity bottle, which is a precisely manufactured flask, made

Figure 5.61 Specific gravity bottle

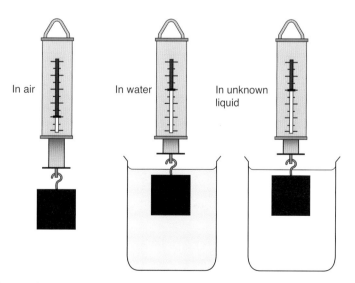

Figure 5.60 Immersion method

to hold a known volume of liquid at a specified temperature, as shown in Figure 5.61 (see page 245). The method also requires the use of precision measuring equipment and should be carried out under controlled laboratory conditions.

The following steps should be taken:

- the mass of the sample is recorded (m_a)
- the mass of the gravity bottle and stopper is recorded (m_b)
- the gravity bottle is filled with water and the mass is recorded (m_c)
- the sample is immersed in the bottle and the mass of the combination is recorded (m_d)
- the mass is removed and the bottle emptied, carefully dried and refilled with the unknown liquid; this mass is recorded (m_e).

From the recorded data, the following can be deduced:

- relative density of unknown liquid:
$$\left(\frac{m_e}{m_c}\right)$$

- relative density of unknown sample:
$$\frac{m_a}{(m_a + m_c - m_d)}$$

Make the Grade D2

Grading criterion D2 requires you to evaluate the methods that might be used to determine the density of a solid material and the density of a liquid. This criterion is best accessed by performing experiments to determine the unknown densities. By performing two experiments, you should be able to compare values and discuss the limitations, sources of error and advantages/disadvantages of each method.

Worked Example

A metal sample is to be checked for density, as is a sample of a given oil.
Two methods are used, with the following results.

Immersion method:

mass of sample in air (m_1)	= 110 g
mass of sample immersed in water (m_2)	= 95 g
mass of sample immersed in oil (m_3)	= 90 g
relative density of unknown sample	$= \dfrac{m_1}{(m_1 - m_2)} = \dfrac{110}{(110 - 95)}$
relative density of unknown sample	= 7.33
actual density of sample	= 7330 kg/m³
relative density of oil	$= \dfrac{(m_1 - m_2)}{(m_1 - m_3)} = \dfrac{(110 - 95)}{(110 - 90)}$
relative density of oil	= 0.75
actual density of oil	= 750 kg/m³

Specific gravity bottle method:

the mass of the sample (m_a)	= 110 g
the mass of the gravity bottle and stopper (m_b)	= 200 g
the mass of the gravity bottle filled with water (m_c)	= 280 g
the mass of the gravity bottle filled with water and the sample (m_d)	= 375 g
the mass of the gravity bottle filled with oil (m_e)	= 210 g

$$\text{relative density of unknown sample} \left(\frac{m_a}{(m_e + m_c - m_d)}\right) = \frac{110}{(110 + 280 - 375)}$$

relative density of sample $= 7.33$

actual density of sample $= 7{,}330\,\text{kg/m}^3$

relative density of oil $= (m_e/m_c)$ $= 210/280$

relative density of oil $= 0.75$

actual density of oil $= 750\,\text{kg/m}^3$

Flow characteristics of a gradually tapering pipe

By the end of this section you should have developed a knowledge and understanding of how the mass and volume flow rates vary as fluid flows through a gradually tapering pipe.

Fluid systems can be static or dynamic. While dynamic fluid systems are generally systems that involve the pressure and forces that involve a fairly consistent body of fluid, such as oil in a tank or the water in a reservoir/lock, we do need to consider the forces, pressure and characteristics of fluid flow – for example, pumping fluids through pipelines, hydraulic systems or domestic water supply.

Key Points

Rate of flow refers to how much of a fluid is delivered in kg/s or m³/s. It is not, however, the velocity of flow in m/s.

1 tonne = 1000 kg

1 m³ = 1000 litres

Make the Grade   P7

Grading criterion P7 requires you to use the continuity of volume and mass flow for an incompressible fluid to determine the design characteristics of a gradually tapering pipe.
This will require you to use given data to determine the cross-sectional area (A) and velocity of flow (v).

Rates of flow

Flow rate is normally defined as how much fluid is delivered per second. If you were filling a tank which had a volume of $100\,\text{m}^3$, you would want to know how long it would take. The volumetric flow rate (m^3/s) would help you to determine this time.

Similarly, you might know that a counterbalance system needs 3 tonnes ($3000\,\text{kg}$) to cause it to activate. If you were pumping water, you would want to know the mass flow rate in (kg/s).

Key Terms

V = volume (m³)	$\dot{Q}$ = volumetric flow rate (m³/s)
m = mass (kg)	$\dot{m}$ = mass flow rate (kg/s)
ρ = density (kg/m³)	$\rho = \dfrac{\dot{m}}{\dot{Q}}$ (kg/m³)

Worked Example

A tank is to be filled with water. If the tank has a volume of 3 m³ and takes two hours to fill, determine the volumetric flow rate and mass flow rate of the supply.

Time = $2 \times 60 \times 60 = 7200$ seconds

$\dot{Q} = \dfrac{3}{7200} = 416.67 \times 10^{-6}$ (m³/s)

$\dot{m} = \rho\,\dot{Q} = 1{,}000 \times 416.67 \times 10^{-6}$
$= 416.67 \times 10^{-3}\,$kg/s

Key Term

Incompressible flow – Although both liquid and gases are considered fluids, we know that gases compress and expand when under pressure. Although this is also true of liquids, because the compressibility is very small, we can safely ignore it and consider liquids to be incompressible.

Flow velocities

Consider the sectional view of a tapering pipe shown in Figure 5.62.

To maintain a steady rate of flow, the flow velocity will increase as fluid moves along the pipe. This is because the cross-sectional area decreases.

The cross-sectional area (A) is measured in m². Flow velocity (v) is measured in m/s.

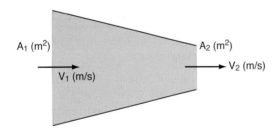

Figure 5.62 Tapering pipe

Continuity of flow

To determine flow rate, you need the cross-sectional area and velocity at any point in the pipe.

$\dot{Q} = Av$ (m³/s)

From this it follows that, if the flow rate $\dot{Q}$ is to remain constant:

$\dot{Q} = A_1 v_1 = A_2 v_2$

Worked Example

A horizontal pipe tapers from a diameter of 100 mm to a diameter of 80 mm, where the flow velocity is 12 m/s. Determine:

• the initial velocity of the flow

• the volumetric flow rate

• the mass flow rate.

$$A_1 = \frac{\pi 0.1^2}{4} = 7.85 \times 10^{-3}\,\text{m}^2$$

$$A_2 = \frac{\pi (8 \times 10^{-3})^2}{4} = 5.03 \times 10^{-3}\,\text{m}^2$$

$$A_1 v_1 = A_2 v_2$$

$$v_1 = 5.03 \times 10^{-3} \times \frac{12}{(7.85 \times 10^{-3})}$$

$$v_1 = 7.68\,\text{m/s}$$

$\dot{Q} = A_2 v_2$

$\dot{Q} = 5.03 \times 10^{-3} \times 12 = 91.80 \times 10^{-3}\,\text{m}^3/\text{s}$

Assuming the density of water is 1000 kg/m³

$\dot{m} = \rho\dot{Q} = 1000 \times 91.80 \times 10^{-3}$

$\dot{M} = 91.8\,\text{kg/s}$

Activity 5.11

1. Oil of density 815 kg/m³ is pumped along a horizontal pipe that is reducing in diameter. The pump flow rate is 30 l/min. If the initial diameter is 50 mm and the flow velocity leaving the pipe is 5 m/s, determine:
   - the initial velocity
   - the final diameter
   - the mass flow rate.

2. Water of density 1000 kg/m³ is pumped up a vertical pipe that is reducing in diameter. The initial diameter of the pipe is 70 mm and the final diameter is 40 mm. For an initial velocity of 14 m/s, determine:
   - the final velocity
   - the volumetric flow rate
   - the mass flow rate.

5.4 Determining the effects of energy transfer in thermodynamic systems

This section will cover the following grading criteria:

Thermodynamic systems are those systems that involve heat and energy transfer. Thermodynamics can involve the study of gases and liquids, but can also consider how solids become liquids or vapours, due to heat or energy transfer occurring.

Having previously studied fluids and, particularly, liquids, the gas laws allow us to consider what happens to gases when temperature, pressure and/or volume changes.

Heat transfer

By the end of this section you should have developed a knowledge and understanding of heat transfer parameters, the different phases that substances can adopt and the principles of heat transfer. You will also be able to determine what happens to materials as they are heated and cooled, and the impact on engineering applications of this principle.

When heat is applied to solid substances, they can either expand or melt. Similarly, if you keep applying heat, you can cause a liquid to vaporise. At what temperature materials change their form depends upon their properties. A good example is if you consider a steel tray full of ice. When you allow heat to be added, the ice melts, turning into water. If sufficient heat is added – in a domestic oven, for example – the water will turn into steam. However, the steel tray will still be a solid. It will, however, have expanded in size very slightly, due to the addition of heat energy.

Key Points

It is important to remember that heat transfer refers to heating and cooling and is a transfer of energy measured in joules (J).

Make the Grade — P8

Grading criterion P8 requires you to calculate dimensional change when a solid material undergoes a change in temperature, and the heat transfer that accompanies a change of temperature and phase. To complete this activity, you will need to demonstrate the use of appropriate formulae which allow you to determine the change in size of a component, or the stress induced if the component is prevented from increasing in size. You will also have to use formulae to determine the amount of energy required to raise the temperature of a substance (sensible heat) and the heat required to effect a change of phase (latent heat).

Heat transfer parameters

There are a variety of different parameters that you need to consider when studying heat transfer concepts.

Temperature

Temperature is the degree of 'hotness' or 'coldness' of a substance. It relates to molecular activity, so when no molecular activity is present, the theoretical temperature of absolute zero has been reached. The most common temperature scale used is the Celsius scale (degrees Celsius). The Celsius scale is based on the freezing and boiling points of water. However, scientific principles often require us to use the Kelvin temperature scale (K), which is based on absolute zero.

Key Points

A temperature change of, say, 10°C is equivalent to a temperature change of 10 K.

However, 0 K = −273°C, so to convert °C to K, add 273.

Pressure

An increase in pressure causes an increase in temperature.

Mass

Mass is a key concept in that the greater the mass of a substance, the more heat energy is required to raise its temperature.

Linear dimensions

Dimensions are key measures of an increase in thermal energy; the increase in size of a component is directly related to the heat energy applied.

Time

The amount of time an object is heated for will have a direct impact on the heat energy absorbed.

Specific heat capacity

Specific heat capacity is a material property and a measure of how much energy is required to raise the temperature of 1 kg of the material by 1 °C. Each material has a unique value. The units are J/kgK.

Specific latent heat of fusion

Specific latent heat of fusion is a material property that determines the amount of heat required to change the state of a material from liquid to solid or vice versa. It is a measure of how much energy is required to convert 1 kg of the substance.

Specific latent heat of vaporisation

Specific latent heat of vaporisation is a material property that determines the amount of heat required to change the state of a material from liquid to vapour or vice versa. It is a measure of how much energy is required to convert 1 kg of the substance.

Linear expansivity

Linear expansivity is a material property that determines the amount of heat required to increase the length of a given sample in terms of the original length. The units are /K.

Phase

The phase of a substance refers to whether it is liquid, solid or gas. It is important to understand that energy is required to change the phase of a substance as well as raising its temperature.

Heat transfer principles

When a kettle is used to heat water, it takes a few minutes to raise the temperature to boiling point. However, if the kettle is allowed to continue heating the water, it takes considerably longer to turn all of the water into steam; while the water is being turned to steam, the temperature remains at 100 °C. The same principle applies when turning ice into water. This is shown in Figure 5.63.

For sensible heat $Q = mc\Delta T$ (J)

For latent heat $Q = mL$ (J)

where:

Q = heat energy (J)
M = mass (kg)
c = specific heat capacity (J/kgK)
ΔT = change in temperature (K)
L = specific latent heat (J/kg)

> **Key Points**
>
> As the water begins to turn to steam, this is known as *wet steam*; when it has turned completely to steam, with no remaining water vapour, it is known as *dry steam*. If it is partly water vapour and partly steam, the *dryness fraction* determines the relative amounts.

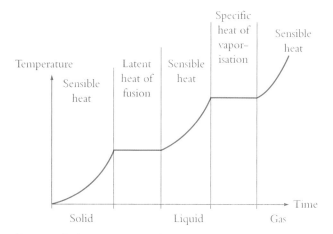

Figure 5.63 Temperature/time graph showing sensible/latent heat

Worked Example

0.22 kg of ice at −5°C is heated to turn it into water, then continually heated until it has turned to dry steam. Determine the total heat supplied. Assume that:

specific heat capacity of water = 4.12 KJ/kgK
specific latent heat of fusion (L_f) = 335 KJ/kg
specific latent heat of vaporisation (L_v) = 2.25 MJ/kg

to turn ice into water $Q = mL_f = 0.2 \times 335 \times 10^3$
$$= 73.7 \times 10^3\,J$$
to raise temperature of water to 100 °C $\qquad Q = mC\Delta T = 0.22 \times 4.19 \times 10^3 \times 100$
$$= 92.18 \times 10^3\,J$$
to turn water into steam $Q = mL_v = 0.22 \times 2.25 \times 10^6$
$$= 495 \times 10^3\,J$$
Total heat supplied $= (73.7 \times 10^3) + (92.18 \times 10^3) + (495 \times 10^3)$
$$= 660.9 \times 10^3\,J$$
Total heat supplied $= 660.9\,kJ$

The heat required to raise the temperature of the water is called *sensible* heat.

The heat required to change the phase of a substance is called *latent* heat.

The efficiency of a device at heating, cooling or changing the phase of a substance is called the thermal efficiency, measured as a percentage.

Worked Example

A heat exchanger is rated at 10 MW and is supplied with water at 20°C. The exchanger heats the water to 100°C at which point it is turned into dry steam. The heat exchanger supplies dry steam at 2 kg/s. If the specific heat capacity of water is 4.19 KJ/kgK and the specific latent heat of vaporisation is 2.25 MJ/kg, determine:

- the heat supplied/kg
- the thermal efficiency.

Sensible heat:

$Q = mc\Delta T$ for 1 kg of steam, remember temperature *change* is the same in Kelvin or °C

$$Q = 1 \times 4.19 \times 10^3 \times 80$$

$$Q = 335.2 \times 10^3\,J$$

Latent heat

$$Q = mL = 1 \times 2.25 \times 10^6$$

$$Q = 2.25 \times 10^6\,J$$

Total heat supplied/kg $= (335.2 \times 10^3) + (2.25 \times 10^6)$

Total heat supplied $= 2.59 \times 10^6\,J$

Rate of heat energy supplied = output power
$= \dot{m}Q = 2 \times 2.59 \times 10^6$
$$= 5.18\,MJ/s$$

Total input power $= 5.18 \times 10^6\,MW$ (as 1 W = 1 J/s)

Thermal efficiency $= \dfrac{\text{output power}}{\text{input power}} = \dfrac{5.18 \times 10^6}{10 \times 10^6}$ $\times 100$

Efficiency = 51.8 per cent

Activity 5.12

1. 2.5 kg of ice at $-3\,°C$ is heated to turn it into water, then continually heated until it has all turned to dry steam. Determine the total heat supplied. Assume that:

specific heat capacity of water	= 4.19 KJ/kgK
specific latent heat of fusion (L_f)	= 335 KJ/kg
specific latent heat of vaporisation (L_v)	= 2.25 MJ/kg

2. A heat exchanger is rated as being 75 per cent efficient and is supplied with water at $40\,°C$. The exchanger heats the water to $100\,°C$, at which point it is turned into dry steam. The heat exchanger supplies dry steam at 1.7 kg/s. If the specific heat capacity of water is 4.19 KJ/kgK and the specific latent heat of vaporisation is 2.25 MJ/K, determine:

   - the heat supplied/kg
   - power input in watts.

Make the Grade — M3

Grading criterion M3 requires you to determine the thermal efficiency of a heat transfer process from given values of flow rate, temperature change and input power. You will be given specific information and will need to determine the heat input in terms of sensible heat and latent heat. The rate of heat input to perform the process can be compared with the power supplied to determine the efficiency.

Make the Grade — M4

Grading criterion M4 requires you to determine the force induced in a rigidly held component that undergoes a change in temperature. To achieve this, you will need to use the appropriate material property to determine the theoretical linear expansion. This value can be used to determine what the strain would have been if the component was not constrained. Using previous formulae, which relate stress to strain, allows the force induced to be determined.

Linear expansion

The linear expansion of a component can be determined, provided that the temperature change, original length and linear expansivity is known:

$$\Delta L = \alpha L \Delta T \ (m)$$

where:

ΔL = change in length (m)
α = linear expansivity (/K)
ΔT = change in temperature (K)
L = original length (m)

Worked Example

A steel rod 3 m long, of diameter 10 mm is heated from 20 °C to 500 °C. Determine the change in length and strain induced, assuming the linear expansivity $\alpha = \dfrac{13 \times 10^{-6}}{K}$.

$$\Delta L = \alpha L \Delta T = 13 \times 10^{-6} \times 3 \times (500 - 20)$$

$$\Delta L = 18.72 \times 10^{-3}\,m$$

Therefore, the change in length is 18.72 mm.

From previous studies, we know that $\epsilon = \dfrac{\Delta L}{L}$
$= \dfrac{18.72 \times 10^{-3}}{3}$

$$\epsilon = 6.24 \times 10^{-3}$$

253

Worked Example

In the previous example, the steel rod was allowed to extend freely. Determine the stress induced if the rod was constrained and the equivalent force required. Take E for steel to be $207\,\text{GN/m}^2$.

$$\epsilon = 6.24 \times 10^{-3}$$

$$E = 207 \times 10^9\,\text{N/m}^2$$

From previous studies, we know that $E = \dfrac{\sigma}{\epsilon}$

$$\sigma = E\epsilon = 6.24 \times 10^{-3} \times 207 \times 10^9$$

$$\sigma = 1.29 \times 10^9\,\text{N/m}^2$$

$$A = \frac{\pi \times (10 \times 10^{-3})^2}{4}$$

$$A = 78.54 \times 10^{-6}\,\text{m}^2$$

$$\sigma = \frac{F}{A}$$

$$F = 78.54 \times 10^{-6} \times 1.29 \times 10^9$$

$$F = 101.32 \times 10^3\,\text{N}$$

$$F = 101.32\,\text{kN}$$

Activity 5.13

An aluminium rod 0.6 m long is heated from 20°C to 240°C. If the rod is constrained, determine the stress induced. Assume E for aluminium $= 70 \times 10^9\,\text{N/m}^2$ and the linear expansivity for aluminium $\alpha = \dfrac{22 \times 10^{-6}}{\text{K}}$.

Thermodynamic process equations

By the end of this section you should have developed a knowledge and understanding of the parameters used in thermodynamic equations and formulae. In addition, you will develop an understanding of how these values are related and how they vary when gases are involved in thermodynamic processes.

Gases are compressible, and the relationship between the increase in pressure that occurs in compression and the changes in volume and temperature are key to our understanding of how thermodynamic systems are designed and how they operate.

Make the Grade

Grading criterion P9 requires you to solve two or more problems that require application of thermodynamic process equations for a perfect gas to determine unknown parameters of the problems. You will need to use given values, being careful to ensure that the correct units are being used, and manipulate the formulae to allow you to determine the required values.

Key Points

When performing calculations in thermodynamics, it is important to use absolute values, as these are what the relationships are developed on.

Process parameters

Absolute temperature

As previously mentioned, absolute temperature is measured in (K).

Absolute pressure

Absolute pressure is not the pressure you would read off a pressure gauge. This is because when a pressure gauge is reading 0 bar, it is not allowing for atmospheric pressure. This is the pressure due to the weight of the atmosphere acting down on us.

Atmospheric pressure is approximately 1.01 bar at sea level; however, this value varies, depending on where you are in the world and your altitude.

Boyle's law

Boyle's law describes the relationship between the pressure and volume of a gas at constant temperature. For example, if gas is contained in a cylinder and the pressure is increased, the volume of the gas in the cylinder will decrease as it is compressed.

Boyle's law tells us that this relationship is inversely proportional, so:

$$pV = \text{constant}$$

Therefore, for constant temperature

$$p_1 V_1 = p_2 V_2$$

Worked Example

A container containing gas at absolute pressure of 1.1 bar is compressed until the gas inside reaches a pressure of 1.8 bar. Determine the reduced volume if the original volume = 400 mm³.

$$p_1 = 1.1 \times 10^5 \, \text{Pa}$$

$$p_2 = 1.8 \times 10^5 \, \text{Pa}$$

$$V_1 = 400 \times 10^{-9} \, \text{m}^3$$

$$p_1 V_1 = p_2 V_2$$

$$V_2 = \frac{1.1 \times 10^5 \times 400 \times 10^{-9}}{1.8 \times 10^5}$$

$$V_2 = 244.44 \times 10^{-9} \, \text{m}^3$$

$$V = 244.44 \, \text{mm}^3$$

Charles' law

Charles' law describes the relationship between the temperature and volume of a gas at constant pressure. For example, if gas is contained in a cylinder and the temperature is increased, the volume of the gas in the cylinder will increase as it is heated.

Charles' law tells us that this relationship is directly proportional, so:

$$\frac{V}{T} = \text{constant}$$

Therefore, for constant pressure

$$\frac{V_1}{T_1} = \frac{V_2}{T_2}$$

Worked Example

A container containing gas at a temperature of 20 °C is heated until the gas reaches a temperature of 90 °C. Determine the increased volume if the original volume = 400 mm³.

$$T_1 = 293 \, \text{K}$$

$$T_2 = 363 \, \text{K}$$

$$V_1 = 400 \times 10^{-9} \, \text{m}^3$$

$$\frac{V_1}{T_1} = \frac{V_2}{T_2}$$

$$V_2 = \frac{363 \times 400 \times 10^{-9}}{293}$$

$$V_2 = 495.56 \times 10^{-9} \, \text{m}^3$$

$$V = 495.56 \, \text{mm}^3$$

General gas equation

Many thermodynamic processes involve changes in pressure, temperature and volume. The general gas equation is a combination of Charles' law and Boyle's law:

$$\frac{pV}{T} = \text{constant}$$

$$\frac{p_1 V_1}{T_1} = \frac{p_2 V_2}{T_2}$$

Worked Example

A cylinder containing gas at a temperature of 70 °C, volume of 80 mm³ and pressure of 2 bar is heated until the gas reaches a temperature of 120 °C. Determine the change in pressure if the final volume = 100 mm³.

$$T_1 = 343\,\text{K}$$

$$p_1 = 2 \times 10^5\,\text{Pa}$$

$$V_1 = 80 \times 10^{-9}\,\text{m}^3$$

$$T_2 = 393\,\text{K}$$

$$V_2 = 100 \times 10^{-9}\,\text{m}^3$$

$$\frac{p_1 V_1}{T_1} = \frac{p_2 V_2}{T_2}$$

$$p_2 = \frac{(2 \times 10^5 \times 80 \times 10^{-9} \times 393)}{(100 \times 10^{-9} \times 343)}$$

$$p_2 = 1.83 \times 10^5\,\text{Pa}$$

$\therefore$ The change in pressure $= p_1 - p_2$

$$= 2 \times 10^5 - 1.83 \times 10^5$$

$$= 0.17 \times 10^5\,\text{Pa}$$

$$= 0.17\,\text{bar}$$

Key Points

Using the general gas equation is a good example of why we change the units to their absolute values. If you try to use the values given in the example, without converting them, the answer will be incorrect.

Characteristic gas equation

Using the general gas equation is very useful when we have the conditions before and after a thermodynamic process. However, for easier analysis, we use a constant to formulate a characteristic equation:

$$pV = mRT$$

where:

p = pressure (Pa)
V = volume (m³)
m = mass (kg)
R = specific gas constant (J/kgK)
T = temperature (K)

Worked Example

A cylinder of volume 85 000 cm³ is filled with 0.3 kg of air at 30 °C. A pressure gauge is attached which reads 2 bar. Because there is some doubt about the accuracy and precision of the gauge, determine the theoretical pressure and compare this with the given value. The specific gas constant for air is 287 J/kgK and the atmospheric pressure can be taken to be 1.01 bar.

$$V = 85 \times 10^{-3}\,\text{m}^3$$

$$T = 303\,\text{K}$$

$$m = 0.3\,\text{kg}$$

$$R = 287\,\text{J/kgK}$$

$$pV = mRT$$

$$p = \frac{(0.3 \times 287 \times 303)}{85 \times 10^{-3}}$$

$$p = 3.07 \times 10^5\,\text{Pa}$$

$$p = 3.07\,\text{bar (absolute)}$$

The pressure gauge reads 2 bar, but we have to add atmospheric pressure to this:

$$2 + 1.01 = 3.01\,\text{bar}$$

The discrepancy is 0.06 bar, or approximately 20 per cent.

Activity 5.14

Nitrogen in a cylinder of volume $500 \times 10^3\,\text{cm}^3$ is compressed from an absolute pressure of 2.2 bar to an absolute pressure of 3.5 bar. If the initial temperature is 25 °C and the final volume is 60 cm³, determine:

- the final temperature.

To achieve a pass grade you will have:

P1 calculated the magnitude, direction and position of the line of action of the resultant and equilibrant of a non-concurrent coplanar force system containing a minimum of four forces acting in different directions

P2 calculated the support reactions of a simply supported beam carrying at least two concentrated loads and a uniformly distributed load

P3 calculated the induced direct stress, strain and dimensional change in a component subjected to direct uniaxial loading and the shear stress and strain in a component subjected to shear loading

P4 solved three or more problems that require the application of kinetic and dynamic principles to determine unknown system parameters

P5 calculated the resultant thrust and overturning moment on a vertical rectangular retaining surface with one edge in the free surface of a liquid

P6 determined the up-thrust on an immersed body

P7 used the continuity of volume and mass flow for an incompressible fluid to determined the design characteristics of a gradually tapering pipe

P8 calculated dimensional change when a solid material undergoes a change in temperature and the heat transfer that accompanies a change of temperature and phase

P9 solved two or more problems that require application of thermodynamic process equations for a perfect gas to determine unknown parameters of the problems.

To achieve a merit grade you will also have:

M1 calculated the factor of safety in operation for a component subjected to combined direct and shear loading against given failure criteria

M2 determined the retarding force on a freely falling body when it impacts upon a stationary object and is brought to rest without rebound, in a given distance

M3 determined the thermal efficiency of a heat transfer process from given values of flow rate, temperature change and input power

M4 determined the force induced in a rigidly held component that undergoes a change in temperature.

To achieve a distinction grade you will also have:

D1 compared and contrasted the use of d'Alembert's principle with the principle of conservation of energy to solve an engineering problem

D2 evaluated the methods that might be used to determine the density of a solid material and the density of a liquid.

By the end of this unit you should be able to:

- use circuit theory to determine voltage, current and resistance in direct current (d.c.) circuits
- understand the concepts of capacitance and determine capacitance values in d.c. circuits
- know the principles and properties of magnetism
- use single-phase alternating current (a.c.) theory.

....

In order to pass this unit, the evidence you present for assessment needs to demonstrate that you can meet all of the above learning outcomes for this unit. The criteria below show the levels of achievement required to pass this unit.

To achieve a pass grade you need to:	To achieve a merit grade you also need to:	To achieve a distinction grade you also need to:
P1 use d.c. circuit theory to calculate current, voltage and resistance in d.c. networks	**M1** use Kirchhoff's laws to determine the current in various parts of a network having four nodes and the power dissipated in a load resistor containing two voltage sources	**D1** analyse the operation and the effects of varying component parameters of a power supply circuit that includes a transformer, diodes and capacitors
P2 use a multimeter to carry out circuit measurements in a d.c. network	**M2** evaluate capacitance, charge, voltage and energy in a network containing a series-parallel combination of three capacitors	**D2** evaluate the performance of a motor and a generator by reference to electrical theory.
P3 compare the forward and reverse characteristics of two different types of semi-conductor diode	**M3** compare the results of adding and subtracting two sinusoidal a.c. waveforms graphically and by phasor diagram.	
P4 describe the types and function of capacitors		
P5 carry out an experiment to determine the relationship between the voltage and current for a charging and discharging capacitor		
P6 calculate the charge, voltage and energy values in a d.c. network for both three capacitors in series and three capacitors in parallel		
P7 describe the characteristics of a magnetic field		
P8 describe the relationship between flux density (B) and field strength (H)		
P9 describe the principles and applications of electromagnetic induction		
P10 use single-phase a.c. circuit theory to determine the characteristics of a sinusoidal a.c. waveform		
P11 use an oscilloscope to measure and determine the inputs and outputs of a single-phase a.c. circuit.		

Introduction

Electrical and electronic principles underpin the modern world; the technology is used in many applications, such as communications, transportation and the media. This technology has had an enormous impact on the way we live today.

The unit will help learners to develop the necessary skills, attitude and knowledge for achieving passes (P), merits (M) and distinctions (D) in this core unit. The unit starts with the fundamental principles of electrical and electronic technology – electron theory – and how this is applied to direct current (d.c.) circuits. This theory develops into electromagnetism, alternating current (a.c.) and electrical machines. The unit concludes with a practical review of circuit measurements.

For learners wishing to follow an electrical/electronic programme, this unit is an essential building block that will provide the underpinning knowledge required for further study of electrical and electronic applications.

This section of the book is organised as follows; each of the subsections can be readily linked to the learning outcome (LO), pass (P), merit (M) and distinction (D) criteria.

Section/content	LO	P	M	D
Electrical and electronic fundamentals				
Safety				
6.1 Electrostatics	2	4		
6.2 Direct current (d.c.)	1	1, 6	1, 2	
6.3 Electromagnetism	3	7, 8		
6.4 Alternating current (a.c.)	4	10	3	
6.5 Electrical machines	3	9		2
6.6 Diodes	1	3		1
6.7 Circuit measurements	1, 2, 4	2, 5, 11		

Electrical and electronic fundamentals

Electrical and electronic fundamentals are based on the scientific model of an atom. The physical nature of all materials can be illustrated by the model of an atom, which consists of a central nucleus with orbiting electrons. Electrical and electronic principles are founded on the electrons, their characteristics and how they behave.

Atomic theory

Figure 6.1 is one illustration of the model of a single atom. Note that this a scientific model used in physics, chemistry and engineering to be able to describe and understand the behaviour of materials, and electrical/electronic circuits. Unlike the solar system, it is not practical to be able to view the elements of an atom, even with powerful optical devices.

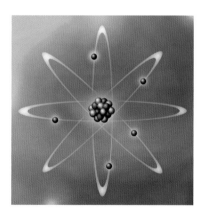

Figure 6.1 Model of an atom

Referring to the model in Figure 6.1, the nucleus of this atom comprises protons and neutrons; orbiting this nucleus are electrons. Electrons are thousands of times lighter than the protons and neutrons in the nucleus. These elements of an atom are retained in position as a result of electrical charge:

- protons are positively charged
- neutrons are neutral
- electrons are negatively charged.

The nucleus is consequently positively charged, and the total charge of the nucleus is equal in magnitude (size) to the sum of all charges on the protons. When two or more atoms combine,

they are linked together to form a molecule (see Figure 6.2). The molecular structure is the basis of a given substance or material that we would recognise – for example, oxygen, water, copper, silver, rubber or plastic.

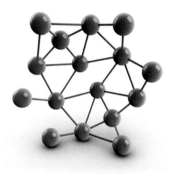

Figure 6.2 Molecules – formed by atoms bonded into a structure

In a stable atom, the number of protons and electrons are in equilibrium, so that overall the atom is neutral, which is to say, it has no charge. Depending on the molecular structure of a material, it is possible to add or remove electrons from an atom. For example, if two particular materials are rubbed together, electrons may be transferred from one material to the other. This alters the stability of the atom, leaving it with a net positive or negative charge. When an atom within a material loses electrons, it becomes positively charged and is known as a positive ion; when an atom gains an electron, it has a surplus negative charge, so is referred to as a negative ion.

These differences in charge give rise to a phenomenon called an electrostatic effect. This can be demonstrated by observing the effect of unrolling a section of cling film – the material is attracted to a person's hand. There is a difference in charge between the cling film and the person, causing an attraction of electrical charges.

Key Term

Periodic table – The method of listing elements in terms of increasing atomic number, i.e. the number of electrons in an atom.

The number of electrons occupying a given orbit within an atom is established from an extension of the simple model previously discussed. These are all characterised in the periodic table. The electrons in specific atoms are located in a particular orbit, or shell, dependent on their energy level. It is important to note that different atoms have different numbers of shells; each of these shells is occupied by a specific number of electrons. Certain atoms have numerous shells, as shown in Figure 6.3. In this example, the innermost shell can have up to two electrons; the second shell can have up to eight; and the third (in certain atoms) up to 18. This outer shell is of particular importance to electrical and electronic theory.

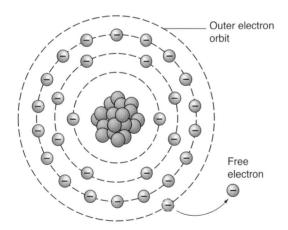

Figure 6.3 An atom with a loosely bound free electron in its outer shell

Outer electron orbit

Free electron

Electrical charge

All electrons and protons carry an electrostatic charge; this is given the unit of the coulomb. The unit is named after the French physicist Charles-Augustin de Coulomb (1736–1806). His work on electrical charge led to the development of Coulomb's law; this defines the electrostatic forces of attraction and repulsion.

Key Term

One coulomb (C) – The total amount of the charge carried by 6.21×10^{18} electrons; a single electron has a very small charge of 1.61×10^{-19} C.

Figure 6.3 shows an atom with one loosely bound outer electron that can easily become detached from the parent atom. Only a small amount of energy is required to overcome the attraction of the nucleus. Sources of such energy may include heat, light or electrostatic fields. The electron, once detached from the atom, is able to move freely around the structure of the material and is called a free electron. These free electrons become the charge carriers within a material. Materials that have large numbers of free electrons make good conductors of electrical energy and heat. A material which has many free electrons that move easily between atoms is known as a conductor. Metals are the best conductors, since they have a very large number of free electrons available to act as charge carriers. Examples of good conductors include silver, aluminium, copper, gold and iron.

Materials that do not conduct charge are called insulators; their electrons are tightly bound to the nuclei of their atoms. Examples of insulators include plastics, glass, rubber and ceramic materials.

In a material containing free electrons, their direction of motion is random, as shown in Figure 6.4(a), but if an external force is applied that causes the free electrons to move in a uniform manner (Figure 6.4(b)), an electric current is said to flow.

The effects of electrical current flow can be observed when electrical energy is converted into other forms of energy. For example, heat is produced when an electric current is passed through a resistive heating element. Light is produced when an electric current flows through the filament wire of an incandescent lamp.

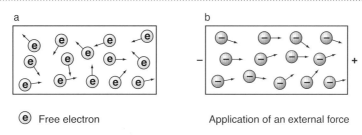

(e) Free electron   Application of an external force

Figure 6.4 Free electrons and the application of an external force

Activity 6.1

Assess your knowledge of electrical and electronic fundamentals by answering these questions.

1 Do electrons carry a very small amount of negative or positive electrical charge?

2 Give examples of materials that readily support the flow of electric current. What general term is given to these materials?

3 Give examples of materials that do not support the flow of electric current. What general term is given to these materials?

4 Explain the following terms: electron, ion, charge.

Temperature coefficient

All materials offer some opposition, or resistance, to current flow. In conductors, the free electrons collide with the relatively large and solid nuclei of the atoms. As the temperature of the material increases, the nuclei vibrate more energetically, further obstructing the path of the free electrons, causing more frequent collisions. The result is that the resistance of a conductor increases with temperature.

Due to the nature of the bonding in insulators, there are only a small number of free electrons. When thermal energy increases as a result of a temperature increase, a few electrons manage to break free from their fixed positions and act as charge carriers. The result is that the resistance of an insulator decreases as temperature increases. By producing special alloys, such as eureka and manganin, which combine the effects of insulators and conductors, it is possible to produce a material where the resistance remains constant with increase in temperature. Figure 6.5 shows how the resistance of insulators, semiconductors and conductors changes with temperature.

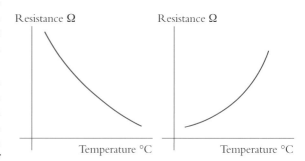

Figure 6.5 Variation of resistance with temperature for negative/positive temperature coefficient materials

Safety

Safety (both yours and that of those around you) is paramount when working with electrical/electronic equipment and systems. The effects of electricity act very quickly and you must always think carefully before working on circuits where voltages in excess of 50 V are present. Failure to observe this simple precaution can result in the risk of an electric shock.

Hazards and risks

Hazards and risks can exist within many circuits, even those that may appear to be totally safe. They include:

- incorrect connection of a power supply
- incorrect connection of components
- incorrect fitting of components.

These can all result in serious risks to personal safety as a result of:

- fire
- explosion
- generation of toxic fumes
- burns
- electric shock.

Voltages in many items of electronic equipment, including all items that derive their power from the domestic power supply, are at a level which can cause sufficient current flow in the body to disrupt the normal operation of muscles and the heart. The threshold will be even lower for anyone with a defective heart.

The most critical path for electric current within the body is when current passes through the heart; this occurs when both hands complete the electrical circuit. (The hand-to-foot route can also have this effect.) Therefore, in preparation for working on an item of electronic equipment, think about what voltages are present from the power supply and/or other sources of stored energy – for example, capacitors. If you have to make measurements or carry out adjustments on a piece of working (or 'live') equipment, a useful precaution is using one hand only to perform the adjustment or to make the measurement. Your 'spare' hand should be placed safely away from contact with anything metal (including the chassis of the equipment, which may or may not be earthed).

Electric shock

The severity of electric shock depends upon several factors, including:

- the magnitude of the current
- whether it is alternating or direct current
- the path of current through the body.

The magnitude of the current through a person depends upon the voltage which is applied and the resistance of the skin. The electrical energy developed in the body will depend on the time for which the current flows. The duration of contact is also crucial in determining the eventual physiological effects of the shock. As an approximation, the effects in Table 6.1 are typical.

The figures quoted in Table 6.1 are provided as a guide for comparison and information; there have been cases of lethal shocks resulting from contact with much lower voltages and at relatively small values of current. Any potential in excess of 50 V must be considered dangerous, and safety precautions should be taken. Get into the habit of treating all electrical and electronic systems with great care.

Current (mA)	Physiological effect	Physical sensation
1	Not usually noticeable	None
1–2	Threshold of perception	Slight tingle may be felt
2–4	Mild shock	Effects of current flow are felt
4–10	Serious shock	Shock is felt as pain
10–20	Nerve paralysis may occur	Unable to let go
20–50	Respiratory control inhibited	Breathing impaired
>50	Ventricular fibrillation of heart muscle	Heart failure

Table 6.1 Effects of current and physiological effect

Key Points

- Safety (both yours and that of those around you) is paramount when working with electrical/electronic equipment and systems.

- It is essential to remove electrical power from equipment before removing or installing components.

- Failure to observe this precaution can result in electric shock and/or damage to components and equipment.

- The most critical path for electric current within the body is when current passes through the heart.

- The magnitude of the current through a person depends upon the voltage which is applied and the resistance of the skin.

- Any potential difference in excess of 50 V must be considered dangerous.

6.1 Electrostatics

This section will cover the following grading criteria:

By the end of this section you should be able to describe the types and functions of capacitors.

Electrostatics builds on atomic theory, in particular the nature and behaviour of electrons. If a material has a deficit of electrons, it will exhibit a net positive charge; if it has a surplus of electrons, it will exhibit a net negative charge. Note that there is no current actually flowing, hence the term 'electrostatics'.

This imbalance in charge can be produced by a variety of methods, the most common being friction. In the extreme case, friction between molecules in the atmosphere gives rise to a massive build-up and subsequent discharge of electrical energy – that is, lightning.

If two bodies are charged in the same way – that is, either both positively or both negatively charged – the two bodies will move apart, indicating that a force of repulsion exists between them. If, on the other hand, the charges on the two bodies are unlike – that is, one positively charged and one negatively charged – the two bodies will move together, indicating that a force of attraction exists between them. From this we can conclude that like charges repel and unlike charges attract.

In the case of friction, electrons and protons in an insulator can be separated from their atoms by rubbing two materials together. This causes electrons to leave one atom for another, thereby creating atoms with an excess of electrons in some atoms, and a deficiency in the others. This creates atoms in the two materials with opposite charges. These charges will remain separated for some time, until they eventually dissipate due to losses in the insulating material or in the air surrounding the materials.

> **Activity 6.2**
> Complete the missing words in this sentence: Charge bodies with the same charge _____ one another, while charges with opposite charge will _____ one another.

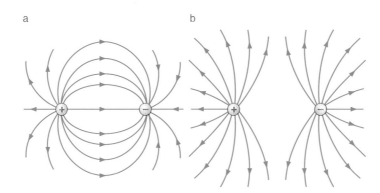

Figure 6.6 (a) Electric field between isolated unlike charges; (b) electric field between isolated like charges

Electric fields

The force exerted on a charged particle is established by the existence of an electric field. The electric field defines the direction and magnitude of a force on a charged object. The field itself is invisible to the human eye, but can be illustrated by constructing lines which indicate the motion of a free positive charge within the field – the number of field lines in a particular region being used to indicate the relative strength of the field at the point in question.

Figure 6.6 illustrates the electric fields between isolated unlike and like charges, while Figure 6.7 shows the field that exists between two charged parallel metal plates.

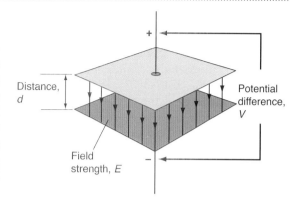

Figure 6.8 Electric field strength between two charged conducting surfaces

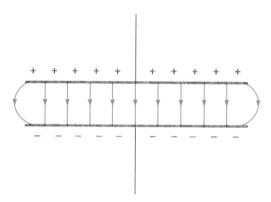

Figure 6.7 Electric field between the two charged parallel metal plates of a capacitor

The strength of an electric field (E) is proportional to the applied potential difference, and inversely proportional to the distance between the two conducting surfaces (see Figure 6.8). The electric field strength (for an infinitely large pair of parallel plates) is given by the relationship:

$$E = \frac{V}{d}$$

where E is the electric field strength (in volts per metre $\frac{V}{m}$), V is the applied potential difference (in volts) and d is the distance (in metres).

These charged conducting surfaces are the basis for a very important component used in electrical and electronic engineering: the capacitor. The amount of charge that can be stored by a capacitor is given by the relationship:

$$Q = C \times V$$

where Q is the unit of charge in coulombs; C is the capacitance expressed in units of farads, and V is the applied voltage.

Key Term

Unit of capacitance (farad) – Named after Michael Faraday, a British chemist and physicist (1791–1867). Faraday was an eminent scientist, contributing significantly to the fields of electromagnetism and electrochemistry.

Mathematical multipliers

The farad is a very large unit. In many branches of engineering, including electrical and electronic, we use mathematical multipliers to express low or high values, written with a prefix name. The International System of Units specifies prefixes used with units of measure (and associated symbol) to form a decimal multiple or sub-multiple. In electrical and electronic engineering, we often deal with very large or very small units of current, voltage and resistance.

0.000001 = 10^{-6} prefix *micro* (μ)
0.001 = 10^{-3} prefix *milli* (m)
1000 = 10^{3} prefix *kilo* (k)
1 000 000 = 10^{6} prefix *mega* (M)

Hence a capacitance of 0.000001 F is written as 1 μF.

Key Term

Unit of voltage – Named after the Italian physicist Alessandro Giuseppe Antonio Anastasio Volta (1745–1827). His work led to the development of the first electric cell in 1800. The volt is the unit of electromotive force, also referred to as voltage; it is also the unit for the related quantity of electrical potential difference.

The relationship $Q = C \times V$ can be rearranged to make C or V the subject, as follows:

$$V = \frac{Q}{C} \text{ and } C = \frac{Q}{V}$$

Worked Example

Two parallel conductors are separated by a distance of 25 mm. Determine the electric field strength if they are fed from a 750 V supply.

The electric field strength will be given by:

$$E = \frac{V}{d}$$

The electric field strength is therefore: $\frac{750}{0.025} = 30\,\text{kV/m}$

Worked Example

The electric field strength between two parallel plates is 20 kV/m. If the plates are separated by a distance of 10 mm, determine the potential difference that exists between the plates.

The electric field strength will be given by:

$$E = \frac{V}{d}$$

Rearranging this formula to make V the subject gives:

$$V = E \times d$$

The potential difference $(V) = 20{,}000 \times 0.01 = 200\,\text{V}$

Worked Example

A potential difference of 100 V appears across two plates with a capacitance of 1 μF. What charge is stored?

The charge can be calculated from:

$$Q = C \times V$$

The stored charge $(Q) = 1 \times 10^{-6} \times 100 = 100\,\mu\text{C}$.

Capacitors

Referring to electrostatic theory, and in particular the accumulation of charge on two conducting surfaces, the definition of capacitance is the ability of a body to hold an electrical charge. The capacitor is a passive electronic device, comprising two conductors separated by an insulating material known as the dielectric. Capacitors are normally formed into cylindrical or rectangular shapes; capacitor construction and circuit symbols are shown in Figure 6.9.

When a voltage potential difference is applied between the plates, an electric field is present in the dielectric. This field stores energy between the plates. Capacitors have many uses in electrical and electronic circuits, based on the properties of energy storage.

Several solid dielectrics are available, including:

- paper
- plastic
- glass
- mica
- ceramic materials.

One widely used type of construction is the electrolytic capacitor; this uses an aluminium or tantalum plate, with the dielectric formed by an oxide layer. The second electrode is a substance containing free ions that is a conductive material; this is connected to the circuit by another foil plate. Electrolytic capacitors offer very high capacitance, but have high tolerances and are generally not stable. They lose capacitance over time; this is accelerated by heat and leakage current. Electrolytic capacitors can only be used in d.c. applications; they can only be connected with the correct terminations.

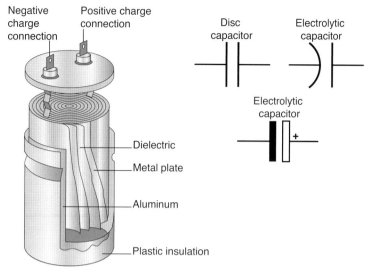

Figure 6.9 Capacitor construction and circuit symbols for capacitors

Safety points

1. When working with high-voltage capacitors, it is essential to ensure that the capacitor is fully discharged before attempting to replace the component.
2. When replacing a capacitor, it is essential to ensure that the replacement component is correctly rated in terms of type, value, working voltage and temperature. Capacitors are prone to failure if their maximum working voltage is exceeded, and they should be de-rated when operated at a relatively high ambient temperature, according to manufacturers' specifications.
3. It is also essential to observe the correct polarity when replacing an electrolytic (polarised) component. This is usually clearly marked on the external casing.

Make the Grade P4

Grading criterion P4 requires that you describe the types and function of capacitors. The type of capacitor can be broadly described by its method and materials used for construction. Applications for capacitors include timing circuits and rectifier output smoothing.

Charging/discharging

When electrical charge is added to the plates of a capacitor, it is said to be charging. When electrical charge is removed from the plates of a capacitor, it is said to be discharging.

Practical charging/discharging is normally achieved in combination with a resistor network; further details are given in section 4.2 (Direct current (d.c.)) of this unit.

Activity 6.3

1. The two plates of a parallel plate capacitor are separated by a distance of 25 mm. If the potential difference between the plates is 100 V, what will the electric field strength be?
2. The electric field between two conducting surfaces is 250 V/m. If the plates are separated by a distance of 25 mm, determine the potential difference between the plates.

Summarising the quantities, symbols, units and abbreviations used in electrostatics, note that the abbreviation is used after writing the value of the unit – for example, a potential difference of 5 volts can be written as 5 V.

Quantity	Unit	Symbol	Abbreviation
Charge	coulomb	Q	C
Electromotive force	volt	V	V
Potential difference	volt	V	V
Electric field	volt/metre	E	V/m

Table 6.2

6.2 Direct current (d.c.)

This section will cover the following grading criteria:

By the end of this section you should be able to use d.c. circuit theory to calculate current, voltage and resistance in d.c. networks; calculate the charge, voltage and energy values in a d.c. network for both three capacitors in series and three capacitors in parallel; use Kirchhoff's laws to determine the current in various parts of a network having four nodes and the power dissipated in a load resistor containing two voltage sources; and evaluate capacitance, charge, voltage and energy in a network containing a series-parallel combination of three capacitors.

d.c. refers to the flow of electrical energy in one direction only. d.c. circuits are found in many practical applications – for example, torches and starter motors. This section focuses on the three basic quantities used in electrical and electronic circuits:

- voltage
- current
- resistance.

The force that creates the flow of current (or rate of flow of charge carriers) in a circuit is the electromotive force (emf). Current is defined as the rate of flow of electrical charge. All materials oppose the movement of electric charge through them; this opposition to the flow of the charge carriers is known as the resistance of the material.

Current/voltage/resistance

The principle of electrical technology is based on the model of an atom as described earlier. Because of their negative charge, electrons will be attracted to positive potentials (remember that like charges repel and unlike charges attract). This model of an atom was developed by scientists in the early twentieth century; prior to this, the early pioneers of electrical engineering believed that the direction of current flow in a circuit was from the positive potential to the negative potential; this is termed conventional current.

Further scientific research, however, established that the charge carriers were electrons; these actually flow in the opposite direction – that is, from negative to positive. The convention of defining current flow from positive to negative remained, however. Electrical engineers continue to refer to conventional current; electron flow is the preference in electronics. In this unit, we will refer to conventional current unless otherwise stated.

The most commonly used method of generating d.c. is the electrochemical cell, invented by Volta.

Key Term

Cell – A device that produces a charge when a chemical reaction takes place. When several cells are connected together, they form a battery (see Figure 6.10). Circuit symbols for cells and batteries are shown in Figure 6.11.

Key Points

- d.c. refers to the flow of electrical energy in one direction only.
- Current is defined as the rate of flow of electrical charge.
- Conventional current flows from positive to negative.
- Electrons travel in the opposite direction, from negative to positive.
- Opposition to the flow of current is known as the resistance of the material.

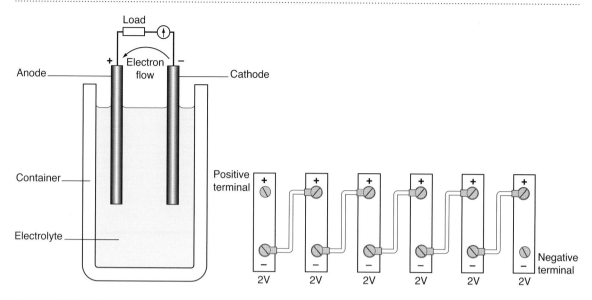

Figure 6.10 Cells and batteries

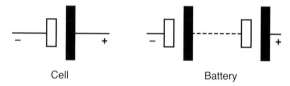

Figure 6.11 Circuit symbols for cells and batteries

A simple d.c. circuit containing a battery, switch and lamp is shown in Figure 6.12.

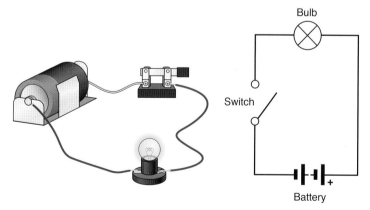

Figure 6.12 Simple d.c. circuit: components and diagram

If a current of 1 A flows for a time (t) of 10 seconds, then the amount of charge (Q) transferred will be:

$$Q = I \times t = 1 \times 10 = 10 \text{ coulombs}$$

Key Points

- Since current is the rate of flow of charge, if more charge moves in a given time, more current will be flowing.
- If no charge moves, then no current is flowing.
- One ampere is equal to a flow of one coulomb per second.

The work done by a source of energy to separate electrical charges and cause current to flow is the emf, measured in V. A typical source of energy is the electrical cell; this transfers energy to the circuit by doing work to raise the electrical potential. (Note that the word 'force' used in the term 'emf' is in fact a misnomer; electromotive force is a property of the energy source.)

Key Term

Potential difference (p.d.) – The voltage difference, or voltage drop, between two given points in a circuit created by the energy source and giving rise to current flow.

All materials oppose the movement of electric charge through them; this opposition to the flow of the charge carriers is known as the resistance (R) of the material. This resistance is due to collisions between the charge carriers (electrons) and the atoms of the material.

Key Term

Ohm, with the symbol Ω – The unit of resistance. This unit is named after the German physicist Georg Simon Ohm (1789–1854).

Key Points

Ohm experimented with the electrochemical cell, recently invented by Volta. Using equipment of his own creation, he determined that there is a direct proportionality between the potential difference (voltage) applied across a conductor and the resultant electric current – this is now known as Ohm's law. The application of Ohm's law is detailed later in this unit.

Note that 1 V is the electromotive force required to move 6.21×10^{18} electrons (one coulomb) through a resistance of one ohm in one second. Hence:

$$V = \left(\frac{Q}{t}\right)R$$

Power stations generate very high voltages, typically in the order of 50,000 V. The integrated circuits used in computers operate with very low currents, typically one-millionth of an ampere. Resistors can be rated in tens, thousands or millions of ohms. For convenience, and to avoid introducing errors, we use mathematical multipliers as before; for example:

- 50 000 volts is written as 50 kV
- 0.001 amperes is written as 1 mA
- 1,000 ohms is written as 1 kΩ.

Resistors

Key Term

Resistor – A component used in electrical networks and electronic circuits.

Resistors are ever-present in most electronic equipment. They can be made of various compounds, chemical films and wire. The primary characteristics of a resistor are the:

- resistance value
- tolerance
- power rating
- temperature coefficient.

Resistors are often miniaturised and formed as part of integrated circuits and printed circuit boards.

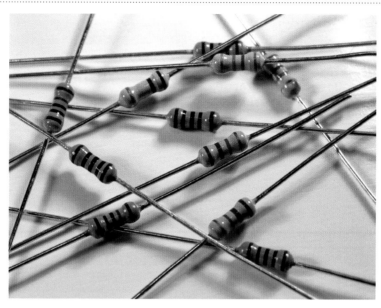

Figure 6.13 Resistors

Fixed-value resistors

Whether designing circuits or investigating problems, it is important to determine the value of a particular resistor component. One method of identifying resistor values is with a colour-coding scheme. Coloured bands correspond to the value and tolerance of the resistor (see Figure 6.13). The first two bands represent the value, the third band is the multiplier and the fourth band is the tolerance.

Band colour	Significant figures	Multiplier	Tolerance (%)
Black	0	$\times 10^0$	–
Brown	1	$\times 10^1$	±1
Red	2	$\times 10^2$	±2
Orange	3	$\times 10^3$	–
Yellow	4	$\times 10^4$	–
Green	5	$\times 10^5$	±0.5
Blue	6	$\times 10^6$	±0.25
Violet	7	$\times 10^7$	±0.1
Grey	8	$\times 10^8$	±0.05
White	9	$\times 10^9$	–
Gold	–	$\times 10^{-1}$	±5
Silver	–	$\times 10^{-2}$	±10
None	–	–	±20

Table 6.3 Resistor colour coding

Worked Example

A resistor is marked with the following coloured bands: yellow, white, orange, silver. What is the value of the resistor?

- First digit: yellow (4)
- Second digit: white (9)
- Multiplier: orange (1000, or 10^3)
- Tolerance: silver ($+/-$ 10 per cent)

Hence the resistor value is 49,000 ohms (or $49\,k\Omega$) $+/-$ 10 per cent.

Variable resistors

Variable resistors are mounted on a linear or rotary shaft to provide a user-adjustable resistance; typical applications include the control of lighting or audio volume in a music player. They are sometimes combined with micro-switches to provide an on/off control function. Variable resistors are produced as one of the following:

- potentiometers
- preset resistors
- rheostats.

Figure 6.14 provides examples of symbols used for variable resistors. When the intention is for the user to adjust the circuit resistance for control purposes – for example, audio volume or lighting intensity – the variable resistor device is used. If the circuit resistance is only intended to be adjusted in the workshop, a preset device is used.

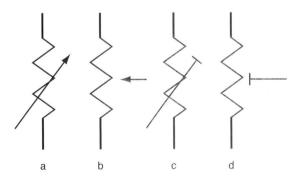

a   b   c   d

Figure 6.14 Circuit symbols for variable resistors

Voltage divider

This is a circuit used to supply a portion of the power supply voltage from a resistive contact. The potentiometer is typically a three-terminal resistor with a sliding centre contact (the wiper). If all three terminals are used, it can be used in the voltage divider application. If only two terminals are used (one end of the resistor and the wiper), it acts as a variable resistor.

A rheostat performs the same function as the potentiometer, but is physically much larger, being designed to operate with much higher voltage and/or current. Rheostats are constructed with a resistive wire formed as a toroidal coil, with the centre contact/wiper moving over the surface of the windings.

Resistivity

Several types of material are used to make wiring, connectors and resistors for electrical circuits. Aluminium conductors are sometimes used for wiring – for example, in aircraft, to save weight; however, the majority of installations are made from copper. The choice of conductor material is a trade-off between the material's resistance over a given length, cost and (in certain applications, such as aircraft) the material's weight.

The relationship between resistance and length is the quantity of resistivity (symbol ρ) measured in units of ohm-metres (Ωm). Annealed copper at $20\,°C$ has a resistivity of $1.725 \times 10^{-8}\,\Omega m$; copper is more ductile than aluminium, and can be easily soldered. Aluminium has a resistivity value of $2.8 \times 10^{-8}\,\Omega m$; it is 60 per cent lighter than copper, but it is more expensive and requires specialised methods of forming joints between wires.

Thermistors

Unlike conventional resistors, the resistance of a thermistor is intended to change with temperature. Thermistors are employed in a wide variety of temperature-sensing and temperature-compensating applications. Two basic types of thermistor are available: negative temperature coefficient (NTC) and positive temperature coefficient (PTC). Typical NTC thermistors have resistances that vary from a few hundred (or thousand) ohms at 25 °C, to a few tens (or hundreds) of ohms at 100 °C. PTC thermistors, on the other hand, usually have a resistance-temperature characteristic that remains substantially flat (usually at around 100 Ω) over the range 0 °C to around 75 °C. Above this, and at a critical threshold temperature (usually in the range 80 °C to 120 °C), their resistances increase rapidly, up to and beyond 10 kΩ (see Figure 6.15).

Quantity	Unit	Symbol	Abbreviation
Current	ampere	I	A
Electromotive force	volt	V	V
Resistance	ohm	R	Ω
Resistivity	ohm-meter	ρ	Ωm

Table 6.4

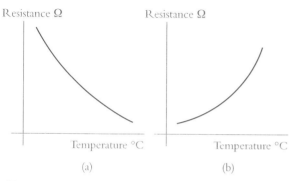

Figure 6.15 (a) Negative temperature coefficient characteristics; (b) positive temperature coefficient characteristics

The resistance of conductors increases with temperature; the resistance of insulators decreases with temperature. This can be summarised as:

- conductors have a positive temperature coefficient
- insulators have a negative temperature coefficient.

Table 6.4 provides a summary of quantities, units, symbols and abbreviations used in this section. Note that the abbreviation is used after writing the value of the unit – for example, a resistance of 5 ohms can be written as 5 Ω.

d.c. circuit theory

A simple d.c. circuit requires two components: a source of electrical energy (or applied voltage) and a resistor (or load) through which a current is passing. These two components are connected together with wire conductors in order to form a completely closed circuit, as shown in Figure 6.16.

Engineers need an understanding of the fundamentals of electrical and electronic principles to analyse simple d.c. circuits. As previously stated, the three electrical quantities used in simple d.c. circuits are:

- voltage
- current
- resistance.

Two eminent pioneers of electrical engineering, Georg Ohm and Gustav Kirchhoff, developed and created fundamental laws of electrical engineering that characterise the relationship between these quantities.

Ohm's law

Georg Ohm carried out research with electrochemical cells. Ohm determined the relationship between the potential difference (voltage) applied across a conductor and the resulting electric current; this is the basis of Ohm's law. As a result of his experimental research, Ohm defined the fundamental relationship between voltage, current and resistance; this represents the basis of electrical circuit theory.

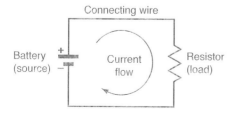

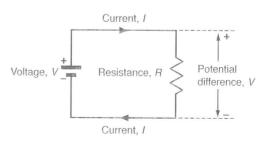

275

Figure 6.16 Simple d.c. circuit diagrams

For any conductor, the current flowing will be dependent on the resistivity and physical dimensions of the conductor (length and cross-sectional area). The amount of current that flows in the conductor when a given emf is applied is inversely proportional to its resistance. Resistance, therefore, may be thought of as an 'opposition to current flow' – the higher the resistance, the lower the current that will flow (assuming that the applied emf and temperature remain constant).

Provided that the temperature does not vary, the ratio of p.d. and current flowing in the conductor is a constant. This relationship is known as Ohm's law and it leads to the relationship:

$$\frac{V}{I} = \text{constant} = R$$

where V is the potential difference (or voltage drop) in volts, I is the current in amperes, and R is the resistance in ohms. This formula may be arranged to make V, I or R the subject, as follows:

$$V = IR$$

$$I = \frac{V}{R}$$

$$R = \frac{V}{I}$$

The triangle shown in Figure 6.17 helps to visualise these three important relationships.

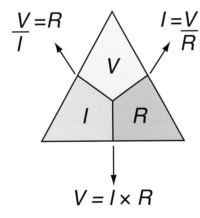

Figure 6.17 Relationship between V, I and R

Another way of linking the relationships between voltage, current and resistance is from the phrase:

Very Important Relationship ($V = IR$)

Kirchhoff's Laws

Gustav Robert Kirchhoff (1824–87) was a German physicist who developed two fundamental laws for the analysis of complex circuits:

- current law
- voltage law.

Key Term

Kirchhoff's current law – This states that the algebraic sum of the currents present at a junction (or node) in a circuit is zero (see Figure 6.18).

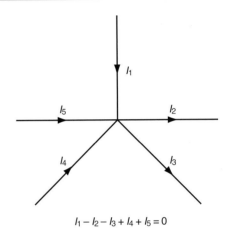

$$I_1 - I_2 - I_3 + I_4 + I_5 = 0$$

Convention:
Current flowing towards the junction is positive (+)
Current flowing away from the junction is negative (−)

Figure 6.18 Kirchhoff's current law

Key Term

Kirchhoff's voltage law – This states that the algebraic sum of the potential drops present in a closed network (or mesh) is zero (see Figure 6.19).

Make the Grade M1

Grading criterion M1 requires that you are able to use Kirchhoff's laws to determine the current in various parts of a network having four nodes and the power dissipated in a load resistor containing two voltage sources.

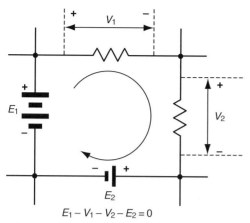

$$E_1 - V_1 - V_2 - E_2 = 0$$

Convention:
Move clockwise around the circuit starting with the positive terminal of the largest e.m.f.
Voltages acting in the same sense are positive (+)
Voltages acting in the opposite sense are negative (−)

Figure 6.19 Kirchhoff's voltage law

Series and parallel resistor circuits

Ohm's law and Kirchhoff's laws can be used in combination to determine voltage, current and resistance in more complex circuits. At this point we need understand the terms 'series' and 'parallel' circuits.

Figure 6.20 shows three circuits, each containing three resistors, R_1, R_2 and R_3.

In Figure 6.20(a), the three resistors are connected in a sequence; the resistors are said to be connected in series. In the series circuit, the same current flows through each resistor. The total value of resistance (R_t) is calculated from:

$$R_t = R_1 + R_2 + R_3$$

This can be expanded for any number of resistors.

In Figure 6.20(b), the three resistors are connected across one another; the resistors are said to be connected in parallel. In the parallel circuit, the same voltage appears across each resistor. The total value of resistance (R_t) is calculated from:

$$\frac{1}{R_t} = \frac{1}{R_1} + \frac{1}{R_2} + \frac{1}{R_3}$$

This can be expanded for any number of resistors.

There is a special case for two resistors in parallel where the following can be used as an alternative:

$$\frac{R_1 \times R_2}{R_1 + R_2}$$

Activity 6.4
Compare the results of calculating resistance values of two resistors in parallel using two different methods.

In Figure 6.20(c), we have a combination of these two types of circuit. Resistor R_1 is connected in series with the parallel combination of R_2 and R_3. In other words, R_2 and R_3 are connected in parallel, and R_1 is connected in series with the parallel combination.

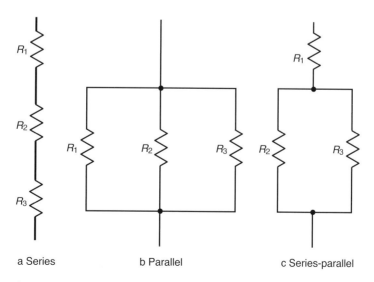

a Series   b Parallel   c Series-parallel

Figure 6.20 Series and parallel circuits

277

Worked Example

Figure 6.21 shows a simple battery test circuit, which is designed to draw a current of 2.0 A from a 24 V d.c. supply. The two test points, A and B, are designed for connecting a meter. Determine:

- **the voltage that appears between terminals A and B (without the meter connected)**
- **the value of resistor, R.**

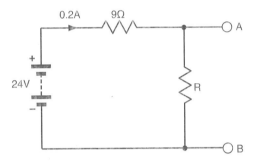

Figure 6.21 Battery test circuit

This problem needs to be solved in several stages. Since we know that the circuit draws 2.0 A from the 24 V supply, we know that this current must flow through both the 9 Ω resistor and R (since these two components are connected in series).

Referring to Figure 6.22, we can determine the voltage drop across the 9 Ω resistor by applying Ohm's law:

$$V = I \times R = 2 \times 9 = 18\,\text{V}$$

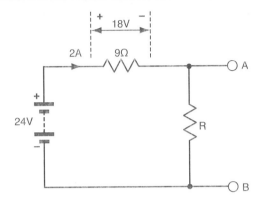

Figure 6.22 Using Ohm's law to find the voltage dropped across a resistor

We can then apply Kirchhoff's voltage law (Figure 6.23) to determine the voltage drop, V, across R – that is, the potential drop between terminals A and B.

$$+\,24\text{V} - 18\text{V} - V_{AB} = 0$$

from which

$$V = 24 - 18 = 6\,\text{V}$$

Now that we know the voltage and current, we can apply Ohm's law (Figure 6.24) to determine the value of R.

The voltage that appears between A and B will be 6 V, and the value of R is 3 Ω.

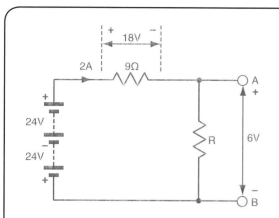

Figure 6.23 Using Kirchhoff's law to determine a voltage

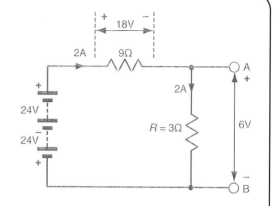

Figure 6.24 Using Ohm's law to find the value of resistance

Make the Grade

Grading criterion P1 requires the use of d.c. circuit theory to calculate current, voltage and resistance in d.c. networks. A thorough understanding of Ohm's law will be required to achieve this. You will need to be able to rearrange the formula associated with this law. Make sure that you can calculate values for simple circuits before attempting more complex networks.

Worked Example

A current of 0.25 A flows in a 100 Ω resistor. What voltage drop (potential difference) will be developed across the resistor?

$$V = IR = 0.25 \times 100 = 25\,V$$

A resistor of 48 Ω is connected to a 12 V battery. What current will flow in the resistor?

$$I = \frac{V}{R} = \frac{12}{48} = 0.25\,A$$

A voltage drop of 50 V is developed across a resistor in which a current of 0.1 A flows. What is the value of the resistor?

$$R = \frac{V}{I} = \frac{50}{0.1} = 500\,\Omega$$

Superposition theorem

In a linear network containing more than one source of emf (see Figure 6.25) the resultant current in any branch is the algebraic sum of the currents that would have been produced by each emf acting alone. The superposition theorem eliminates all but one source of emf within a network.

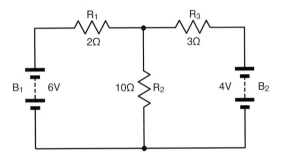

Figure 6.25 Linear network containing two sources of emf

To analyse this circuit, we effectively remove the second source of emf (battery B_2) (see Figure 6.26).

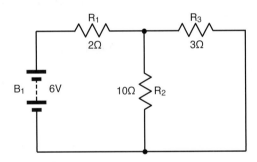

Figure 6.26 Linear network: one source of emf removed

Resistors R_2 and R_3 are in parallel; their combined resistance is calculated using the theory from Figure 6.20.

$$\frac{R_2 \times R_3}{R_2 + R_3} = \frac{10 \times 3}{10 + 3} = \frac{30}{13} = 2.31\,\Omega$$

The total circuit resistance is therefore $2 + 2.31$ $= 4.31\,\Omega$ (these resistors are in series). The total current (I_1) flowing in this network is therefore:

$$I_1 = \frac{V}{R} = \frac{6}{4.31} = 1.39\,\text{A}$$

From Ohm's law, the voltage drop across R_1 is $1.39 \times 2 = 2.78\,\text{V}$

the voltage drop across R_2 and R_3 is therefore $6 - 2.78 = 3.22\,\text{V}$

the value of I_2 is $\frac{3.22}{10} = 0.322\,\text{A}$ (through R_2)

the value of I_3 is $\frac{3.22}{3} = 1.07\,\text{A}$ (through R_3)

Now consider the other source of emf being removed, as illustrated in Figure 6.27. The equivalent resistance of R_1 and R_2 is

$$\frac{R_1 \times R_2}{R_1 + R_2} = \frac{20}{12} = 1.67\,\Omega$$

The total value of resistance is therefore $3 + 1.67$ $= 4.67\,\Omega$ (these resistors are in series).

To summarise the above, once the two parallel resistors are treated as single values, they are, in series with the resistor, connected to the output of the battery. Two parallel and two series calculations are required.

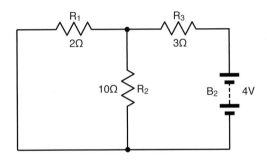

Figure 6.27 Linear network: second source of emf removed

The total current (I_4) flowing out of B_2 in this network is therefore:

$$I_4 = \frac{V}{R} = \frac{4}{4.67} = 0.86\,\text{A}$$

From Ohm's law, the voltage drop across R_3 is $0.86 \times 3 = 2.58\,\text{V}$

the voltage drop across R_1 and R_2 is therefore $4 - 2.58 = 1.42\,\text{V}$

the value of I_5 is $\frac{1.42}{10} = 0.142\,\text{A}$ (through R_2)

the value of I_6 is $\frac{1.42}{2} = 0.71\,\text{A}$ (through R_1)

Superimposing the results from Figure 6.26 on those for Figure 6.27, the resulting current in the left-hand side of the circuit is:

$$I_1 - I_6 = 1.39 - 0.71 = 0.68\,\text{A}$$

The resulting current in the right-hand side of the circuit is:

$$I_4 - I_3 = 0.86 - 1.07 = -0.21\,\text{A}$$

(The minus sign indicates that battery B2 is being charged by 0.21 A.)

The resultant current through R_2 is therefore $I_2 + I_5 = 0.322 + 0.142 = 0.464\,\text{A}$.

Circuits containing resistance and capacitance

Practical circuits used for charging and discharging capacitors contain resistors to control the rate of increasing or decreasing the charge on the plates. The time taken for the capacitor to fully charge or discharge is the time constant for the circuit.

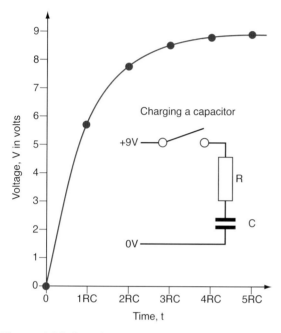

Figure 6.28 Capacitor charging

Referring to Figure 6.28, when a capacitor is charging, the current flow builds up rapidly with an in-rush of current; it then levels off as the capacitor becomes fully charged. At this point in time, the voltage across the capacitor equals the supply voltage. When a capacitor is discharging (Figure 6.29), the current flow is initially very high; it then levels off as the capacitor becomes fully discharged.

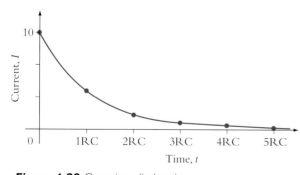

Figure 6.29 Capacitor discharging

Time constant – The time taken for a given capacitor to fully charge or discharge through a resistor, being the same in both cases.

This time constant can be varied with different values of capacitance and/or resistance; the time constant is given by the expression:

$$t = 5CR$$

Networks that contain both capacitance and resistance (CR networks) are used in timing and delay circuits. Further mathematical analysis of a simple charging circuit can be used to show that at $t = 5CR$, the voltage across the capacitor is in fact 99.33 per cent of the supply voltage, and that a very small current (0.007 per cent of the maximum) is still flowing – that is, the capacitor is not fully discharged. From a practical perspective, the time constant is a very useful design feature in electrical/electronic applications. Timing circuits can be used in many practical circuits – for example, security lights.

Capacitors in series/parallel

When two or more capacitors are connected in series, the overall capacitance of the circuit reduces (see Figure 6.30).

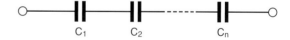

Figure 6.30 Capacitors in series

The total capacitance (C_t) for multiple capacitors connected in series is calculated using the following formula:

$$\frac{1}{C_t} = \frac{1}{C_1} + \frac{1}{C_2} + \frac{1}{C_3}$$

This can be expanded for any number of capacitors.

When two or more capacitors are connected in parallel, the overall capacitance of the circuit increases (see Figure 6.31).

281

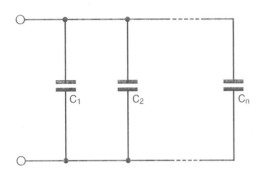

Figure 6.31 Capacitors in parallel

The total capacitance (C_t) for multiple capacitors connected in parallel is calculated using the following formula:

$$C_t = C_1 + C_2 + C_3$$

(This can be expanded for any number of capacitors.)

Energy and power

Energy is the capacity to do work; power is the rate of converting energy from one form to another. An electric fire, for example, converts electrical energy into heat; a filament lamp converts electrical energy into light. The unit of power is the watt, named after James Watt (1736–1819), a Scottish inventor and mechanical engineer. The larger the amount of power, the greater the amount of energy that is converted in a given period of time. The relationship between power, energy and time is given as:

1 watt = 1 joule per second, or:

$$\text{power} = \frac{\text{energy}}{\text{time}}$$

The unit of energy is the joule, named after James Prescott Joule (1818–89), an English physicist. Joule studied the nature of heat, and discovered its relationship to mechanical work. From the definition of power:

1 joule = 1 watt × 1 second

Hence

energy (W) = (power, P) × (time, t) with units of (watts × seconds)

thus

$$W = P\,t\,\text{J}$$

Joules can therefore be expressed in watt-seconds. If the power was to be measured in kilowatts and the time in hours, then the unit of electrical energy would be the kilowatt-hour (kWh, commonly known as a unit of electricity in domestic and industrial power supplies). The electricity meter in homes and buildings records the amount of energy that has been used, expressed in kilowatt-hours.

The power in an electrical circuit is equivalent to the product of voltage and current. Hence:

$$P = I \times V$$

where P is the power, I is the current and V is the voltage. The relationship shown in Figure 6.32 illustrates these three relationships.

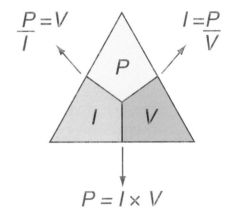

Figure 6.32 Relationship between P, I and V

This triangle can be applied in a similar way to the Ohm's law expressions in Figure 6.17. For the power equations:

$$P = IV, I = \frac{P}{V} \text{ and } V = \frac{P}{I}$$

We can combine Ohm's law with the above to develop two expressions for use in power calculations:

$$P = I \times V = I \times (IR) = I^2R$$
$$P = I \times V = \left(\frac{V}{R}\right) \times V = \frac{V^2}{R}$$

To summarise: power is the rate of using (or converting) energy; one watt corresponds to energy being used at the rate of one joule per second.

282

Worked Example

An electrical heating element provides an output of 1.5 kW for 20 minutes. How much energy has it supplied?

Recalling that $W = P\,t$

where $P = 1.5\,\text{kW} = 1{,}500\,\text{W}$ and $t = 20$ minutes $= 20 \times 60 = 1{,}200\,\text{s}$.

Thus: $W = 1{,}500 \times 1{,}200 = 1{,}800{,}000\,\text{J} = 1.8\,\text{MJ}$

Worked Example

The capacitor in a power supply is required to store 20 J of energy. How much power is required to store this energy in a time interval of 0.5 s?

Rearranging $W = P\,t$ to make P the subject gives:

$$P = \frac{W}{t}$$

Thus

$$P = \frac{20}{0.5} = 40\,\text{W}$$

Worked Example

An aircraft battery is used to start a gas turbine engine. If the starter demands a current of 1000 A for 30 s and the battery voltage remains at 12 V during this period, determine the amount of electrical energy required to start the engine.

First, we need to find the power delivered to the starter from:

$$P = I \times V = 1000 \times 12 = 12\,000\,\text{W} = 12\,\text{kW}$$

Next, we need to find the energy from:

$$W = P\,t = 12\,000 \times 30 = 360\,\text{kJ}$$

Worked Example

A 24 V workshop power unit is to be tested at its rated load current of 3 A. What value of load resistor is required and what should its minimum power rating be?

The value of load resistance required can be calculated using Ohm's law, as follows:

$$R = \frac{V}{I} = \frac{24}{3} = 8\,\Omega$$

The minimum power rating for the resistor will be given by:

$$P = IV = 3 \times 24 = 72\,\text{W}$$

Energy storage

Energy can be stored in a variety of ways:

- chemically – for example, in fuel
- mechanically – for example, in a spring
- electrically – for example, in a capacitor.

For a capacitor, there is a direct relationship between the charge and applied voltage (remember that $Q = CV$); Figure 6.33 depicts this graphically.

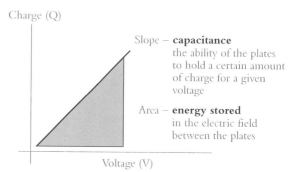

Figure 6.33 Energy stored in a capacitor

The energy stored in the capacitor is the shaded area in Figure 6.33, given by the formula:

$$\text{energy } (W) = 0.5\,QV$$

Combining this with the formula $Q = CV$ gives:

$$W = 0.5(CV)\,V = 0.5\,CV^2$$

283

Make the Grade
P6

Grading criterion P6 requires that you calculate the charge, voltage and energy values in a d.c. network for both three capacitors in series and three capacitors in parallel. Make sure that you can calculate charge and voltage with three capacitors in series, using your knowledge of electrostatics and stored energy.

Worked Example

Referring to Figure 6.34, series and parallel capacitor networks can be reduced into a single (or total) value of capacitance (C_T) using the formulae:

total capacitance (C_T) for the series circuit:

$$\frac{1}{C_T} = \frac{1}{C_1} + \frac{1}{C_2} + \frac{1}{C_3}$$

total capacitance (C_T) for the parallel circuit:

$$C_T = C_1 + C_2 + C_3$$

If all three capacitors have the same value of $1{,}000\,\mu F$:

the total capacitance (C_T) for the series circuit
$$= \frac{1}{C_1} + \frac{1}{C_2} + \frac{1}{C_3} = \frac{3}{1{,}000 \times 10^{-6}} = 333\,\mu F$$

the total capacitance (C_T) for the parallel circuit
$$= C_1 + C_2 + C_3 = 3{,}000\,\mu F$$

The energy values depend on the applied voltages, so for a 24 V d.c. supply, the energy stored in each network:

for the series circuit:

$$\text{energy } (W) = 0.5\,CV^2 = 0.5 \times 333 \times 10^{-6} \times 24^2$$
$$= 0.5 \times 333 \times 10^{-6} \times 576 = 0.1\,J$$

and for the parallel circuit:

$$\text{energy } (W) = 0.5\,CV^2 = 0.5 \times 3{,}000 \times 10^{-6} \times 24^2 = 0.86\,J$$

Figure 6.34 Single (or total) value of capacitance (C_T)

Make the Grade
M1

Grading criterion M1 requires the use of Kirchhoff's laws to determine (i) the current in various parts of a network having four nodes and (ii) the power dissipated in a load resistor containing two voltage sources. For part (i) of the criterion, it is essential that you know the difference between Kirchhoff's two laws. Part (ii) of the criterion requires that you can apply the power equations.

Worked Example

The circuit shown in Figure 6.35 has been assigned four nodes (where more than two connections are made). Determine the total current and current through each resistor.

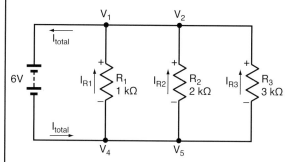

Figure 6.35 Assigning node voltages

The three resistors are in parallel, so the total resistance can be calculated as follows:

$$\frac{1}{R_t} = \frac{1}{R_1} + \frac{1}{R_2} + \frac{1}{R_3}$$

$$\frac{1}{R_t} = \frac{1}{1000} + \frac{1}{3000} + \frac{1}{2,000}$$

$$\frac{1}{R_t} = \frac{1}{1.8 \times 10^{-3}} = 545\,\Omega$$

The total current can now be calculated: $I = \frac{V}{R} = \frac{6}{545} = 11\,\text{mA}$

The current through each resistor is calculated from:

$$I_{R1} = \frac{V}{R_1} = \frac{6}{1000} = 6\,\text{mA}$$

$$I_{R2} = \frac{V}{R_2} = \frac{6}{3000} = 2\,\text{mA}$$

$$I_{R3} = \frac{V}{R_3} = \frac{6}{2000} = 3\,\text{mA}$$

To confirm these calculations, the sum of all three currents through the resistors equals the total current:

$$6\,\text{mA} + 2\,\text{mA} + 3\,\text{mA} = 11\,\text{mA}$$

Worked Example

From the superposition theorem (illustrated in Figures 6.25, 6.26 and 6.27 on page 280) we derived a current of 0.464 A through the load resistor of 10 Ω (represented by R_2). The power dissipated by this resistor is calculated from:

$$P = I^2R = 0.464^2 \times 10 = 2.15\,\text{W}$$

Make the Grade   M2

Grading criterion M2 requires that you evaluate capacitance, charge, voltage and energy in a network containing a series-parallel combination of three capacitors. In the first instance, you will need to be able to reduce a series-parallel network of capacitors into one value. You can then apply your knowledge of charge, voltage and energy to calculate the required values in a given circuit.

Worked Example

Referring to Figure 6.36, three capacitors are connected to a power supply such as a battery. Capacitor C_1 is in series with C_2 and C_3; the latter are in parallel.

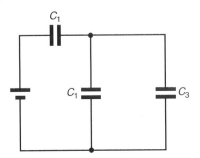

Figure 6.36 Series-parallel combination of three capacitors

Assuming that all three capacitors have the same value of $1000\,\mu F$, the first task is to calculate the total capacitance of the circuit. This has to be broken down in two stages:

The total capacitance (C_t) of C_2 and C_3 is calculated:

$$C_t = C_2 + C_3$$

$$C_t = 10^{-3} + 10^{-3}$$

$$C_t = 2 \times 10^{-3} = 2000\,\mu F$$

So the total value of the circuit's capacitance (C_T) is now calculated by C_t and C_1 in series:

$$\frac{1}{C_T} = \frac{1}{C_t} + \frac{1}{C_1}$$

$$\frac{1}{C_T} = \frac{1}{(2 \times 10^{-3})} + \frac{1}{(1 \times 10^{-3})}$$

$$\frac{1}{C_T} = 1.5 \times 10^3$$

$$C_T = \frac{1}{(1.5 \times 10^3)}$$

$$C_T = 666\,\mu F$$

To calculate the charge: assuming a 1.5 V battery, the charge can be calculated from:

$$Q = C \times V = 666 \times 10^{-6} \times 1.5 = 1\,mC$$

so the energy stored in the capacitor network is:

$$\text{energy } (W) = 0.5\,QV = 0.5 \times 10^{-3} \times 1.5$$
$$= 0.75\,mJ$$

Table 6.5 gives a summary of quantities, units, symbols and abbreviations used in this section. Note that the abbreviation is used after writing the value of the unit – for example, the power output of 5 watts can be written as 5 W.

Quantity	Unit	Symbol	Abbreviation
Power	watt	P	W
Energy	joule	W	J

Table 6.5

6.3 Electromagnetism

This section will cover the following grading criteria:

By the end of this section you should be able to describe the characteristics of a magnetic field; describe the relationship between flux density (B) and field strength (H); and describe the principles and applications of electromagnetic induction.

Key Term

Electromagnetism – The physical relationship and effect between electrical energy and the phenomenon of magnetism.

When current flows through a conductor, it generates a magnetic field. This magnetic field has the same effect that attracts certain materials to permanent magnets. When a conductor is exposed to a magnetic field, a current flow is created. The effects of electromagnetism are very important subjects within electrical engineering.

Magnetic fields

Key Terms

Magnetism – A force that causes objects (usually metals such as iron) to be attracted or repelled to one another.

Poles – Two regions of the magnet where the magnetic strength is concentrated.

When a magnet is suspended and allowed to rotate (see Figure 6.37), the magnet lines up with the Earth's magnetic field – that is, in a north–south direction. The poles are therefore called north-seeking and south-seeking poles, north (N) and south (S) for short.

If you place the magnet on a table, cover it with a sheet of paper or clear plastic and sprinkle the sheet with iron filings, when the sheet is lightly tapped, the filings take up a curved pattern between the poles (see Figure 6.38 on page 288). The pattern formed by the filings indicates the presence of the magnetic force. (The filings were moved by this force to form this pattern.) It can be seen that these patterns are in the form of lines that are concentrated at the poles. The area (or space) in which these lines are formed is called the magnetic field; the individual lines are referred to as lines of flux – that is, the total amount of magnetism.

(a) Earth's magnetic field

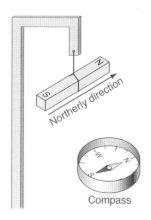

(b) North- and South-seeking poles

Figure 6.37 North- and south-seeking poles

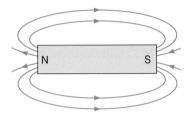

Figure 6.38 Magnetic field and lines of flux

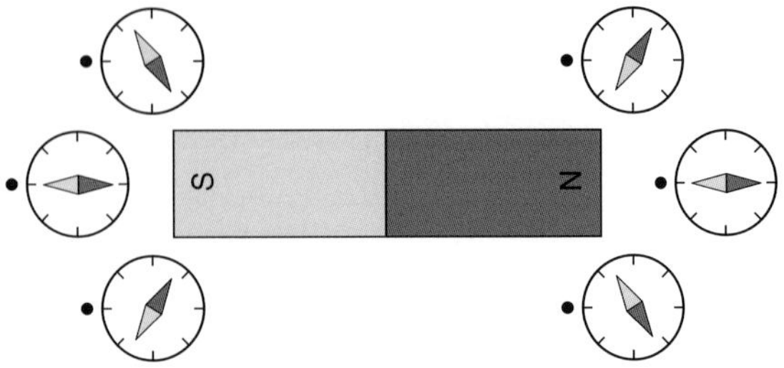

Figure 6.39 Field and flux directions for a magnet

Direction of magnetic field

The flux lines from the permanent magnet exit at a north pole and enter at a south pole, as indicated by the direction arrows (see Figure 6.39).

This effect can be confirmed by placing a small compass within different areas of the field and noting the position that the needle takes. While the needle aligns with the same pattern established by the iron filings, the needle (which is itself a weak magnet) points towards the poles of the magnet. One end of the compass's needle will always point to the north pole of our magnet. It can be imagined that the lines of flux also pass internally through the magnet, in the same direction that continues from the external field.

Behaviour of magnetic flux

Although lines of magnetic flux do not have any physical substance, they provide a means of illustrating a number of observable facts:

1. They form lines that create a pattern which takes up a direction that aligns with the poles of the magnet.
2. Each line of flux is continuous: externally through the field and internally through the magnet.
3. Lines of magnetic flux never cross each other.
4. Like poles of a magnet repel each other; opposite poles attract (see Figure 6.40).

Magnetic fields and electrical current

Whenever an electric current flows in a conductor, a magnetic field is created around the conductor in the form of concentric circles, as shown in Figure 6.41. The field is present along the whole length of the conductor and is strongest nearest to the conductor. For a straight conductor, the lines of flux form concentric circles that radiate from the conductor. As with permanent magnets, this field also has direction. The direction of the magnetic field is dependent on the direction of the current passing through the conductor.

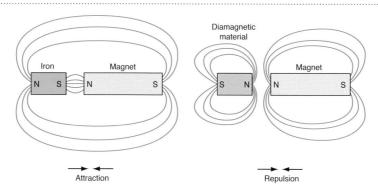

Figure **6.40** Behaviour of magnetic flux

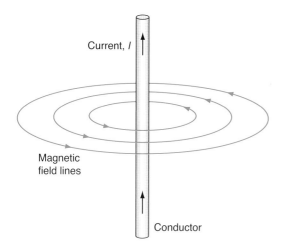

Figure 6.41 Magnetic field around a current-carrying conductor

The direction of the current passing through the conductor creates the field that has a specific direction. This can be represented by a corkscrew being driven into a piece of material (see Figure 6.42 on page 290).

Key Points

In electromagnetism, the right-hand screw rule depicts the direction of the magnetic field, created by the current in the conductor; this is clockwise, as shown in Figure 6.42 (see page 290).

If we place a current-carrying conductor in a magnetic field, the conductor has a force exerted on it. Consider the arrangement shown in Figure 6.43 (see page 290), in which a current-carrying conductor is placed between two magnetic poles.

The net effect of combining these two magnetic force fields is that at position A, they both travel in the same direction and reinforce one another, while at position B, they travel in the opposite direction and tend to cancel one another. With a stronger force field at position A and a weaker force at position B, the conductor is forced upwards out of the magnetic field. If the direction of the current was reversed – that is, if it was to travel 'out of the page' – then the direction of the magnetic field in the current-carrying conductor would be reversed and, therefore, so would the direction of motion of the conductor.

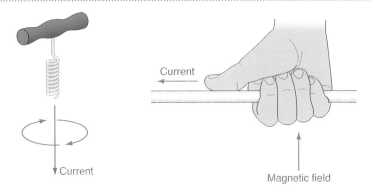

(a) Right-hand screw rule   (b) Right-hand grip rule

Figure 6.42 Right-hand screw rule and right-hand grip rule

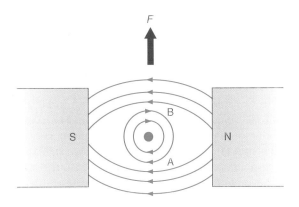

Figure 6.43 Current-carrying conductor in a magnetic field

Key Term

Magnetic field of flux – The region in which the forces created by the magnet have influence. This field surrounds a magnet in all directions and is concentrated at the north and south poles of the magnet.

Make the Grade   P7

Grading criterion P7 requires that you describe the characteristics of a magnetic field. Referring to Figure 6.44, you could prepare for this by using an experimental approach – that is, using a magnet, a sheet of paper, iron filings and a compass. Compare your actual results with the theoretical predictions.

The field radiates out around the conductor in concentric circles, with the greatest density of magnetic flux nearest to the conductor. The magnitude of the force acting on the conductor depends on the:

- current flowing in the conductor
- length of the conductor in the field
- strength of the magnetic flux.

The magnetic flux is expressed in terms of its flux density; the size of the force will be given by the expression:

$$F = BIl$$

where F is the force in newton (N), B is the flux density in tesla (T), I is the current (A) and λ is the length (m).

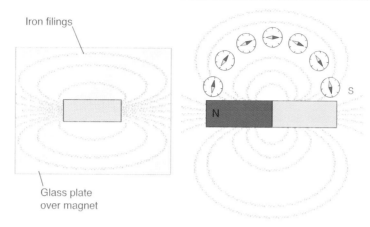

Figure 6.44 Characteristics of a magnetic field: experimental approach

Key Terms

The unit of force, newton – Named after Sir Isaac Newton (1643–1727), an English physicist, mathematician and astronomer.

The unit of flux density, tesla – Named after Nikola Tesla (1856–1943), an Austrian mechanical and electrical engineer often credited with developing the fundamentals of electrical engineering for commercial purposes.

Flux density

The total flux present in a magnetic field is a measure of the total magnetic intensity present in the field; it is measured in webers (Wb) and represented by the Greek symbol, ϕ. The flux density, B, is simply the total flux, ϕ, divided by the area over which the flux acts, A. Hence:

$$B = \frac{\phi}{A}$$

where B is the flux density (T), ϕ is the total flux present (Wb) and A is the area (m²).

Key Term

The unit of magnetic flux intensity, weber – Named after Wilhelm Eduard Weber (1804–91), a German physicist and, together with Carl Friedrich Gauss, inventor of the first electromagnetic telegraph communication system.

Worked Example

A flux density of 0.25 T is developed in free space over an area of 20 cm². Determine the total flux.

Rearranging the formula to make ϕ the subject gives:

$$\phi = B A = 0.25 \times 0.002 = 0.0005 \text{ or } 0.5 \text{ mWb.}$$

Activity 6.5

Complete the missing words in the following sentence.

If we place a current-carrying conductor in a magnetic field, the conductor has a _____ exerted on it. If the conductor is free to move, this force will produce _____.

Key Points

Flux density is determined by dividing the total flux present by the area over which the flux acts.

Magnetic field strength

The strength, or intensity, of the magnetic field is directly proportional to the current and inversely proportional to the distance from the conductor. Magnetic field strength is equivalent to electric field strength (see page 267).

Electromagnetic induction

Electromagnetic induction is the phenomenon of creating current flow in a conductor by magnetism. This current flow can be created in one of two ways:

- when the conductor is stationary in a changing magnetic field
- when the conductor is moving through a stationary magnetic field.

The manner in which electrons flow in a conductor is the opposite to that which produces the force. In order to create current flow, we require movement in to get electricity out. The components required to create (or generate) electricity are:

- closed conductor
- magnetic field
- relative movement.

Whenever relative motion occurs between a magnetic field and a conductor acting at right angles to the field, an emf is induced or generated in the conductor. The manner in which this emf is generated is the principle of electromagnetic induction. In order to increase the strength of the field, a conductor may be shaped into a loop (Figure 6.45) or coiled to form a coil (Figure 6.46).

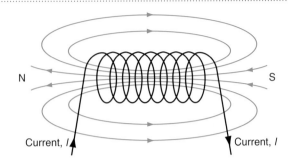

Figure 6.46 Magnetic field around a coil

Magnetic circuits

The flux density within a coil can be increased by using a material such as iron in the centre of the coil to form a core (see Figure 6.47). Not only will the flux density increase, the core will also become magnetised.

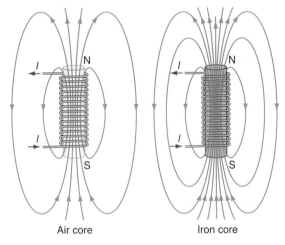

Air core   Iron core

Inserting an iron core may give a magnetic field several hundred times that of the equivalent air core solenoid

Figure 6.47 Concentrated electromagnetic field using a core

The concentration of flux is formed by a magnetic circuit (see Figure 6.48).

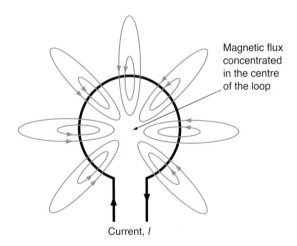

Magnetic flux concentrated in the centre of the loop

Current, *I*

Figure 6.45 Magnetic field around a single-turn loop

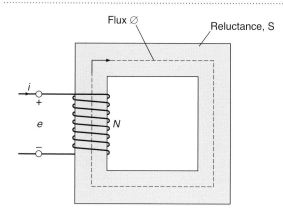

Figure 6.48 Magnetic circuit

In order to produce a magnetic field, we need to pass a current through the coil; this creates a magnetomotive force (mmf), causing flux to circulate within the core. This magnetic circuit is defined by its own terminology, which can be compared to an electrical circuit:

Magnetic circuit	Electrical circuit
Magnetomotive force (mmf)	Electromotive force (emf)
Reluctance (S)	Resistance (R)
Flux (ϕ)	Current (I)
Permeability (μ)	Resistivity (ρ)

Table 6.6

The amount of mmf distributed over the length of the electromagnet determines the field intensity. As with the electrical circuit, the type of material used for the core will have an effect on the magnetic circuit. There is a mathematical relationship between flux density and field intensity for any particular material, which can be depicted in a graph, as shown in Figure 6.49. This is called the normal magnetisation curve, or B–H curve, for any particular material. As the field intensity increases, the core material becomes magnetised. It can be seen that the flux density levels off with increased field intensity, an effect known as saturation. At this point, the core cannot be further magnetised, despite the increasing field intensity.

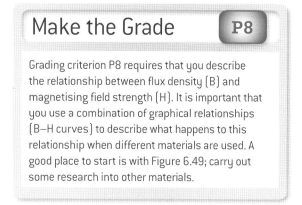

Make the Grade   P8

Grading criterion P8 requires that you describe the relationship between flux density (B) and magnetising field strength (H). It is important that you use a combination of graphical relationships (B–H curves) to describe what happens to this relationship when different materials are used. A good place to start is with Figure 6.49; carry out some research into other materials.

Faraday and Lenz

Consider Figure 6.50 (see page 294), which shows relative movement between a magnet and a closed coil of wire. An emf will be induced in the coil whenever there is relative movement of a magnetic field and conductor – for example, when the magnet is moved in or out of the coil (or the magnet is held stationary and the coil moved).

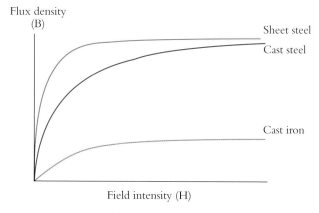

Figure 6.49 Normal magnetisation curve, or B–H curve

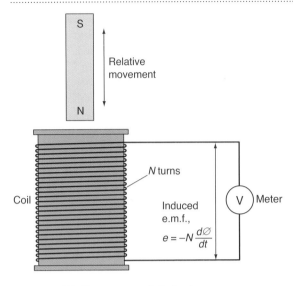

Figure 6.50 Electromagnetic induction

At any given time, the instantaneous magnitude of the induced emf (e), depends on the number of turns (N) and the rate at which the flux changes in the coil.

Key Term

Faraday's law – The magnitude of the induced emf is dependent on the relative velocity with which the conductor cuts the lines of magnetic flux.

This can be expressed for the instantaneous emf (e), given by the relationship:

$$e = -N \frac{d\phi}{dt}$$

where N is the number of turns and $\frac{d\phi}{dt}$ is the rate of change of flux. The minus sign indicates that the polarity of the generated emf opposes the change.

The number of turns (N) is directly related to the length (l) of the conductor, moving through a magnetic field. The velocity with which the conductor moves through the field determines the rate at which the flux changes in the coil as it cuts the flux field. Thus, the magnitude of the induced (generated) emf is proportional to the flux density, length of conductor and relative velocity (v) between the field and the conductor. Expressed mathematically:

$$e \propto \frac{B}{v}$$

(where α is the mathematical sign for proportionality).

In order to generate an emf, the conductor must cut the lines of magnetic flux. If the conductor cuts the lines of flux at right angles (Figure 6.51(a)), then the maximum emf is generated; cutting them at any other angle θ (Figure 6.51(b)) reduces this value until $\theta = 0°$, at which point, the lines of flux are not being cut at all and no emf is induced or generated in the conductor. So the magnitude of the induced emf is dependent on $\sin\theta$; therefore:

$$e = B \, l \, v \, \sin\theta$$

When a magnetic flux through a coil is made to vary, an emf is induced. The magnitude of this emf is proportional to the rate of change of magnetic flux. In effect, the relative movement between the

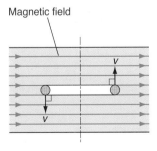

(a) Cutting lines of flux at 90°

$E = B \, l \, v$

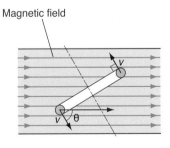

(b) Cutting lines of flux at angle θ

$E = B \, l \, v \, \sin\theta$

Figure 6.51 Cutting lines of flux and the generated emf

magnetic flux and the conductor is essential to generate an emf. The voltmeter shown in Figure 6.50 indicates the induced emf; if the direction of motion changes, the polarity of the induced emf in the conductor changes.

Heinrich Lenz (1804–65) was an Estonian physicist who made significant contributions to electrodynamics. His work included the development of a simple rule known as Lenz's law.

Key Term

Lenz's law – The current induced in a conductor opposes the changing field that produces it. (This is the reason for the negative sign in the expression $e = -N \dfrac{d\phi}{dt}$)

Key Points

- When a magnetic flux through a coil is made to vary, an emf is induced.
- The induced emf opposes any change of current; this is referred to as back emf.
- An emf will be induced in the coil whenever there is relative movement of a magnetic field and conductor.
- The magnitude of this emf is proportional to the rate of change of magnetic flux.

Self- and mutual inductance

We have already seen how an induced emf is produced by a flux change in an inductor. The back emf is proportional to the rate of change of current (from Lenz's law), as illustrated in Figure 6.52.

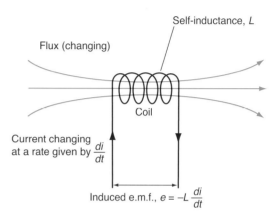

NB: Induced e.m.f. opposes current change

Figure 6.52 Self-inductance

This effect is called self-inductance (or just inductance), which has the symbol L. Self-inductance is measured in henries (H) and is calculated from:

$$e = -L \frac{di}{dt}$$

where L is the self-inductance, $\frac{di}{dt}$ is the rate of change of current and the minus sign indicates that the polarity of the generated emf opposes the change.

Worked Example

A closed conductor of length 15 cm cuts the magnetic flux field of 1.25 T with a velocity of 25 m/s. Determine the induced emf when the:

- **angle between the conductor and field lines is 60°**
- **angle between the conductor and field lines is 90°.**

The induced emf is found using $e = B l V \sin\theta$; hence:

$e = 1.25 \times 0.15 \times 25 \times \sin 60° = 4.688 \times 0.866$
$$= 4.06\,\text{V}$$

The maximum induced emf occurs when the lines of flux are cut at 90°. In this case,

$e = Blv$ (recall that $\sin 90° = 1$),
hence $e = 1.25 \times 0.15 \times 25 = 4.69\,\text{V}$

Key Term

The henry (H) – Unit of inductance named after Joseph Henry (1797–1878), an American scientist who discovered the phenomenon of electromagnetic self-inductance.

A coil is said to have an inductance of 1 H if a voltage of 1 V is induced across it when a current changing at the rate of 1 A/s is flowing in it.

Key Term

Mutual inductance – The effect when two inductors are placed close to one another, and the flux generated from a changing current flow in the first inductor will cut through the other inductor (see Figure 6.53). This changing flux will, in turn, induce a current in the second inductor. This effect occurs whenever two inductors are inductively coupled.

Worked Example

A coil has a self-inductance of 15 mH and is subject to a current that changes at a rate of 450 A/s. What emf is produced?

The calculation is as follows:

$$e = -L\frac{di}{dt} = -15 \times 10^{-3} \times 450 = -6.75\,\text{V}$$

Note the minus sign, indicating that a back emf of 6.75 V is induced.

Inductors

Inductors provide a means of storing electrical energy in the form of a magnetic field. The electrical characteristics of an inductor are determined by a number of factors, including the material of the core (if any), the number of turns and the physical dimensions of the coil.

In practice, every coil comprises both inductance and resistance, and the circuit of Figure 6.54 shows these as two discrete components. In reality, the inductance, L, and resistance, R, are both distributed throughout the component, but it is convenient to treat the inductance and resistance as separate components in the analysis of the circuit. Typical inductors are shown in Figure 6.55.

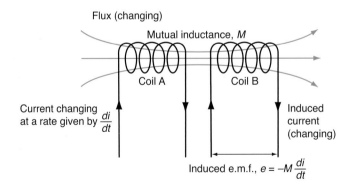

Figure 6.53 Mutual inductance

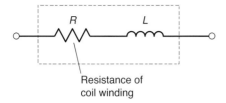

Resistance of
coil winding

Figure 6.54 Practical inductor with resistance and inductance

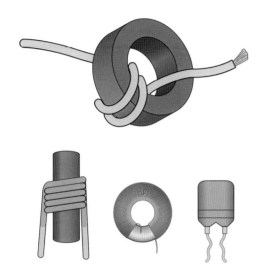

Figure 6.55 A selection of inductors

Activity 6.6

1. A 1.5 m length of wire moves perpendicular to a magnetic flux field of 0.75 T. Determine the emf that will be induced across the ends of the wire if it moves at 10 m/s.
2. An emf of 30 V is developed across the terminals of an inductor when the current flowing in it changes from zero to 10 A in half a second. What is the value of inductance?

Table 6.7 gives a summary of quantities, units, symbols and abbreviations used in this section. Note that the abbreviation is used after writing the value of the unit – for example, an inductance of five henries can be written as 5 H.

Quantity	Unit	Symbol	Abbreviation
Inductance	henry	L	H
Force	newton	F	N
Flux density	tesla	B	T
Flux	weber	Φ	Wb
Magnetic force	amperes per metre	H	A/m
Reluctance	amperes per weber	S	A/Wb

Table 6.7

6.4 Alternating current (a.c.)

This section will cover the following grading criteria:

By the end of this section you should be able to use single-phase a.c. circuit theory to determine the characteristics of a sinusoidal a.c. waveform; and compare the results of adding and subtracting two sinusoidal a.c. waveforms graphically and by phasor diagram.

Alternating currents are bidirectional and continuously reverse their direction of flow. Referring to Figure 6.56, d.c. flows in one direction only; a.c. flows in two directions.

Waveforms

The variation of voltage (or current) in a circuit follows a profile referred to as a waveform. There are many common types of waveform encountered in electrical circuits, as shown in Figure 6.57, including:

- sinusoidal
- square
- triangle
- ramp/saw-tooth (positive and negative)
- pulse.

Complex waveforms (Figure 6.58) comprise many components at different frequencies; these are typically produced from electronic devices – for example, microphones.

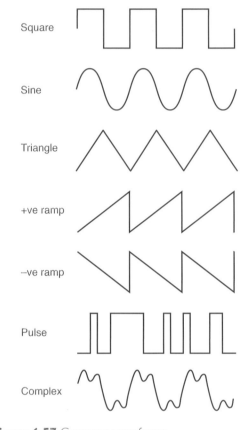

Figure 6.57 Common waveforms

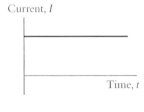

(a) Direct current

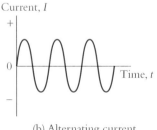

(b) Alternating current

Figure 6.56 Comparison of (a) d.c. and (b) a.c.

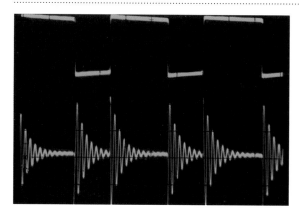

Figure 6.58 Complex waveforms

a.c. waveform terminology

Frequency

The unit is named after Heinrich Rudolf Hertz (1857–94), a German physicist associated with electromagnetic theory, who was the first to categorically demonstrate the existence of electromagnetic waves. A frequency of 1 Hz is equivalent to one cycle per second; five cycles per second is 5 Hz; ten cycles per second is 10 Hz etc. (Figure 6.59).

The relationship between periodic time and frequency is thus:

$$t = \frac{1}{f}$$

where t is the periodic time (in seconds) and f is the frequency (in Hz).

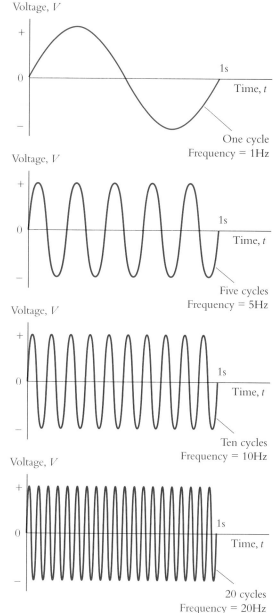

Figure 6.59 Waveforms with different frequencies

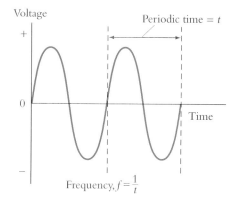

Figure 6.60 Periodic time

299

An aircraft generator operates at a frequency of 400 Hz. What is the periodic time of the voltage generated?

The voltage will have a periodic time of $t = \frac{1}{f}$ $= \frac{1}{400} = 0.0025$ ms

A bench power supply has a periodic time of 20 ms. What is the frequency of the supply?

The supply has a frequency of $f = \frac{1}{t} = \frac{1}{0.02} =$ 50 Hz.

Average values

The average value of an alternating current which oscillates symmetrically above and below zero will be zero when measured over time. Hence, average values of currents and voltages are invariably taken over one complete half-cycle (either positive or negative) rather than over one complete full cycle (which would result in an average value of zero).

Peak values

The peak-to-peak value for a wave that is symmetrical about its resting value is twice its peak value.

Root mean square value

The root mean square (r.m.s.) of an alternating voltage or current is the value that would produce the same heat energy from a resistor (or light energy in a lamp) as a d.c. of the same magnitude. This can be demonstrated with a simple circuit experiment (see Figure 6.61). Since the r.m.s. value of a waveform is very much dependent upon its shape, values are only meaningful when dealing with a waveform of known profile. Where the profile of a waveform is not specified, r.m.s. values are normally based on sinusoidal conditions. (r.m.s. is also referred to as the effective value.)

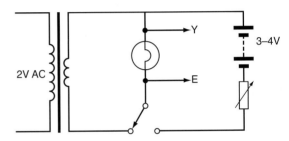

Figure 6.61 Experiment to demonstrate an r.m.s. value

For a given waveform (Figure 6.62), a set of fixed relationships exist between average, peak, peak–peak and r.m.s. values. The required multiplying factors for sinusoidal voltages and currents are summarised in Table 6.8.

Wanted quantity:					
		average	peak	peak–peak	r.m.s.
Given quantity:	average	1.0	1.57	3.14	1.11
	peak	0.636	1.0	2.0	0.707
	peak–peak	0.318	0.5	1.0	0.353
	r.m.s.	0.9	1.414	2.828	1.0

Table 6.8

From the table, we can conclude that:

$$V_{av} = 0.636 \times V_{pk}$$
$$V_{pk-pk} = 2 \times V_{pk}$$
$$V_{r.m.s} = 0.707 \times V_{pk}$$

Similar relationships apply to the corresponding a.c.'s, thus:

$$I_{av} = 0.636 \times I_{pk}$$
$$I_{pk-pk} = 2 \times I_{peak}$$
$$I_{r.m.s} = 0.707 \times I_{pk}$$

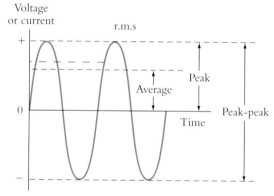

Figure 6.62 r.m.s average, peak and peak–peak values of a sine wave

Worked Example

A sinusoidal current of 40 A peak–peak flows in a circuit. What is the r.m.s. value of the current?

$$I_{r.m.s.} = 0.353 \times I_{pk-pk} = 0.353 \times 40 = 14.12\,A$$

Worked Example

A generator produces an r.m.s. sine wave output of 110 VAC. What is the peak value of the voltage?

From the stated fixed relationships,
$$V_{pk} = 1.414 \times V_{r.m.s.} = 1.414 \times 110 = 155.5\,VAC$$

Hence the voltage has a peak value of 311 V.

Make the Grade   P10

Grading criterion P10 requires that you use single-phase a.c. circuit theory to determine the characteristics of a sinusoidal a.c. waveform. A good starting point would be to produce your own sketch of a sine wave on graph paper and plotting amplitudes at 30° intervals over one complete cycle.

Reactance

Reactance is the opposition to current caused by capacitance or inductance.

$$I = \frac{V}{X}$$

where X is the reactance, V is the alternating potential difference and I is the alternating current.

In the case of capacitive reactance, we use the suffix, C, so that the reactance equation becomes:

$$X_C = \frac{V_C}{I_C}$$

Similarly, in the case of inductive reactance, we use the suffix, L, so that the reactance equation becomes:

$$X_L = \frac{V_L}{I_L}$$

The voltage and current in a circuit containing pure reactance (either capacitive or inductive) will be out of phase by 90°. In the case of a circuit containing pure capacitance, the current will lead the voltage by 90° (alternatively, we can say that the voltage lags the current by 90°). This relationship is illustrated by the waveforms shown in Figure 6.63.

In the case of a circuit containing pure inductance, the voltage will lead the current by 90° (alternatively, we can say that the current lags the voltage by 90°). This relationship is illustrated by the waveforms shown in Figure 6.64.

A useful way of remembering leading and lagging phase relationships is to recall the word *CIVIL*, as shown in Figure 6.65. In the case of a circuit containing pure capacitance (C), the

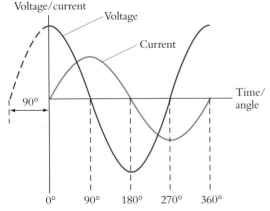

Figure 6.63 Voltage and current in a purely capacitive circuit

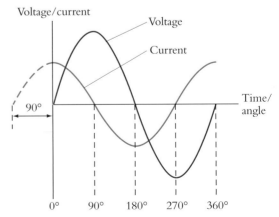

Figure 6.64 Voltage and current waveforms for a purely inductive circuit

302

current (I) will lead the voltage (V) by 90°, while in the case of a circuit containing pure inductance (L), the voltage (V) will lead the current (I) by 90°.

The reactance of an inductor (inductive reactance)

In a capacitor (C), current (I) leads voltage (V)

In a conductor (C), current (I) lags voltage (V)

Figure 6.65 Using *CIVIL* to determine phase relationships in circuits containing capacitance and inductance

is directly proportional to the frequency of the applied a.c. and can be determined from the following formula:

$$X_L = 2\pi f L$$

where X_L is the reactance in ohms, f is the frequency in hertz and L is the inductance in henries.

Since inductive reactance is directly proportional to frequency, the graph of inductive reactance plotted against frequency takes the form of a straight line (see Figure 6.66).

Worked Example

Determine the reactance of a 0.1 H inductor at (a) 100 Hz and (b) 10 kHz.

- at 100 Hz, $X_L = 2\pi f L = 2 \times \pi \times 100 \times 0.1$
  $= 62.8\,\Omega$
- at 10 kHz, $X_L = 2\pi f L = 2 \times \pi \times 10{,}000 \times 0.1 = 6.28\,k\Omega$

The reactance of a capacitor (capacitive reactance) is inversely proportional to the frequency of the applied a.c. and can be determined from the following formula:

$$X_C = \frac{1}{2\pi f C}$$

where X_C is the reactance in ohms, f is the frequency in hertz and C is the capacitance in farads.

Since capacitive reactance is inversely proportional to frequency, the graph of inductive reactance plotted against frequency takes the form of a rectangular hyperbola (see Figure 6.67).

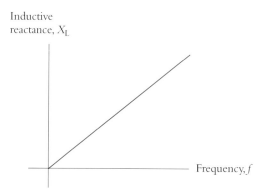

Figure 6.66 Variation of inductive reactance with frequency

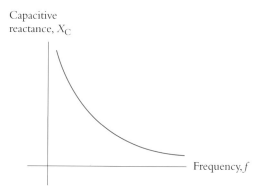

Figure 6.67 Variation of capacitive reactance with frequency

Worked Example

Determine the reactance of a 1 μF capacitor at (a) 100 Hz and (b) 10 kHz.

At 100 Hz, $X_C = \dfrac{1}{2\pi f C} = \dfrac{1}{2\pi \times 100 \times 10^{-6}} = \dfrac{0.159}{10^{-4}} = 1.59\,\text{k}\Omega$

At 10 kHz, $X_C = \dfrac{1}{2\pi f C} = \dfrac{1}{2\pi \times 10\,000 \times 10^{-6}} = \dfrac{0.159}{10^{-2}} = 15.9\,\Omega$

To summarise: when alternating voltages are applied to capacitors or inductors, the magnitude of the current flowing will depend on both the value of capacitance or inductance and on the frequency of the voltage. In effect, capacitors, inductors and resistors oppose the flow of current, albeit in a different way, the important difference being that the reactance of the component varies with frequency (unlike the case of a conventional resistor, where the magnitude of the current does not change with frequency).

Phasor diagrams

One technique used to describe and analyse the phase relationship between voltage and current in a.c. circuits is through the use of a rotating vector called a phasor. In most branches of engineering, vectors are used to represent physical quantities that have both magnitude and direction, such as the magnitude and direction force in mechanical engineering. The term 'phasor' is used for the graphic representation of the magnitude and phase of a sinusoidal current or voltage.

Phasors can be used to represent sinusoidal a.c. waveforms. An a.c. waveform can be thought of as being generated by a rotating phasor. Referring to Figure 6.68, any specific point on the sine wave can be projected to corresponding points on the circle to the left. (By convention, phasors rotate anticlockwise.) Manipulation of these vectors allows both the amplitudes and the phase angles of current currents and voltage to be determined.

The phasor's length (the radius of the circle) represents the quantity's peak value; its angle (θ) represents the quantity's phase angle. To illustrate this, consider the point on the sine wave when the instantaneous value is zero and the phase angle is zero degrees; the phasor indicates an instantaneous value of:

$$A\sin\theta_0$$

The sine of $0°$ has a value of zero; therefore the phasor's value at this point is zero.

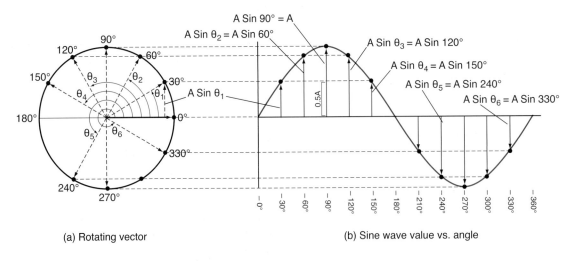

(a) Rotating vector

(b) Sine wave value vs. angle

Figure 6.68 Phasor representation for a sine wave

A phase angle of 30° (θ_1) corresponds to 30° of phasor rotation on the circular representation; this indicates an instantaneous value of:

$$A\sin\theta_1$$

The sine of 30° has a value of 0.5, so the phasor's value at this point is 0.5 A, as can be seen on the sine wave. This pattern continues for any phase angle from zero through to 360°.

> **Key Points**
>
> - Vectors are used to represent physical quantities that have both magnitude and direction.
> - Phasors are used for the graphic representation of the magnitude and phase of a sinusoidal a.c. or voltage.
> - The phasor's length represents the quantity's *peak* value (not the instantaneous value) at any given time.

Analysing sine waves

Some a.c. circuits can have more than one voltage source; these can be connected in series, as shown in Figure 6.69.

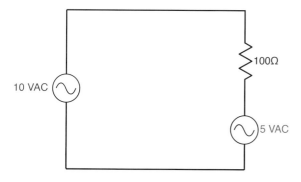

Figure 6.69 Multiple sources of alternating voltage

If the two voltages are in phase, they can be added together by summing the amplitudes of each wave at each point in time; Figure 6.70 demonstrates this principle. In this example, the red and blue sine waves represent the two sources of alternating voltage.

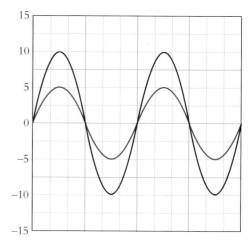

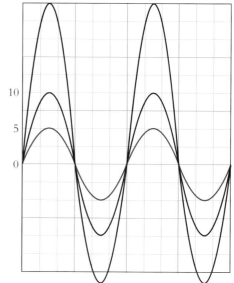

Figure 6.70 Addition of sine waves

The resulting sine wave (the voltage developed across the resistor) can be plotted as illustrated in Figure 6.70.

The addition of sine waves can also be achieved by using phasors; consider the two sources of alternating voltage shown in Figure 6.71 (see page 306).

The two phasors are simply 'joined' to form one composite phasor, giving a total peak output of 7 VAC.

In these examples, we stated that the two voltages were in phase. If they were out of phase by 180°, the two voltages would be subtracted using the same principles.

305

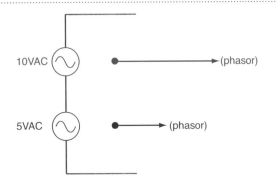

Figure 6.71 Phasor representation of voltage sources

Make the Grade   M3

Grading criterion M3 requires that you compare the results of adding and subtracting two sinusoidal a.c. waveforms graphically and by phasor diagram. First of all, sketch your own a.c. circuit, with two sources of alternating voltage and a resistor (refer to Figures 6.68–6.71 if you need reminding of any details). There are two scenarios to address: the voltages being (i) in phase and (ii) out of phase by 180°.

Scenario (i) requires that the two sine waves are added; scenario (ii) requires that they are subtracted. Plot the two scenarios graphically and by phasor diagrams.

Key Points

Phasors give us a convenient method of adding sinusoidal waves of the same frequency. When two sinusoidal a.c. waveforms of the same frequency are added or subtracted, the resulting waveform is another sinusoid of the same frequency.

Activity 6.8

Referring to Figure 6.72, explain what each of the following phasor diagrams represents.

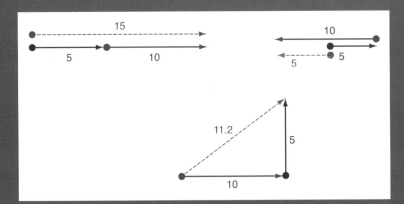

Figure 6.72 Representative phasor diagrams

Impedance

Circuits that combine both resistance and reactance (either capacitive, inductive or both) are said to exhibit impedance.

Key Term

Impedance, like resistance and reactance – The ratio of applied voltage to the current flowing.

Thus

$$Z = \frac{V}{I}$$

where Z is the impedance in ohms, V is the (alternating) potential difference in volts and I is the alternating current in amps.

Because the voltage and current in a pure reactance are at 90° to one another (we say that they are in quadrature), we cannot simply add up the resistance and reactance present in a circuit in order to find its impedance. Instead, we can use the impedance triangle shown in Figure 6.73. The impedance triangle takes into account the 90° phase angle, and from it we can infer that the impedance of a series circuit (R in series with X) is given by:

$$Z = \sqrt{R^2 + x^2}$$

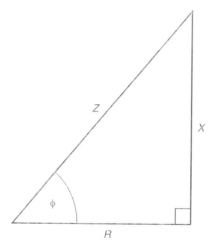

Figure 6.73 The impedance triangle

Worked Example

A resistor of 30 Ω is connected in series with a capacitive reactance of 40 Ω. Determine the impedance of the circuit and the current flowing when the circuit is connected to a 115 V a.c. supply.

First, we must find the impedance of the C R series circuit:

$$Z = \sqrt{(R^2 + X^2)} = \sqrt{(30^2 + 40^2)} = \sqrt{2500} = 50\,\Omega$$

The current taken from the supply can now be found:

$$I = \frac{V}{Z} = \frac{115}{50} = 2.3\,\text{A}$$

Worked Example

A coil is connected to a 50 V a.c. supply at 400 Hz. If the current through the coil is 200 mA and the coil has a resistance of 60 Ω, determine the value of inductance.

Like most practical forms of inductor, the coil in this example has both resistance and reactance (see Figure 6.54 on page 297). We can find the impedance of the coil from:

$$Z = \frac{V}{I} = \frac{50}{0.2} = 250\,\Omega$$

Since $Z = \sqrt{(R^2 + X^2)}$, the formula can be rearranged as $Z^2 = R^2 + X^2$

from which: $X^2 = 250^2 - 60^2 = 62\,500 - 3600 = 58\,900\,\Omega$

thus

$$X_{\text{L}} = \sqrt{58\,900} = 243\,\Omega$$

since $X_{\text{L}} = 2\pi f L$

hence

$$L = \frac{X_{\text{L}}}{2}\pi f = \frac{243}{6.28} \times 400 = 97\,\text{mH}$$

307

Resistance and reactance combine to form impedance in a.c. circuits. In other words, impedance is the result of combining resistance and reactance in the impedance triangle. Because of the quadrature relationship between voltage and current in a pure capacitor or inductor, the angle between resistance and reactance in the impedance triangle is always 90°.

Resonance

A special case occurs when $X_C = X_L$, in which case the two equal but opposite reactances effectively cancel each other out. The result of this is that the circuit behaves as if only resistance, R, is present – in other words, the impedance of the circuit, $Z = R$. In this condition, the circuit is said to be resonant. The frequency at which resonance occurs is when:

$$X_C = X_L$$

Thus

$$\frac{1}{2\pi f C} = 2\pi f L$$

from which

$$f^2 = \frac{1}{4\pi^2 LC}$$

and thus

$$f = \frac{1}{2\pi\sqrt{LC}}$$

where f is the resonant frequency, L is the inductance and C is the capacitance.

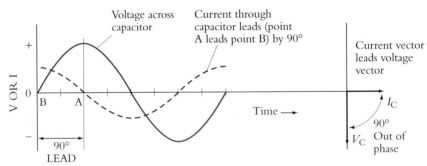

(a) Capacitor current leads the voltage across the capacitor by 90°

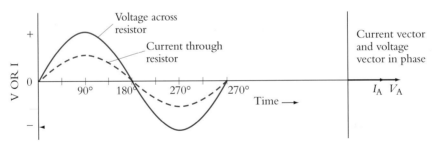

(b) Volgate and current are in-phase in a resistive circuit

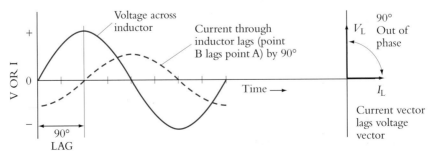

(c) Inductor curret lags the voltage across the inductor by 90°

Figure 6.74 Current and voltage vectors

Worked Example

A series circuit comprises an inductor of 10 H, a resistor of 50 Ω and a capacitor of 40 μF. Determine the frequency at which this circuit is resonant and the current that will flow in it when it is connected to a 50 V a.c. supply at the resonant frequency.

Using $f = \dfrac{1}{2\pi\sqrt{LC}}$

$$f = \frac{1}{6.28\sqrt{10 \times 40 \times 10^{-6}}} = 7.95\,\text{Hz}$$

At the resonant frequency, the circuit will behave as a pure resistance (recall that the two reactances will be equal but opposite) and thus the supply current at resonance can be determined from:

$$I = \frac{V}{Z} = \frac{V}{R} = \frac{50}{50} = 1\,\text{A}$$

Impedance and phasors

Impedance is the opposition to a.c. current flow that capacitance and/or inductance offer to an a.c. current; this is determined by the magnitude of their reactance. We have also noted the phase shift caused between voltage and current in circuits with capacitance and/or inductance. In practical circuits, reactance occurs together with resistance. The effects of this can be illustrated and analysed through phasors (see Figure 6.74).

From the term 'CIVIL', an a.c. voltage applied to a capacitor will lag the current by a full 90°, and less if there is resistance in the circuit; and in a capacitor the current will lag by 90°. For a purely resistive circuit, the voltage and current are in phase. For an inductive circuit, the current lags the voltage by 90°.

These effects can be determined by working with the reactive elements as phasors in a right-angled triangle. Figure 6.75 illustrates how trigonometry is used to calculate the magnitude and phase

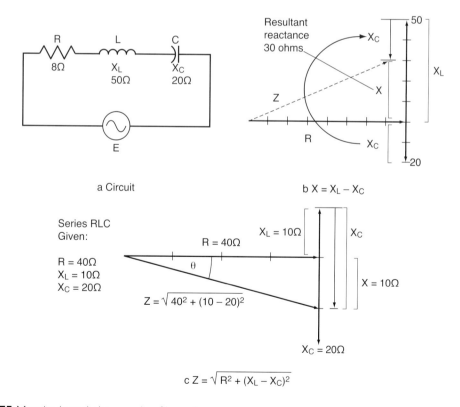

Figure 6.75 Magnitude and phase angle of impedance

angle of impedance in a series circuit containing R, L and C. The magnitude of the inductive reactance is plotted in the positive (Y) direction, capacitive reactance in the negative (Y) direction and the resistance in the positive (X) direction. The impedance is represented by the hypotenuse of the right-angled triangle that is formed. The length of the hypotenuse is the magnitude of the impedance, and the angle formed with the X axis determines the phase angle between the current and the voltage.

For the circuit values given in Figure 6.75(c) (see p. 309), $Z = 17.32\,\Omega$ and has a phase angle θ calculated by:

$$\theta = \tan^{-1}\frac{(X_L - X_C)}{R} = -14°$$

The resulting effective current is the magnitude of the applied voltage divided by the magnitude of the impedance. If the applied voltage is 50 volts, an effective current of 2.9 amperes will flow. Since there is a predominance of capacitive reactance, the current is leading the voltage.

Power factor

For an a.c. circuit containing both resistance and reactance, power factor is the ratio of true power to apparent power:

$$\text{power factor} = \frac{\text{true power}}{\text{apparent power}}$$

The true power in an a.c. circuit is the power that is dissipated as heat in the resistive component:

$$\text{true power} = I^2R$$

where I is the r.m.s. current and R is the resistance; true power is measured in watts (W).

The apparent power in an a.c. circuit is the power that is apparently consumed by the circuit and is the product of the supply current and supply voltage (which may not be in phase). Note that unless the voltage and current are in phase (that is, $\varnothing = 0°$), the apparent power will not be the same as the power that is actually dissipated as heat. Hence:

$$\text{apparent power} = IV$$

where I is the r.m.s. current and V is the supply voltage. To distinguish apparent power from true power, apparent power is measured in volt-amperes (VA).

Now, since $V = IZ$, we can rearrange the apparent power equation as follows:

$$\text{apparent power} = IV = I \times IZ = I^2Z$$

Returning to our original equation:

$$PF = \frac{TP}{AP} = \frac{I^2R}{IV} = \frac{I^2R}{I \times IZ} = \frac{I^2R}{I^2Z} = \frac{R}{Z}$$

From the impedance triangle shown earlier, we can infer that:

$$PF = \frac{R}{Z} = \cos\varnothing$$

Worked Example

An a.c. load has a power factor of 0.8. Determine the true power dissipated in the load if it consumes a current of 2 A at 110 V.

Since: $PF = \cos\varnothing = \dfrac{TP}{AP}$

True power = power factor $\times$ apparent power
= power factor $\times$ VI

thus

true power = $0.8 \times 2 \times 110 = 176\,\text{W}$

Key Terms

From the electrical perspective, the power factor in an a.c. circuit – The ratio of true power to apparent power.

From the mathematical perspective, the power factor in an a.c. circuit – The cosine of the phase angle between the supply current and supply voltage.

Worked Example

A coil having an inductance of 150 mH and resistance of 250 Ω is connected to a 115 V 400 Hz a.c. supply. Determine the:

- **power factor of the coil**

- **current taken from the supply**

- **power dissipated as heat in the coil.**

First, we must find the reactance of the inductor and the impedance of the coil at 400 Hz.

$$X_L = 2\pi f L = 2\pi \times 400 \times 150 \times 10^{-3} = 376 \, \Omega$$

and

$$Z = \sqrt{(R^2 + X^2)} = \sqrt{(250^2 + 376^2)} = 452 \, \Omega$$

We can now determine the power factor from:
$$PF = \frac{R}{Z} = \frac{250}{452} = 0.553$$

The current taken from the supply can be determined from:

$$I = \frac{V}{Z} = \frac{115}{452} = 0.254 \, A$$

The power dissipated as heat can be found from:

$$\text{true power} = \text{power factor} \times VI = 0.553 \times 115 \times 0.254 = 16.15 \, W$$

Activity 6.9

1. Determine the reactance of a 60 mH inductor at (a) 20 Hz and (b) 4 kHz.
2. Determine the reactance of a 220 nF capacitor at (a) 400 Hz and (b) 20 kHz.
3. A 0.5 μF capacitor is connected to a 110 V 400 Hz supply. Determine the current flowing in the capacitor.
4. A resistor of 120 Ω is connected in series with a capacitive reactance of 160 Ω. Determine the impedance of the circuit and the current flowing when the circuit is connected to a 200 V a.c. supply.
5. A capacitor or 2 μF is connected in series with a 100 Ω resistor across a 24 V 400 Hz a.c. supply. Determine the current that will be supplied to the circuit and the voltage that will appear across each component.

Transformers

Transformers provide a means of increasing (stepping up) or decreasing (stepping down) an a.c. voltage. They are formed with two coils (the primary and secondary) onto a common core (see Figure 6.76).

For a step-up transformer, the output (or secondary) voltage will be greater than the input (or primary), while for a step-down transformer, the secondary voltage will be less than the primary voltage. Since the primary and secondary power must be the same (no increase in power is possible), an increase in secondary voltage can only be achieved at the expense of a corresponding reduction in secondary current, and vice versa (in fact, the secondary power will be very slightly less than the primary power, owing to losses within the transformer).

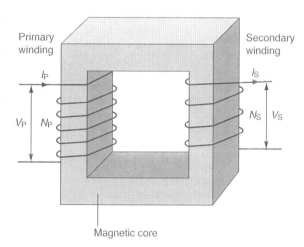

Figure 6.76 Transformer principles

311

The principle of the transformer is illustrated in Figure 6.76 (see page 311). The primary and secondary windings are wound on a common low-reluctance magnetic core consisting of a number of steel laminations. All of the alternating flux generated by the primary winding is coupled into the secondary winding (very little flux escapes due to leakage). A sinusoidal current flowing in the primary winding produces a sinusoidal flux within the transformer core.

The flux in the transformer core will increase with increasing frequency in the primary turns (up to the saturation point of the B–H curve).

Flux variation

From a.c. theory and transformer principles, we know that the flux has to change from a positive a.c._{max} to a negative a.c._{max} in one half of the cycle. This occurs in $0.5\,f$ seconds; therefore, the average rate of flux change is given by:

$$\frac{(2\Phi_{max})}{2f} = 4f\Phi_{max} \text{ webers/second}$$

The average emf induced per turn is given by:

$$4f\Phi_{max} \text{ volts}$$

However, for a sinusoidal wave, the emf (or effective value) is 1.11 times the average value; therefore, the r.m.s. value of induced emf is:

$$1.11 \times 4f\Phi_{max}$$

So, for the primary turns, the r.m.s. value of the primary voltage (V_P) is given by:

$$V_P = 4.44 f N_P \Phi_{max}$$

Similarly, the r.m.s. value of the secondary voltage (V_s) is given by:

$$V_S = 4.44\, f N_S \Phi_{max}$$

From these two relationships (and since the same magnetic flux appears in both the primary and secondary windings), we can infer that:

$$\frac{V_P}{V_S} \approx \frac{N_S}{N_P}$$

Note that losses in the transformer mean that the relationship is not exactly equal, hence the use of the approximation symbol ($\approx$).

Furthermore, assuming that minimal power is lost in the transformer, we can conclude that:

$$\frac{I_P}{I_S} \approx \frac{N_S}{N_P} \approx \frac{V_S}{V_P}$$

The ratio of primary turns to secondary turns ($\frac{N_P}{N_S}$) is known as the turns ratio.

Referring to Figure 6.77, we can further analyse the transformer. Since the ratio of primary voltage to primary turns is the same as the ratio of secondary turns to secondary voltage, we can conclude that, for a particular transformer:

$$\text{turns per volt (t.p.v.)} \approx \frac{V_P}{N_P} \approx \frac{V_S}{N_S}$$

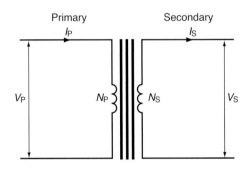

Figure 6.77 Transformer turns and voltages

The t.p.v. rating can be quite useful when it comes to designing transformers with multiple secondary windings.

Worked Example

A transformer has 2000 primary turns and 120 secondary turns. If the primary is connected to a 220 V a.c. mains supply, determine the secondary voltage.

Since $\frac{V_P}{V_S} \approx \frac{N_S}{N_P}$ we can derive: $V_S \approx \frac{V_P N_S}{N_P} \approx 220 \times \frac{120}{2000} = 13.2\,V$ a.c.

Worked Example

A transformer has 1100 primary turns and is designed to operate with a 110 V a.c. supply. If the transformer is required to produce an output of 10 V, determine the number of secondary turns required.

Since $\frac{V_P}{V_S} \approx \frac{N_S}{N_P}$ we can derive: $N_S \approx \frac{V_S N_P}{V_P} \approx 1100$ $\times \frac{10}{110} \approx 100$ turns

Worked Example

A transformer has a turns-per-volt rating of 1.2. How many turns are required to produce secondary outputs of (a) 50 V and (b) 350 V?

Using $N_S \approx$ t.p.v. $\times V_S$

In the case of a 50 V secondary winding:

$$N_S \approx 1.5 \times 50 = 75 \text{ turns}$$

In the case of a 350 V secondary winding:

$$N_S \approx 1.5 \times 350 = 525 \text{ turns}$$

Worked Example

A transformer has 1200 primary turns and 60 secondary turns. Assuming that the transformer is loss-free, determine the primary current when a load current of 20 A is taken from the secondary.

Since $\frac{I_P}{I_S} \approx \frac{N_S}{N_P}$ we can derive: $I_P \approx I_S \frac{N_S}{N_P} \approx \frac{20 \times 60}{1200}$ $= 1 \text{ A}$

Worked Example

A transformer produces an output voltage of 110 V under no-load conditions and an output voltage of 100 V when a full load is applied. Determine the per unit regulation.

$$V_{S \text{ (no load)}} - \frac{V_{S \text{ (full load)}}}{V_{S \text{ (no load)}}}$$

$$\approx 110 - \frac{100}{110} = 0.1 \text{ (10 per cent)}$$

The output voltage produced at the secondary of a transformer reduces progressively as the load imposed on the transformer increases – that is, as the secondary current increases from its no-load value. The voltage regulation of a transformer is a measure of its ability to keep the secondary output voltage constant over the full range of output load currents – that is, from no load to full load at the same power factor. This change, when divided by the no-load output voltage, is referred to as the per unit regulation for the transformer.

Most transformers operate at very high values of efficiency. Despite this, in high-power applications the losses in a transformer cannot be completely neglected. Transformer losses can be categorised into two types of loss:

- losses in the magnetic core (referred to as iron loss)
- losses due to the resistance of the coil windings (referred to as copper loss).

Iron loss can be further divided into hysteresis loss (energy lost in repeatedly cycling the magnet flux in the core backwards and forwards) and eddy current loss (energy lost due to current circulating in the steel core).

Hysteresis losses can be reduced by using magnetic core material that is easily magnetised and has a very high permeability (see Figure 6.78) – note that energy loss is proportional to the area inside the B–H curve). Eddy current loss can be reduced by laminating the core; these laminations and small gaps help to ensure that there is no closed path for current to flow. Copper loss results from the resistance of the coil windings; it can be reduced by using low-resistivity wire material.

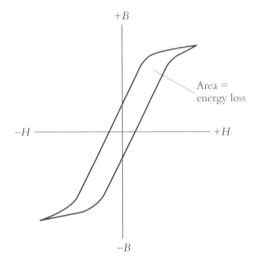

Figure 6.78 Hysteresis curves and energy loss

Since the flux within a transformer varies only slightly between the no-load and full-load conditions, iron loss is substantially constant, regardless of the load actually imposed on a transformer. Copper loss, on the other hand, is zero when a transformer is under no-load conditions and rises to a maximum at full load.

The efficiency of a transformer is given by:

$$\frac{\text{output power}}{\text{input power}} \times 100 \text{ per cent}$$

from which:

$$\text{efficiency} = \frac{\text{input}}{\text{power}} - \frac{\text{losses}}{\text{input power}} \times 100 \text{ per cent}$$

Therefore:

$$\text{efficiency} = (1 - \frac{\text{losses}}{\text{input power}}) \times 100 \text{ per cent}$$

The losses present are attributable to iron loss and copper loss, but the copper loss appears in both the primary and the secondary windings, hence:

$$\text{efficiency} = \frac{(1 - (\text{iron losses} + \text{copper losses}_{\text{PRIMARY}} + \text{copper losses}_{\text{SECONDARY}}))}{\text{input power}} \times 100 \text{ per cent}$$

Worked Example

A transformer rated at 500 VA has an iron loss of 3 W and a full-load copper loss (primary plus secondary) of 7 W. Calculate the efficiency of the transformer at 0.8 power factor.

The input power to the transformer will be given by the product of the apparent power – that is, the transformer's VA rating – and the power factor.

Hence

$$\text{input power} = 0.8 \times 500 = 400 \, \text{W}$$

Now

$$\text{efficiency} = (\frac{1 - (7+3)}{400}) \times 100 \text{ per cent} = 97.5$$

Typical transformers are shown in Figure 6.79.

Figure 6.79 Typical transformers

Activity 6.10

1. A transformer has 480 primary turns and 120 secondary turns. If the primary is connected to a 110 V a.c. supply, determine the secondary voltage.

2. A step-down transformer has a 220 V primary and a 24 V secondary. If the secondary winding has 60 turns; how many turns are there on the primary?

3. A transformer has 440 primary turns and 1800 secondary turns. If the secondary supplies a current of 250 mA, determine the primary current (assume that the transformer is loss-free).

4. A transformer produces an output voltage of 220 V a.c. under no-load conditions, and an output voltage of 208 V a.c. when full load is applied. Determine the per unit regulation.

5. A 1 kVA transformer has an iron loss of 15 W and a full-load copper loss (primary plus secondary) of 20 W. Determine the efficiency of the transformer at 0.9 power factor.

6.5 Electrical machines

This section will cover the following grading criteria:

By the end of this section you should be able to evaluate the performance of a motor and a generator by reference to electrical theory.

Electrical machines are based on the practical application of electromagnetism. The motor and generator are electrical machines that share many common features in terms of their operating principles, construction and operation.

a.c. generators

Earlier in this unit we explored the concept of electromagnetic induction. This can be summarised as the generation of an emf across the ends of a conductor when it cuts through a changing magnetic flux. The action of cutting through the lines of magnetic flux results in a generated emf (see Figure 6.80). The amount of emf (e) induced in the conductor will be directly proportional to the:

- density of the magnetic flux (B)
- effective length of the conductor (l) within the magnetic flux

- speed (v) at which the lines of flux cut through the conductor
- sine of the angle (θ) between the conductor and the lines of flux.

The induced emf is given by the formula:

$$e = Blv\sin\theta$$

When the conductor moves at right angles to the field (as shown in Figure 6.80), maximum emf will be induced. When the conductor moves along the lines of flux, the induced emf will be zero. At any other point in the magnetic field, there is a value of emf dependent on the sine of the angle (θ) between the conductor and the lines of flux.

Moving a conductor through a magnetic field is usually achieved by attaching a coil to a shaft and driving this from a suitable source of power – for example, from an engine gearbox, as shown in Figure 6.81. The loop, or coil, is made to rotate inside a permanent magnetic field, with opposite poles (N and S) on either side of the loop.

Making contact with the loop as it rotates inside the magnetic field is achieved by a pair of carbon brushes and copper slip-rings (see Figure 6.82). The brushes are spring-loaded and held against the rotating slip-rings so that there is a permanent connection for current to flow from the loop to the external load to which it is connected.

The opposite sides of the loop consist of conductors that move through the field. At 0° (with the loop vertical, as shown at A in Figure 6.83 on

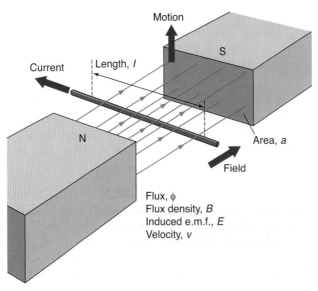

Figure 6.80 A conductor moving inside a magnetic field

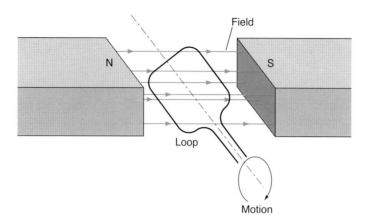

Figure 6.81 A loop rotating within a magnetic field

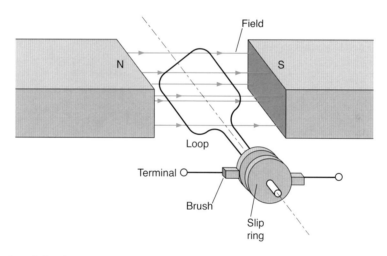

Figure 6.82 Brush and slip-ring arrangement

page 318), the opposite sides of the loop will be moving in the same direction as the lines of flux. At that instant, the angle, θ, at which the field is cut, is 0°, and since the sine of 0° is zero, the generated voltage (from e = Blvsinθ) will be zero.

When the loop is rotated to a position B (which is 90° to the lines of flux), the two conductors will effectively be moving at right angles to the field. At that instant, the generated emf will take a maximum value (since the sine of 90° is unity).

At 180° from the starting position, the generated emf will reduce back to zero, since the conductors are moving along the flux lines (but in the direction opposite to that at 0°, as shown in C).

At 270°, the conductors will be moving perpendicular to the flux lines (but in the direction opposite to that at 90°). At this point (D), a maximum generated emf will be produced. Note that the emf generated at this instant is the opposite polarity to that which was generated at 90°. The relative direction of motion (between the conductors and flux lines) has now been reversed.

Since e = Blvsinθ, the generated emf will take a sinusoidal form, as shown in Figure 6.84 (see page 318). Note that the maximum values of emf occur at 90° and 270°, and that the generated voltage is zero at 0°, 180° and 360°.

In practice, the single loop shown in Figure 6.84 would comprise a coil of wire wound on a

317

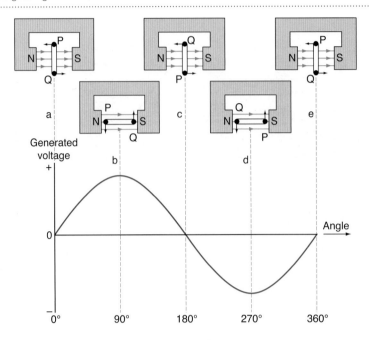

Figure 6.83 emf generated at various angles

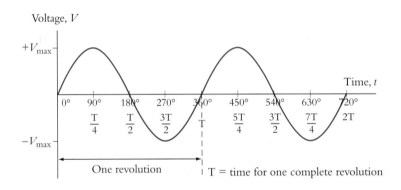

Figure 6.84 Sinusoidal voltage produced by a rotating loop

suitable non-magnetic former. This coil of wire effectively increases the length of the conductor within the magnetic field, and the generated emf will then be directly proportional to the number of turns on the coil.

To summarise, in a simple a.c. generator, a loop of wire rotates inside the magnetic field produced by two opposite magnetic poles. Contact is made to the loop as it rotates by means of slip-rings and brushes.

Alternators

Alternators are based on the principles that relate to the simple a.c. generator, except that it is the magnetic field that is rotated rather than the conductors from which the output is taken. Furthermore, the magnetic field is usually produced by a rotating electromagnet (the rotor) rather than a permanent magnet. There are a number of reasons for doing this:

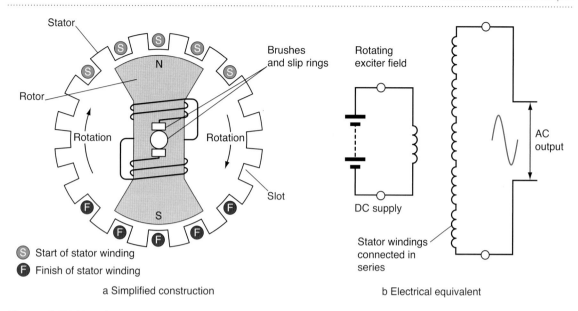

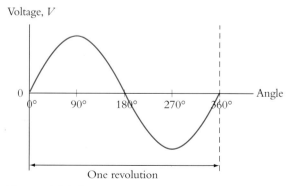

S Start of stator winding
F Finish of stator winding

a Simplified construction

b Electrical equivalent

Figure 6.85 Simplified construction of a single-phase a.c. generator

- the conductors can be lighter in weight, thereby allowing faster rotational speeds
- it is more practical to insulate the conductors if they are stationary
- the currents that produce the rotating magnetic field are very much smaller than those that are produced by the conductors; the slip-rings can be made smaller and more reliable.

Figure 6.85 shows the simplified construction of a single-phase a.c. generator. The stator consists of five coils of insulated heavy-gauge wire located in slots in the high-permeability laminated core. These coils are connected in series to make a single stator winding, from which the output voltage is derived.

The two-pole rotor comprises a field winding that is connected to a d.c. field supply via a set of slip-rings and brushes. As the rotor moves through one complete revolution, the output voltage will complete one full cycle of a sine wave, as shown in Figure 6.86.

By adding more pole pairs to the physical arrangement, it is possible to produce several cycles of output voltage for one single revolution of the rotor. The frequency of the output voltage produced by an a.c. generator is given by:

$$f = \frac{pN}{60}$$

where f is the frequency of the induced emf (in Hz), p is the number of pole pairs, and N is the rotational speed in r.p.m. Note that the '60' is required in this expression to effectively convert r.p.m. into cycles per second.

Figure 6.86 Output voltage produced by the single-phase a.c. generator

Worked Example

An alternator is to produce an output at a frequency of 60 Hz. If it uses a four-pole rotor, determine the shaft speed at which it must be driven.

Rearranging $f = \frac{pN}{60}$ to make N the subject gives: $N = 60f/p$

A four-pole machine has two pairs of poles; thus

$p = 2$, and so and $N = \dfrac{(60 \times 60)}{2} = 1800$ r.p.m.

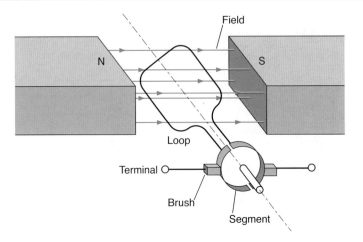

Figure 6.87 Commutator arrangement

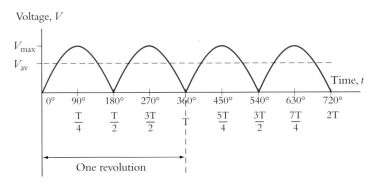

Figure 6.88 d.c.-generated output

d.c. generators

Where the application requires d.c. output, this can be achieved by replacing the brushes and slip-rings with a commutator arrangement, as shown in Figure 6.87. The commutator functions as a rotating/reversing switch, which ensures that the emf generated by the loop is reversed after rotating through 180°. The generated emf for this arrangement is shown in Figure 6.88.

The generated emf shown in Figure 6.88 has only one polarity, either all positive or all negative. This may be acceptable in certain applications; however, in some cases we need a d.c. power source that provides a constant voltage output rather than a series of pulses. One way of achieving this is with a second loop (or coil) at right angles to the first, as shown in Figure 6.89. The commutator is then divided into four (rather than two) segments;

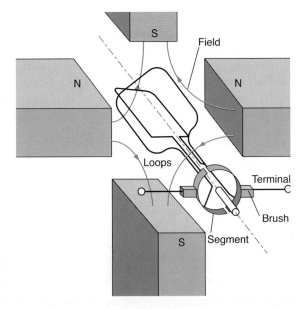

Figure 6.89 Improved d.c. generator construction

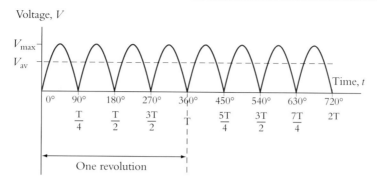

Figure 6.90 Improved d.c. generator output

the generated emf produced by this arrangement is shown in Figure 6.90.

In practical generators, a coil comprising a large number of turns of copper wire replaces the single-turn rotating loop. This arrangement effectively increases the total length of the conductor within the magnetic field and, as a result, also increases the generated output voltage. The output voltage also depends on the density of the magnetic flux through which the current-carrying conductor passes; the denser the field, the greater the output voltage will be.

d.c. motors

A simple d.c. motor consists of a very similar arrangement to the d.c. generator. A loop of wire that is free to rotate on a shaft is placed inside a permanent magnetic field (see Figure 6.91). When a d.c. current is applied to the loop of wire, two equal and opposite forces are set up which act on the conductor in the directions indicated in Figure 6.91.

The direction of the forces acting on each arm of the conductor can be established by using two rules of electromagnetism:

- the right-hand grip rule
- Fleming's left-hand rule (for electric motors).

The right-hand grip rule is a principle applied when electric current flows through a coil, resulting in a magnetic field (see Figure 6.92). With the right hand wrapped around the coil, the fingers point in the direction of the conventional current, and the thumb indicates the coil's magnetic north pole.

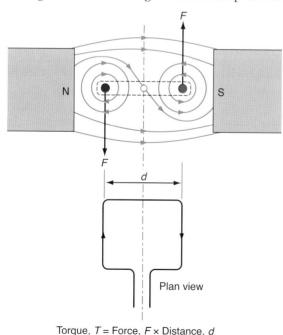

Torque, T = Force, F × Distance, d

Figure 6.91 Torque on a current-carrying loop located within a permanent magnetic field

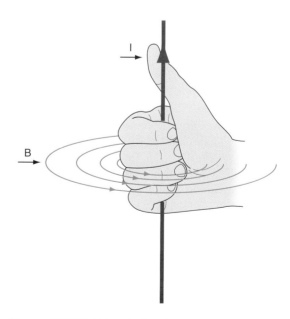

Figure 6.92 Right-hand grip rule

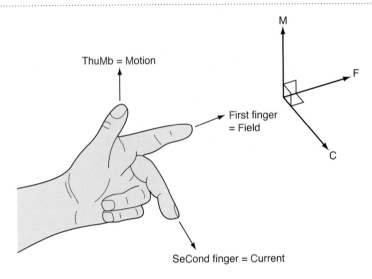

Figure 6.93 Fleming's left-hand rule (for motors)

The right-hand grip rule can also be applied to simple conductors; now the thumb points in the direction of the conventional current (from positive to negative), and the fingers indicate the direction of the magnetic lines of flux.

Fleming's left-hand rule (for electric motors) shows the direction of the thrust on a conductor carrying a current in a magnetic field (see Figure 6.93). This rule was developed by Sir John Ambrose Fleming (1849–1945), an English electrical engineer and physicist.

Since the conductors are equidistant from their pivot point and the forces acting on them are equal and opposite, this forms a couple. The moment of this couple is equal to the magnitude of a single force multiplied by the distance between them; this moment is known as torque (T), expressed as:

$$T = Fd$$

where T is the torque (in newton-metres, Nm), F is the force (N) and d is the distance (m).

We have already seen that the magnitude of the force F is given by $\Phi = B\,I\,\lambda$. Thus the torque expression can be written as:

$$T = B\,I\,l\,d$$

where T is the torque (Nm), B is the flux-density (T), I is the current (A), λ is the length of conductor in the magnetic field (m) and d is the distance (m).

The torque produces a turning moment such that the coil (or loop) rotates within the magnetic field. This rotation continues whenever current is flowing in the coil. A practical form of d.c. motor consists of a rectangular coil of wire (instead of a single-turn loop of wire) mounted on a former and free to rotate about a shaft in a permanent magnetic field, as shown in Figure 6.94.

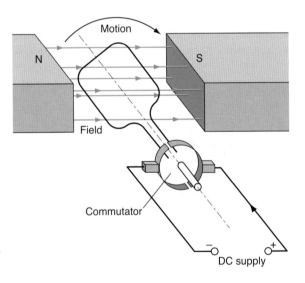

Figure 6.94 Simple motor with commutator

In practical motors, this rotating coil is known as the armature and consists of many hundreds of turns of conducting wire. This arrangement maximises the force imposed on the conductor by introducing the longest possible conductor into the magnetic field. Furthermore (from the relationship $\Phi = B I \lambda$), it can be seen that the force used to derive the torque in a motor is directly proportional to the size of the magnetic flux (B). As an alternative to using a permanent magnet to produce this flux, an electromagnet is used (this is the usual arrangement in most motor applications).

The electromagnetic field is set up using the principle shown in Figure 6.95. This arrangement constitutes a field winding; each of the turns in the field winding assists each of the other turns, in order to produce a strong magnetic field.

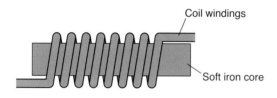

Coil windings

Soft iron core

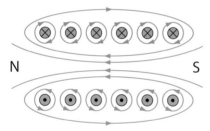

Cross section showing magnetic field lines

Figure 6.95 Magnetic field used as a field winding

This field may be intensified by inserting a ferromagnetic core inside the coil. When current is applied to the conducting coil, the core is magnetised and acts in combination with the coil to produce a permanent magnet, having its own north and south poles.

Referring to the simple motor illustrated in Figure 6.94, we know that when current is supplied to the rotating armature (rotor), a torque is produced. In order to produce continuous rotary motion, this torque (turning moment) must always act in the same direction. The current in each of the armature conductors must be reversed as the conductor passes between the north and south magnetic field poles. The commutator acts like a rotating switch, reversing the current in each armature conductor at the appropriate time to achieve this continuous rotary motion.

Key Points

- A commutator acts like a rotating switch.
- The effect of not having a commutator in a d.c. motor would be that only a half-turn of movement is achieved.
- When current is supplied to the rotating armature (rotor), a torque is produced.
- The torque produced by a d.c. motor is directly proportional to the product of the current flowing in the rotating armature winding.

In Figure 6.96(a), the rotation of the armature conductor is given by Fleming's left-hand rule for motors. When the coil reaches a position midway between the poles (Figure 6.96(b)), no rotational torque is produced in the coil. At this point, the commutator reverses the current in the coil. Finally (Figure 6.96(c)), with the current reversed, the motor torque now continues to rotate the coil in its original direction.

a

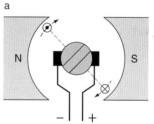

Rotation of each armature conductor given by left-hand rule

b

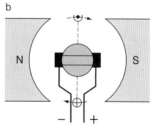

Coil mid-way between poles – no rotational torque produced. Commutator reverses current

c

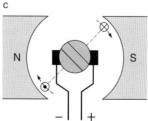

With current reversed motor torque continues to rotate coil in the original direction.

Figure 6.96 Action of the commutator

323

Field connections

The field winding of a d.c. motor can be connected in various ways, depending on the application; the following configurations can be used:

- series-wound
- shunt-wound
- compound-wound (where both series and shunt windings are present).

In the series-wound d.c. motor, the field winding is connected in series with the armature and the full armature current flows through the field winding (see Figure 6.97). This arrangement results in a d.c. motor that produces a large starting torque at slow speeds. This type of motor is ideal for applications where a heavy load is applied from rest. The disadvantage of this type of motor is that, with reduced loads, the motor speed may become excessively high. Typical torque and speed characteristics for a series-wound d.c. motor are shown in Figure 6.98.

In the shunt-wound d.c. motor, the field winding is connected in parallel with the armature; the supply current is divided between the armature

and the field winding (see Figure 6.99). This results in a d.c. motor that runs at a reasonably constant speed over a wide variation of loads, but does not perform well when heavily loaded. Typical torque and speed characteristics for a shunt-wound d.c. motor are shown in Figure 6.100.

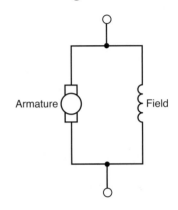

Figure 6.99 Shunt-wound d.c. motor

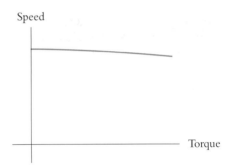

Figure 6.100 Typical torque and speed characteristics for a shunt-wound d.c. motor

The compound-wound d.c. motor combines both series and shunt field windings (see Figure 6.101) and is therefore able to combine some

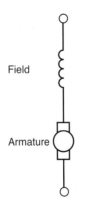

Figure 6.97 Series-wound d.c. motor

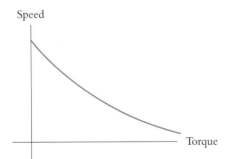

324

Figure 6.98 Typical torque and speed characteristics for a series-wound d.c. motor

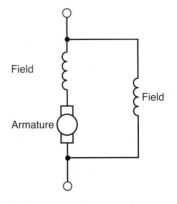

Figure 6.101 Compound-wound d.c. motor

of the advantages of each type of motor. Typical torque and speed characteristics for a compound-wound d.c. motor are shown in Figure 6.102.

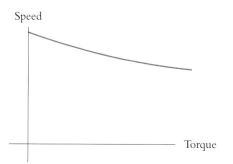

Figure 6.102 Typical torque and speed characteristics for a compound-wound d.c. motor

In order to avoid the need for a large permanent magnet in a d.c. machine – that is, a motor or generator – a separate field winding is used. This winding is energised with direct current. In the case of a d.c. generator, this current can be derived from the output of the generator (in which case it is referred to as a self-excited generator) or it can be energised from a separate d.c. supply.

Make the Grade P9

Grading criterion P9 requires that you describe the principles and applications of electromagnetic induction. This should be approached from both an experimental and a theoretical perspective. Decide what items will be required – for example, bar magnets, coils of wire. Develop some experiments. Compare your results with your knowledge of the theory. Figure 6.103 gives some examples and hints for getting started.

Make the Grade D2

Grading criterion D2 requires that you evaluate the performance of a motor and a generator by reference to electrical theory. This should be approached from both a practical and theoretical perspective. Decide what items will be required – for example, motors, power supplies – and then develop some experiments. (For safety reasons, only use low-voltage power supplies.) You should address a range of performance details for motors, such as power supply voltage and direction of motion. For generators, make notes of the speed of turning the shaft versus output voltage. Compare your results with your knowledge of the theory.

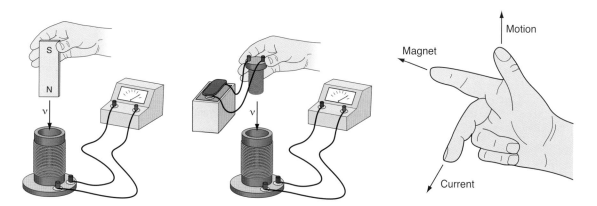

Figure 6.103 Principles of electromagnetic induction

6.6 Diodes

This section will cover the following grading criteria:

By the end of this section you should be able to compare the forward and reverse characteristics of two different types of semiconductor diode, and analyse the operation and the effects of varying component parameters of a power supply circuit that includes a transformer, diodes and capacitors.

Diodes are electronic components, designed to conduct and oppose current, depending on how they are connected into the circuit. They are manufactured from two or more semi-conducting materials that either conduct or oppose the flow of electrons. The physical appearance of a diode used in electronic circuits usually takes the form of a small cylinder; larger diodes are formed into larger metallic packages to dissipate heat.

Semiconductors

Semiconductor materials are formed by changing the molecular structure of an insulating material. The crystalline structure of this base material is effectively modified into one of two types of semiconductor:

- with an increased number of electrons in the crystal structure (N-type)
- with a decreased number of electrons in the crystal structure (P-type).

Typical semiconducting materials include silicon and germanium. Diodes are created by joining P-type and N-type semiconductors to form a junction. When the diode is connected to an electrical circuit, it allows current to flow in one direction (with the diode connected in forward bias), but not the other direction (with the diode connected in reverse bias) (see Figure 6.104).

Diode principles

The mechanical analogy of a diode is the non-return valve (see Figure 6.105). The valve in this example contains a ball and spring. When fluid passes into the valve in one direction, it compresses the spring and the fluid flows through the valve. When fluid is directed into the valve from the opposite direction, the spring and fluid pressure keep the ball against the seat of the valve, preventing the flow of fluid.

The characteristics of a diode can be shown graphically, as in Figure 6.106. Note the point at which current flows in the forward and reverse directions.

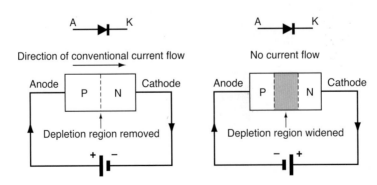

Figure 6.104 Diodes in forward and reverse bias

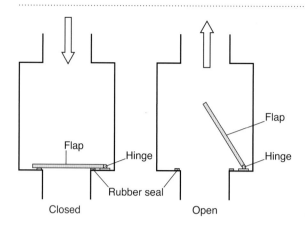

Figure 6.105 Non-return valve

Diode types

There are many types of diode used in everyday applications; these include:

- junction diode
- Zener diode
- light-emitting diode
- photodiode.

Circuit symbols for each of these diodes is shown in Figure 6.107 (see page 328).

Junction diode

The function of a simple junction diode is to conduct current in one direction and to prevent the flow of current in the opposite direction. A typical application of diodes used in this way is called a rectifier (see Figure 6.108 on page 328).

In this application, the diode converts a.c. to d.c., a process known as rectification. With a single diode, either the positive or negative half-cycle of the a.c. wave is conducted, while the other half is not conducted.

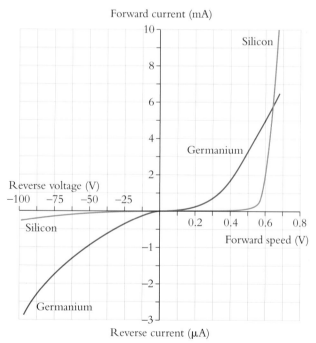

Figure 6.106 Characteristics of a diode

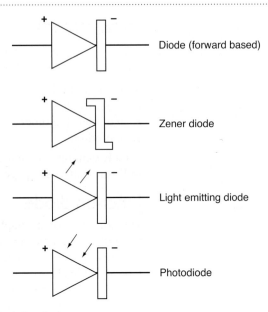

Figure 6.107 Circuit symbols for diodes

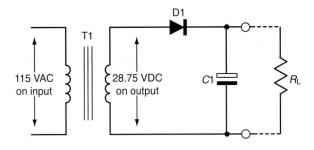

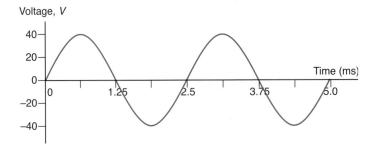

a T1 secondary voltage

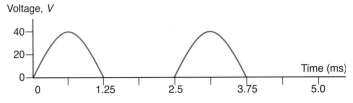

b Voltage developed across R_L

328

Figure 6.108 Diode used as a rectifier

Since only one half of the input a.c. waveform reaches the output, it is a very inefficient method of converting a.c. to d.c. A more efficient circuit uses four diodes, as shown in Figure 6.109. This provides full-wave rectification.

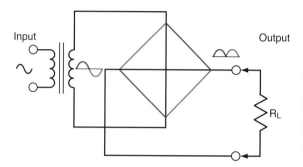

Figure 6.109 Full-wave rectification

A full-wave rectifier converts the whole of the a.c. input to a positive (or negative) output. The output voltage can be smoother by incorporating a capacitor, as shown in Figure 6.110.

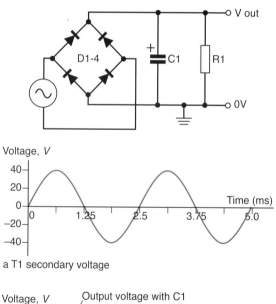

a T1 secondary voltage

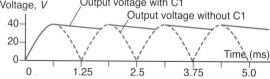

b Voltage developed across R_L

Figure 6.110 Smoothed d.c. output

Light-emitting diode

As their name implies, light-emitting diodes (LEDs) emit light energy when conducting. They are manufactured from a range materials to produce colours from the infrared to the near ultraviolet (see Figure 6.111).

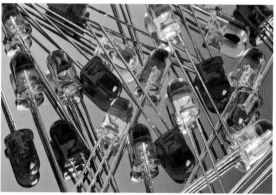

Figure 6.111 LEDs

LEDs require very low power to operate and are also more reliable than filament lamps. Most LEDs will provide a reasonable level of light output when a forward current of between 5 mA and 20 mA is applied. A typical LED circuit is shown in Figure 6.112; the value of resistance is selected to allow the required current through the diode.

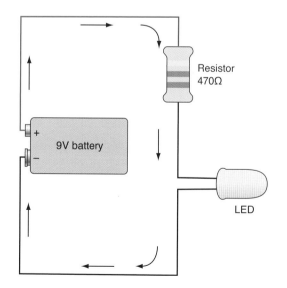

Figure 6.112 LED circuit

Photodiode

Photodiodes (also called photodetectors or photo-cells) are designed to sense light. When sufficient light energy reaches the diode, it energises electrons, thereby creating a free electron and a resultant positive charge. With increasing light energy, electrical current flows through the diode. A typical photodiode circuit is shown in Figure 6.113.

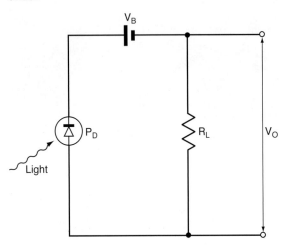

Figure 6.113 Photodiode circuit

Zener diode

A Zener diode conducts current in the forward direction (as a normal diode), but also in the reverse direction when the applied voltage exceeds the breakdown voltage known as the Zener voltage. This effect, called Zener breakdown, occurs at a predetermined and precise voltage (see Figure 6.114); the Zener diode is used as a precision voltage reference in voltage regulators (Figure 6.115).

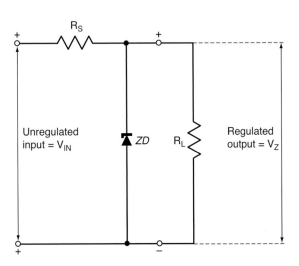

Figure 6.115 Zener diode circuit

Make the Grade   P3

Grading criterion P3 requires you to compare the forward and reverse characteristics of two different types of semiconductor diode. To fully address this grading criterion, you should use a combination of diode theory and practical circuit measurement. Make sure that you can relate the theory to practical results. Figure 6.116 gives some ideas for a circuit layout. Think about the power supply: should it be fixed at 6 V d.c., or would a variable supply be more useful?

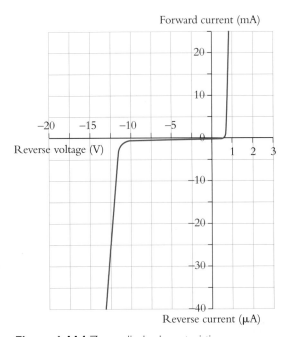

Figure 6.114 Zener diode characteristics

Forward-biased

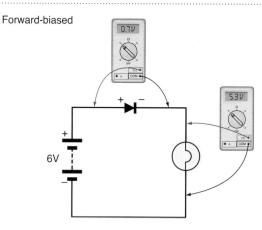

Reverse-biased

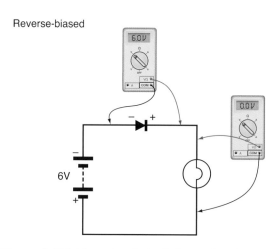

Figure 6.116 Measuring diodes in forward and reverse bias

Activity 6.11
Design a simple circuit that could be used to convert mains power to charge a 12V d.c. battery. Develop this thought process to analyse other input/output voltages – for example, larger aircraft use 115V a.c. from the main generators to charge 28V d.c. batteries.

Full-wave rectifier circuit

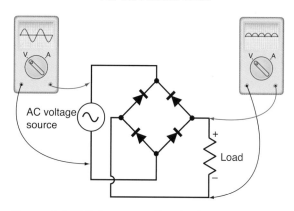

Figure 6.117 Full-wave rectifier circuit

Make the Grade   D1

Grading criterion D1 requires you to analyse the operation and the effects of varying component parameters of a power supply circuit that includes a transformer, diodes and capacitors. Practical power supplies vary in complexity, depending on the application and specification. Figure 6.117 gives an example of a full-wave rectifier circuit; for safety reasons, design and build circuits that operate at low voltage. Make notes of the input/output waveforms and how capacitors can be used to smooth the output.

6.7 Circuit measurements

By the end of this section you should be able to use a multimeter to carry out circuit measurements in a d.c. network; carry out an experiment to determine the relationship between the voltage and current for a charging and discharging capacitor; and use an oscilloscope to measure and determine the inputs and outputs of a single-phase a.c. circuit.

Circuit measurements will be required whether building new circuits or diagnosing existing circuits. There is a variety of equipment available to measure the three basic quantities in electrical and electronic circuits – current, voltage and resistance.

Multimeters

The most popular and convenient item of test equipment used for simple measurements is the multimeter; these can be either analogue or digital instruments (see Figure 6.118). Analogue instruments measure current through a coil/armature arrangement using electromagnetic principles; a pointer moves across the display to indicate the quantity being measured. Digital instruments use electronic circuits and displays. Multimeters combine several functions into a single item of test equipment; the basic functions of simple multimeters include the measurement of current, voltage and resistance.

Analogue instruments

A popular analogue instrument used in the aircraft industry is the Avometer® (registered trade mark of Megger Group Limited). This instrument is often referred to simply as an AVO, deriving its name from the words amperes, volts, ohms. It has been in widespread use in the UK from the 1930s, and can still be found in many hangars, workshops and repair stations. Referring to Figure 6.119, AVO features include the measurement of:

- a.c. up to 10 A
- voltages up to 1000 V
- resistance from 0.1 Ω up to 200 kΩ.

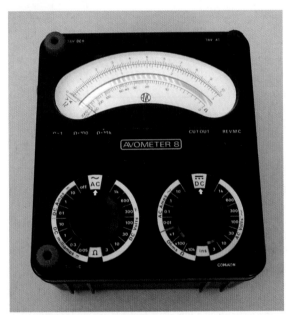

Figure 6.119 Avometer®

The instrument is very accurate, typically ±1 per cent of full-scale deflection (FSD) on d.c. ranges and ±2 per cent on a.c. ranges. Maximum current consumption for voltage measurement is 50 μA (corresponding to a sensitivity of 20 000 ohms per volt), thereby minimising voltage measurement error. Two rotary switches are used to select the function and range to be measured; if the wrong combination of function or range is selected,

Figure 6.118 Analogue and digital multimeters

an overload cut-out switch (similar to a circuit breaker) disconnects the test circuit.

Some analogue multimeters can measure the voltage drop across semiconductor junctions and measure additional circuit/component quantities, such as:

- capacitance
- frequency
- temperature
- conductance
- inductance.

Digital multimeters

Digital multimeters are generally smaller than the analogue type, making them a useful hand-held device for basic diagnostic work. Higher-specification devices can measure to a very high degree of accuracy and will commonly be found in repair shops and calibration laboratories. Digital multimeters offer more features than basic analogue instruments; commonly available measurement features include:

- auto-ranging
- sample and hold
- graphical representation
- data acquisition
- personal computer (PC) interface.

Auto-ranging selects the appropriate range for the quantity under test, so that meaningful digits are shown. For example, if a battery cell terminal voltage of 1.954 V d.c. was being measured, auto-ranging by a four-digit multimeter would automatically display this voltage, instead of 0.019 (range set too high) or 0.999 (range set too low), via manual range selection.

Sample and hold retains the most recent display for evaluation after the instrument is disconnected from the circuit being tested. Graphical representation of the quantity under test can be displayed in a number of ways – for example, as a bar graph; this facilitates observations of trends.

Simple data acquisition features are used in some multimeters to record maximum and minimum readings over a given period of time, or to take a number of sample measurements at fixed time intervals. Higher-specification multimeters feature PC interface, typically achieved by infrared (IR) links or datalink connections – for example, universal serial bus (USB). This interface allows the multimeter to upload measured data into the computer for storage and/or analysis.

Oscilloscopes

The oscilloscope (often referred to as a scope) is an item of electronic test equipment used to measure and display signal voltages as a two-dimensional graph (usually signal voltage on the vertical axis versus time on the horizontal axis). The oscilloscope will be found in most workshops, repair stations and calibration laboratories (see Figure 6.120).

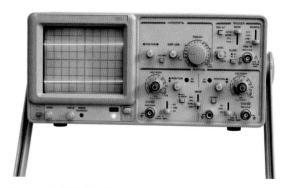

Figure 6.120 Oscilloscope

Original oscilloscope equipment was based on the cathode ray tube (CRT) display, making them heavy and bulky. Use of liquid crystal display (LCD) technology makes the oscilloscope useful as a portable device. The portable oscilloscope has many uses during troubleshooting – for example, checking for electrical noise and measurement of digital signals.

PC-based oscilloscopes can be configured in an existing laptop via a specialised signal acquisition board and suitable hardware interfaces. The PC-based oscilloscope has a number of features:

- lower cost compared to a stand-alone oscilloscope
- efficient exporting of data into standard PC software – for example, spreadsheets
- control of the instrument via custom programmes on the PC
- utilisation of the PC's networking and disc storage functions
- larger-/higher-resolution colour displays
- colours can differentiate between waveforms
- portability.

d.c. circuit measurements

There are three basic measurements that are required for analysing an electrical or electronic circuit:

- voltage
- current
- resistance.

333

When measuring any of these (and taking into account safety precautions), the type of measuring equipment has to be selected. For simple d.c. circuits, the multimeter is ideal. In the first instance, the meter must be set to d.c., together with the required parameter (usually denoted on the meter as V, A or Ω). Second, the meter must be set to the anticipated range to be measured; in some meters, this is set automatically.

Referring to Figure 6.121, when measuring:

- resistance, the component is usually removed from the circuit to avoid measuring the effects of parallel circuits
- current, the meter is connected in series
- voltage, the meter is used across two points.

a.c. circuit measurements

Referring to the typical a.c. waveforms described earlier, voltage and current can also be measured using a multimeter that is designed for the purpose. These items have a selector switch with a graphical a.c. symbol, such as ~, on one or more of the switch positions. Multimeters are usually designed to measure r.m.s. values of sinusoidal a.c. If peak values are required, it would be appropriate to use an oscilloscope.

Measuring r.m.s. values of other waveforms – for example, square waves – with a multimeter will not produce a true value. Once again, it would be more appropriate to use an oscilloscope to determine r.m.s. and/or peak values.

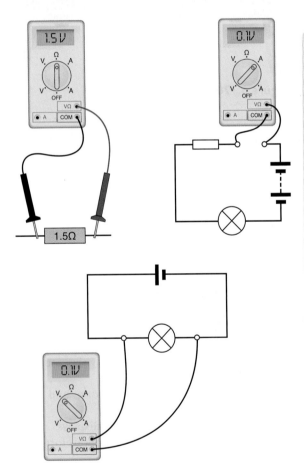

Figure 6.121 Measuring resistance, current and voltage

Make the Grade — P2

Grading criterion P2 requires you to use a multimeter to carry out circuit measurements in a d.c. network. This relates back to what can be calculated – for example, from Ohm's law and/or Kirchoff's laws – and what can be measured. Try designing and building some simple d.c. networks; then use your knowledge to measure current, voltage and resistance. Remember to measure voltages across resistors and power supplies. Resistors should be removed from the circuit if you need to measure their values. Ammeters are connected in series with the circuit.

Make the Grade — P5

Grading criterion P5 requires you to carry out an experiment to determine the relationship between the voltage and current for charging and discharging a capacitor. A simple but effective way of doing this is to use a power supply unit, a variety of electrolytic capacitors/resistors, a stopwatch and a multimeter. Figure 6.122 gives a typical circuit arrangement; think about how/where you connect a multimeter.

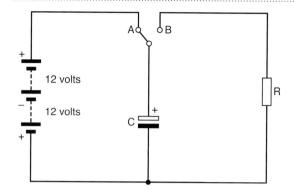

Figure 6.122 Charging and discharging a capacitor

Make the Grade   P11

Grading criterion P11 requires you to use an oscilloscope to measure and determine the inputs and outputs of a single-phase a.c. circuit. (For safety reasons, this should be based on a low-voltage power supply or a function generator.) Build some simple resistor series/parallel networks and experiment with resistor values. Figure 6.123 illustrates a typical experimental arrangement.

Activity 6.12

Charge and discharge the various capacitors using a variety of resistor values. Plot graphs of capacitor voltage against time for each capacitor/resistor network; evaluate the network time constants by comparing the results with theory.

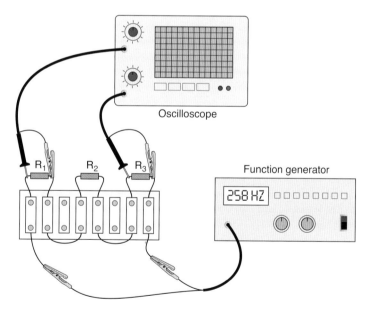

Figure 6.123 Using an oscilloscope in an a.c. circuit

To achieve a pass grade you will have:

P1 used d.c. circuit theory to calculate current, voltage and resistance in d.c. networks

P2 used a multimeter to carry out circuit measurements in a d.c. network

P3 compared the forward and reverse characteristics of two different types of semi-conductor diode

P4 described the types and function of capacitors

P5 carried out an experiment to determine the relationship between the voltage and current for a charging and discharging capacitor

P6 calculated the charge, voltage and energy values in a d.c. network for both three capacitors in series and three capacitors in parallel

P7 described the characteristics of a magnetic field

P8 described the relationship between flux density (B) and field strength (H)

P9 described the principles and applications of electromagnetic induction

P10 used single-phase a.c. circuit theory to determine the characteristics of a sinusoidal a.c. waveform

P11 used an oscilloscope to measure and determine the inputs and outputs of a single-phase a.c. circuit.

To achieve a merit grade you will also have:

M1 used Kirchhoff's laws to determine the current in various parts of a network having four nodes and the power dissipated in a load resistor containing two voltage sources

M2 evaluated capacitance, charge, voltage and energy in a network containing a series-parallel combination of three capacitors

M3 compared the results of adding and subtracting two sinusoidal a.c. waveforms graphically and by phasor diagram.

To achieve a distinction grade you will also have:

D1 analysed the operation and the effects of varying component parameters of a power supply circuit that includes a transformer, diodes and capacitors

D2 evaluated the performance of a motor and a generator by reference to electrical theory.

Exercise and Activity Answers

Unit 4

Exercise 4.2

String of numbers	Simplified using indices (or index notation)
$2 \times 2 \times 2$	2^3
$3 \times 3 \times 3 \times 3$	3^4
$4 \times 4 \times 4 \times 4 \times 4 \times 4$	4^6
$2 \times 2 \times 2 \times 3 \times 3 \times 3 \times 3$	$2^3 \times 3^4$
$3 \times 3 \times 4 \times 4 \times 2 \times 3 \times 4 \times 2 \times 2 \times 4 \times 3$	$2^3 \times 3^4 \times 4^4$
$7 \times 7 \times 4 \times 5 \times 6 \times 4 \times 5 \times 3 \times 9 \times 8 \times 9 \times 3 \times 2 \times 9$	$2 \times 3^2 \times 4^2 \times 5^2 \times 6 \times 7^2 \times 8 \times 9^3$ Or $2^1 \times 3^2 \times 4^2 \times 5^2 \times 6^1 \times 7^2 \times 8^1 \times 9^3$
Can you see any relationship between the indices and the frequency with which each number occurs? Write the relationship in your own words in the box to the right.	You should have something like: the index or power indicates the number of times each digit occurs in the multiplication.
Can you see any advantage in using the index notation instead of the string of numbers in the left-hand column? Write your comments in the box on the right.	It shortens the expression, making it smaller.

Exercise 4.3

String of numbers	Simplified using indices (or index notation)
$2 \times 2 \times 3 \times 3 \times 4 \times 4 + 2 \times 2 + 3 \times 3 \times 3 + 4 \times 4$	$2^2 \times 3^2 \times 4^2 + 2^2 + 3^3 + 4^2$
$3 \times 3 \times 3 - 2 \times 2 \times 2 \times 2 - 4 \times 4 \times 4 \times 4 - 5 \times 2$	$3^2 - 2^4 - 4^4 - 10$
$\dfrac{6 \times 3 \times 6 \times 6 \times 3}{2 \times 3 \times 2 \times 6 \times 3 \times 6}$	$\dfrac{3 \times 6^3}{2^2 \times 3^2 \times 6^2}$

Exercise 4.4

1. 2^{19}
2. 2^{11}
3. x^4

Exercise 4.5

1. 0.11111111111111
2. 0.11111111111111
3. 0.00462962962963
4. 0.00462962962963

Exercise 4.6

1. $3^6 \times 3^3 = 3^9$
2. $6^4 \times 6^3 = 6^7$
3. $24^6 \times 24^8 = 24^{14}$
4. $y^5 \times y^7 = y^{12}$
5. $(p^2)^3 = p^6$
6. $(4^3)^3 \times 4^8 = 4^{17}$
7. $\dfrac{5^4}{5^2} = 5^2$
8. $\dfrac{6^3}{6^5} = 6^{-2}$

Exercise 4.7

1. 3
2. 5
3. 13

Exercise 4.8

Since **2** is the **logarithm** of 100, it follows that **100** is the **antilogarithm** of 2.

Exercise 4.9

1.

Number	Log to base 10
2.3	0.3617
23	1.3617
2 300	3.3617

2. (a) The characteristic is the digit on the left of the decimal point

(b) The mantissa is the decimal part of a logarithm

3.

Number	Log to base 10 (to 4 d.p.)
2 300 000	6.3617
23 000 000 000	10.3617
2 300 000 000 000 000 000 000	21.3617

Exercise 4.10

1. -0.6383 and -2.6383 (to 4 sig. figs.)

2. It is the same for both numbers.

3. They increase to represent the number of zeros which follow the decimal point in the original number.

4. Consider 2.3×10^{-1} meaning the same as 0.23×10^0 (take the index '0' to indicate no zeros after the decimal point). Writing 0.0023 in the same format would give 0.23 (because it is the same value of fraction, hence it gives the same logarithm mantissa) $\times 10^{-2}$ (because there are two zeros after the decimal point).

5. When the **number** is a decimal fraction **less** than 1, the **logarithm** has the same **mantissa**, but the **characteristic** becomes **negative**.

Exercise 4.13

Your answer should be along the lines of:

The log of any number (A) to a power (n) is the same as the log of that number (A) multiplied by the power (n).

Try with numbers, to see it working:

$$\log 2^4 = \log 16, \text{ and the } \log 16 = 1.20$$
$$4\log 2 = 4 \times 0.301029995, \text{ which is } 1.20$$

Exercise 4.14

Take logs

$$\text{Log} 2^3$$

Rearrange this to

$$3\log 2$$

Take logs again

Log3 plus loglog2 (or log3 plus the log of the log of 2)

	Log	Log of log	
3	0.47		
2	0.30	-0.52	
Sum 0.4774$-$0.5214		-0.04	
Antilog		= antilog -0.04	= 0.90
Antilog			7.99 (= 8)

Exercise 4.16

1. $x = 5$

2. $x = 4$

3. $x = 3$

4. $CR\left(\dfrac{\ln I}{i}\right) = t$

5. $\dfrac{1}{\mu}\left(\dfrac{\ln T}{T_0}\right) = \theta$

Exercise 4.18

$v = V(1 - e^{-t/\tau})$

$(1 - v/V) = e^{-t/\tau}$

Taking natural logs:

$$\log_e (1 - v/V) = -t/\tau$$
$$-\tau \times \log_e (1 - v/V) = t$$
$$-20\,\text{k}\Omega \times 33\,\mu\text{F} \times \log_e(1 - 50\,\text{V}/100\,\text{V}) = t$$

$t = 0.457$ seconds for the voltage to reach 40V

Exercise 4.19

$$N = N_i \, (e^{-0.13t})$$
$$N/N_i = e^{-0.13t}$$
$$0.5 = e^{-0.13t}$$

Taking logs to base e

$$\log_e 0.5 = -0.013t$$
$$-0.693 = -0.013t$$
$$t = 50 \text{ years}$$

Exercise 4.23

2. Empty drum = 850 kg (intercept)

    Cable mass = 2.5 kg/m (gradient)

    Equation is $M_t = 2.5l + 850\,\text{kg}$ (or similar following $y = mx + c$ format)

3. Intercept = 23.5 Ω

    Gradient = for example, $5/40 = 0.12\,\Omega/°\text{C}$

    Equation is $R_s = 0.12\text{T} + 23.5\,\Omega$ (or similar in $y = mx + c$ format)

4. Intercept (original mass) = 949 tonnes

    Gradient (in tonnes of fuel per hour) = $-19/5$ = -3.8 tonnes per hour (negative, therefore reducing in mass)

    Equation is $m = -3.8t + 949$ tonnes

    where 't' is the time in flight in hours

5. $R = 0.31\text{T} + 38\,\Omega$

Exercise 4.25

1. Cost per resistor is 1.1p

    Cost per capacitor is 2.7p

2. The bolts cost 39p each and the nuts cost 17p each.

Exercise 4.26

1. $6 + 2x$

2. $3x - 12$

3. $-x^2 + 6x$

4. $2x^2$

5. 18

Exercise 4.27

1. $14x^2 - 2xy$

2. $3y + 3z$

3. $14f - 14R/L$

4. $8x + 4$

Exercise 4.28

1. $x^2 - 4x - 32$

2. $Rx + 7R - 5x - 35$

3. $n^2 - n - 6$

4. $(v + 3)^2 = (v + 3)(v + 3) \rightarrow v^2 + 6v + 9$

Exercise 4.29

2. $(R + 7)(R - 2)$

3. $(x - 2)(x - 5)$

Exercise 4.30

1. $(x + 5)(x - 3)$, therefore $x = -5$ or $+3$

2. $(y - 7)(y - 4)$, therefore $y - + 7$ or $+4$

3. $(z - 1.5)(z + 2)$, therefore $z = +1.5$ or -2

4. $(2x + 3)(x + 3)$, therefore $x = -1.5$ or -3

5. $A = 2\theta r^2 + 2\theta rh$

   $4.5 = 2\theta r^2 + 2\theta r(1.8)$

   $4.5 = 2\theta r^2 + 3.6\theta r$

   $2\theta r^2 + 3.6\theta r - 4.5 = 0$

   $(a \qquad\quad b \qquad\quad c)$

   Using the quadratic formula

   $$\frac{-(3.6\pi) \pm \sqrt{(3.6\pi)^2 - 4(2\pi)(-4.5)}}{2(2\pi)}$$

   $$\frac{-11.31 \pm \sqrt{127.91 + 113.10}}{12.57}$$

   $$\frac{-11.31 \pm 15.52}{812.57}$$

   $$\frac{-11.31 \pm 15.52}{12.57}$$

   Hence $r = -2.31$ m (and a radius cannot be negative, so reject this result)

   or the radius $(r) = 0.33$ m

Exercise 4.31

1. $3a^2 - 2ab + 6a^2 - 6ab \rightarrow 9a^2 - 8ab \rightarrow$ $a(9a - 8b)$

2. $8a + 12b - 6a - 6b - 3a + 2a^2 \rightarrow 2a^2 - a - 6b$

3. $C - q(q - 3p)$

Exercise 4.32

1. $(a + 3)(x + 1)$

2. $(3q - 2r)(p + s)$

3. One option is $2(A + B) + B(3 + C) + C(4 + D)$. No common factor

Exercise 4.33

1. Answer = about 30.86 m

   Circumference is about 40 km = 40 000 000 m

   Divide by 360 → 111,111 m for a degree of arc

   Divide by 60 → 1,851.85 m for a minute of arc (this is 1 nautical mile, apart from a slight error by rounding the Earth's circumference to 40,000 km)

   Divide by 60 → 30.86 m for a second of arc.

2. The value is transcendental and is actually equal to 2π

   $360°/2\pi = 57.29°$ − usually taken as $57.3°$

Exercise 4.34

Degrees	Radians as a decimal fraction	Radians as multiples of π	Pronounced as
360	6.28	2π	2 pi
270	4.71	1.5π or $3\pi/2$	3 pi by 2
180	3.14	π	pi
90	1.57	$\pi/2$	pi by 2
60	1.047	$\pi/3$	pi by 3
45	0.78	$\pi/4$	pi by 4
30	0.52	$\pi/6$	pi by 6
15	0.26	$\pi/12$	pi by 12

Exercise 4.35

1. The angle through which the pulley and belt are in contact with each other is $360° - 140° = 220°$.

   $220° = 220° \, \pi/180$ rad (the abbreviation for radians is rad)

   $\theta = 3.8397$ rad

   using $s = \theta r$

   the distance $s = 0.24 \times 3.83$

   distance $s = 0.92$ m

2. Approximately 62cm and 20cm, making a total of 82cm of the belt being in contact with the pulleys. Using the formula will make it more accurate.

Exercise 4.36

1. $\pi/2$ radians is a quarter of a turn, so it will take 4 seconds to do a complete rotation. Hence there will be 15 revolutions per minute.

2. 120° is twice 60° so it will be twice as many radians. 240° is double that again.

Hence the line voltages would be written as follows:

$$V_1 = V\sin(\theta)$$
$$V_2 = V\sin(\theta + 2\pi/3)$$
$$V_3 = V\sin(\theta + 4\pi/3)$$

Exercise 4.38

Ratio	Values and calculation	Final value
sine	$\frac{opp}{hyp} = \frac{3}{5}$	0.6000
cosine	$\frac{adj}{hyp} = \frac{4}{5}$	0.8000
tangent	$\frac{opp}{adj} = \frac{3}{4}$	0.7500

Exercise 4.39

They should all be 0.5 (within the limits of error from drawing and measuring).

Exercise 4.40

angleθ°	sinθ°
0	0.00
30	0.50
60	0.86
90	1.00
120	0.86
150	0.50
180	0.00
210	−0.50
240	−0.86
270	−1.00
300	−0.86
330	−0.50
360	0.00

angleθ°	cosθ°
0	1.00
30	0.86
60	0.50
90	0.00
120	−0.50
150	−0.86
180	−1.00
210	−0.86
240	−0.50
270	0.00
300	0.50
330	0.86
360	1.00

angleθ°	tanθ°
0	0.00
30	0.57
60	1.73
90	16,324,552,277,619,100.00
120	−1.73
150	−0.57
180	0.00
210	0.57
240	1.73
270	5,441,517,425,873,020.00
300	−1.73
330	−0.57
360	0.00

Exercise 4.41

Angle (degrees)	Tangent of the angle
89.99	5,729.57
89.9999	572,957.79
89.999999	57,295,779.51
90	
90.000001	−57,295,779.51
90.0001	−572,957.79
90.01	−5,729.57

Exercise 4.42

There are a few ways to work this out, but here's one: $\frac{opp}{hyp}$ divided by $\frac{adj}{hyp}$ is division by a fraction.

Hence, this becomes $\frac{opp}{hyp}$ multiplied by $\frac{hyp}{adj}$.

The 'hyp's cancel each other out, leaving $\frac{opp}{adj}$.

From the right2angled triangle, or SOHCAHTOA,

$\tan\theta = \frac{opp}{adj}$. Hence, $\frac{\sin\theta}{\cos\theta} = \tan\theta$

Exercise 4.43

(with some rounding)

Base length (cm)	Opposite side (cm)	Hypotenuse (cm)	Angle (°)
5	9	10.3	61
2	9	9.22	77.5
3	12	12.4	76
5	35	35.35	81.87
576	852	1 028	56

Exercise 4.44

1. 250V

2. $-47.4°$

3. From $\overline{Z} = \overline{R} + \overline{X}_L$

$X_L = \sqrt{(Z^2 - R^2)}$

$X_L = \sqrt{(130^2 - 50^2)}$

$X_L = 120\,\Omega$

Using the triangle and the impedance values, Z, X_L and R, the angle can be found using any ratio chosen from SOHCAHTOA.

Assuming $\cos\theta = \dfrac{adj}{hyp} = \dfrac{50}{130} = 0.38$

The angle whose cosine is $0.3846 = \cos^{-1} 0.38 = 67.4°$

Exercise 4.45

1. $\dfrac{12}{\sin 50°} = \dfrac{b}{\sin B} = \dfrac{c}{\sin 38°}$

Using and rearranging the first and third ratios,

$c = \dfrac{12\sin 38°}{\sin 50°}$

$c = 12 \times 0.80 = 9.64\,\text{cm}$

B must $= 180° - 38° - 50° = 92°$

Using the first and middle ratios,

$b = \dfrac{12\sin 92°}{\sin 50°} = 15.66\,\text{cm}$

2. $\dfrac{6}{\sin A} = \dfrac{4}{\sin 40°} = \dfrac{c}{\sin C}$

Using and rearranging the first and second ratios to find A,

$\sin A = \dfrac{6}{4} \times \sin 40°$

$A = 74.62°$ or $105.38°$

This can be 'seen' by drawing to scale and setting compass to 4 cm. Put the point at 'C'

and draw an arc that passes through the base at 'A' where $A = 74.62°$ and at a point closer to 'B' where A would equal $105.38°$. The decision on which to use depends on the application, hence the need to understand the application of the sine rule and the problems which can occur by accepting the first answer you arrive at.

Exercise 4.46

For Triangle 1:

$c = \sqrt{(a^2 + b^2 - 2ab\,\cos C)}$

$c = \sqrt{(49 + 36 - 84\cos 58°)}$

$c = \sqrt{40.49)} = 6.36\,\text{cm}$

To find angle B:

$\dfrac{6}{\sin B} = \dfrac{6.36}{\sin 58°}$

$B = \sin^{-1}\left(\dfrac{6\sin 58°}{6.36}\right) = 53.1°$

$A = 180° - 53.1° - 58° = 68.9°$

For Triangle 2:

Selecting any cosine rule equation, such as:

$a^2 = b^2 + c^2 - 2bc\cos A$

Rearrange to obtain 'A':

$A = \cos^{-1}\left(b^2 + c^2 - \dfrac{a^2}{2bc}\right)$

$A = \cos^{-1}\left(\dfrac{-32}{80}\right) = 113.6°$

Now using:

$b^2 = a^2 + c^2 - 2ac\,\cos B$

Then:

$B = \cos^{-1}\left(a^2 + c^2 - \dfrac{b^2}{2ac}\right)$

$B = 41.8°$

And:

$C = 108° - 41.8° - 113.6° = 24.6°$

Activity 4.9 Some suggested answers:

Quantity	Formula	Application at home	Application in engineering
Volume of a cylinder	$\pi r^2 h$	Containers for food or drink — for example, diameter and height of a can holding 440 ml of drink. What is the optimum size?	Pneumatic or hydraulic cylinder or car engine piston displacement. Why is a 3.4 litre V6 engine called 3.4 litre?
Total surface area of a cylinder	$2\pi rh + \pi r^2$	Amount of insulation to cover a hot water cylinder (with only one end — the top) — the formula becomes $\pi rh + \pi r^2$ if cylinder is, for example, 1.4 m high and 0.45 m diameter.	Fume extraction ducts of 2.4 m diameter take exhaust gases from a steel furnace to a filter plant 150 m away. If the duct is to be insulated to a thickness of 12 cm, we could calculate the volume of insulating paste which is required.
Volume of a sphere	$\frac{4}{3}\pi r^3$	Air in a football is pressurised to make it solid enough to use. Find out what the recommended pressure is and determine the volume of air which has to be pumped into a ball to achieve that pressure.	A dome-shaped building is visited by the public and regulations advise that the ventilation must ensure that there are five air changes per hour. What amount of air is this, and what capacity fan is required?

Quantity	Formula	Application at home	Application in engineering
Surface area of a sphere	$4\pi r^2$	Someone makes spherical candles of 150mm diameter and wants to coat them with glitter. The adhesive for the glitter is supplied in tubes of 100ml, and each has a surface coverage of $0.3\,m^2$. How many tubes would be needed to totally cover 24 candles?	A cathedral dome is to be decorated with high-power LEDs, at one LED per square centimetre. If the radius of the dome is 8 m, how many LEDs would be required?
Volume of a cone	$\frac{1}{3}\pi r^2 h$	A fisherman wants to make conical metal fishing weights, 12 cm high and 4 cm diameter at the base. The metal he has to use has a density of $9.5\,gm/cm^3$. What will be the mass of each cone?	A baker scoops flour using a large cone, with base diameter of 22 cm and a height of 0.78 m. How many scoopfuls are needed to load a cubic metre of flour?
Curved surface area of a cone	$\pi r \times$ slant height	A conical funnel has been used to pour paint into a small tin and the paint left on the funnel's surface is 1.2 mm thick. If the cone is 12 cm high and a base width/diameter of 10 cm, what volume of paint has been left on the funnel's surface?	An exhaust pipe is made of sheet metal formed into a cone with a small cone cut off the end. Determine the amount of sheet metal required for a cone-shaped exhaust, 045 m long with diameters of 2 cm at one end and 8cm at the other.

Activity 4.10

Term	Definition, meaning or explanation
Variable	A value or quantity which is not constant. For example, the number 3 is constant, but the mass of different people varies, hence m = a variable.
Mean (to use its full title — the arithmetic mean)	Add up the values of all items, then divide by the number of items.
Mode	The most commonly used value in a series of terms. There may be more than one mode (referred to as bi-modal or tri-modal data, and so on).
Median	When all data values are listed in numerical order, the value which is halfway from each end (the value which is in the middle) is called the median. If there is an even number of values, the median is obtained by taking the arithmetic mean of the two values which are in the middle.
Discrete data (also known as discontinuous variables or discontinuous variants)	Separate items, such as the number of days at work, and so on, number of people in work, and so on — usually obtained by counting things. For example — we cannot have 43.2 people out of work, but we can have 2.46% of people out of work
Continuous data (also known as continuous variables or continuous variants)	Data which can have any value, such as time, size, length, mass, and so on — usually obtained by measuring things.
Grouped data (data put into a 'set' or 'group' or 'class' or 'category')	Data is collected into ranges or bands, to reduce the number of variables requiring consideration. For example, people spend various amounts on travel each week. These could be grouped into £5–£15, £16–£25, £26–£35 and so on.
Bar chart	A diagram which consists of a series of bars, each one representing a set of data. The length of each bar is proportional to the size of the quantity or set of data which it represents. Multiple bar charts are used to compare sets of data, perhaps from different years, countries and so on. Bars generally have gaps between them to represent the 'lumping together' of separate data items.
Pie chart	A circle divided into segments, with the size of each segment (by proportion of 360°) representing the proportion of each item of data.
Histogram	A diagram which is used to represent data in a grouped frequency distribution. The area of each 'bar' is proportional to the frequency; hence the intervals need not be in identical-width steps. When the class intervals are equal, the height of the 'rectangles' represents the frequency. The 'bars' of a histogram are not separated from each other to represent the continuousness of the data.
Frequency distribution	This is the number of times a certain event has occurred or the number of times specific items have been counted. For example, in a road-use survey, the number of different vehicles is counted using a tally chart. The tally chart would show how the frequency of each vehicle is distributed throughout the observation time.

Term	Definition, meaning or explanation
Class boundary	Upper and lower class boundaries are the 'edges' or 'limits' of the groups where values are rounded up or down to fit in them. For example, if we measured the weekly amounts spent on travel (given above) to an accuracy of two decimal places, the £16−£25 group would actually include £15.50 to £25.49, due to rounding. Hence, for that 'class' or group, the lower class boundary is £15.50 and the upper class boundary is £25.49.
Class width	If we subtract the lower class boundary from the upper class boundary, we get the class width. From the above example, the class width for the £16−£25 group is £25.49−£15.50 = £9.99.
Frequency table	A table which contains groups or classes of grouped data and their frequencies of occurrence. The table can contain the different groups or classes of grouped data and their frequency of occurrence. The table contents can be arrived at adding up the occurrences between upper and lower class boundaries on a tally chart.
Variant	A variable that may have a randomly chosen value − something that varies with no obviously determinable pattern.
Cumulative frequency (occasionally referred to as a 'running total graph')	The addition (accumulation) of all frequencies which have occurred prior to the one under consideration. The upper class boundary values are used to plot the curve (see below).
Cumulative frequency curve	If the cumulative frequencies are plotted on a graph, the value increases until 100% of the sample has been included. A line (called an ogive) is created

Exercise 4.48

It seems that 0.03% of the sample of 10 000 died.

One of these ate raw vegetables, and the other two ate cooked vegetables, along with the other 4998. The very small sample that died indicates an almost indemonstrable link, and the fact that the headline took the comparison *only* between those who had died to create the image of eating raw vegetables as being hazardous.

Exercise 4.49

You could have said many things:

- they are all odd numbers
- the largest value is in the middle
- the lowest value is first
- they are all prime numbers.

Exercise 4.50

1. 9 cars
2. 1 cars
3. 4 cars (there are five occurrences of this)
4. 1 cars (there are four occurrences of this)
5. 4 cars
6. 4.4 cars

Exercise 4.51

1. 1, 1, 2, 2, 2, 2, 3, 3, 3, **3, 4**, 4, 5, 5, 5, 6, 7, 7, 8, 9

The 'mid value is the arithmetic mean of 3 and 4 = 3.5

Hence, the median may not be a value that actually exists either.

The modes are 2 and 3, indicating that the most common number of cars is 3 and 4.

2. Mean = 32.3 Ω

Mode = 33.1 Ω

Median = 33.1 Ω

3. (a) 10% of 33 Ω is 3.3 Ω, so the acceptable values would be between 29.7 Ω and 36.3 Ω. All except 29.1 Ω would be acceptable, so 14 out of 15 would be suitable.

(b) 5% of 33 Ω is 1.65 Ω, so the acceptable values would be between 31.35 Ω and 34.65 Ω. All the values from 31.4 Ω would be suitable, which is 12 out of the 15.

Exercise 4.52

1. $\Sigma xf = 48{,}060$ and $\Sigma f = 256$ Hence the mean life of the lamps is 187.7 days or 188 days.

2. The two values would be 182.7 days and 191.8 days, respectively. Both are valid answers to the valid problem.

3. If the 256 lamps were all changed after 188 days of use, half of them would have blown (on average) so the workplace may be too dark to work safely and with the required quality.

343

If the lamps are changed after about every 160 to 170 days, only a few would have blown and the light level may still be acceptable. If not, more frequent change of lamps would be required.

There are endless explanations and interpretations.

Exercise 4.53

Motor scooters mileage

Days	a	b	c	d	e
1	234	201	35	77	128
2	52	211	197	194	152
3	176	179	145	140	134
4	38	134	179	201	166
5	145	88	130	117	193
6	132	149	82	172	15
7	201	160	170	126	69
8	98	52	201	159	129
9	108	93	77	72	163
10	211	106	29	35	231
11	78	168	107	9	149
12	156	202	170	125	192
13	143	183	140	172	169
14	187	55	162	169	56
15	123	238	94	110	170
16	99	174	157	218	147
17	75	196	129	57	184
18	43	121	185	179	156
19	199	153	63	183	43
20	165	104	221	147	147
21	135	218	184	125	204
22	123	93	149	165	77
23	109	42	182	121	149
24	62	12	62	135	129
25	26	160	129	53	174
26	8	23	55	178	162
27	193	182	158	93	79
28	49	111	21	124	92
29	108	77	126	201	111
30	45	93	53	42	44
Total	3,521	3,978	3,792	3,899	4,014
Mean	117.4	132.6	126.4	130.0	133.8

Motor scooters mileage

a	b	c	d	e
8	12	21	9	15
26	23	29	35	43
38	42	35	42	44
43	52	53	53	56
45	55	55	57	69
49	77	62	72	77
52	88	63	77	79
62	93	77	93	92
75	93	82	110	111
78	93	94	117	128
98	104	107	121	129
99	106	126	124	129
108	111	129	125	134
108	121	129	125	147
109	**134**	**130**	**126**	**147**
123	**149**	**140**	**135**	**149**
123	153	145	140	149
132	160	149	147	152
135	160	157	159	156
143	168	158	165	162
145	174	162	169	163
156	179	170	172	166
165	182	170	172	169
176	183	179	178	170
187	196	182	179	174
193	201	184	183	184
199	202	185	194	192
201	211	197	201	193
211	218	201	201	204
234	238	221	218	231

	a	b	c	d	e
Median	116	142	135	131	148

Motor scooters mileage

a	b	c	d	e
8	12	21	9	15
26	23	29	35	43
38	42	35	42	44
43	52	53	53	56
45	55	55	57	69
49	77	62	72	77
52	88	63	77	79
62	**93**	77	93	92
75	**93**	82	110	111
78	**93**	94	117	128
98	104	107	121	**129**
99	106	126	124	**129**
108	111	**129**	125	134
108	121	**129**	125	147
109	134	130	126	147
123	149	140	135	**149**
123	153	145	140	**149**
132	160	149	147	152
135	160	157	159	156
143	168	158	165	162
145	174	162	169	163
156	179	**170**	**172**	166
165	182	**170**	**172**	169
176	183	179	178	170
187	196	182	179	174
193	201	184	183	184
199	202	185	194	192
201	211	197	**201**	193
211	218	201	**201**	204
234	238	221	218	231

Mode(s)	108	93	129	125	129
	123		170	172	147
				201	149
	bi-modal		bi-modal	tri-modal	tri-modal

Exercise 4.54

You should obtain values like this — apart from the odd minor error (yours or mine)

1.

Grouped mileage data	Number of occurrences for *all* riders taken together
1−25	6
26−50	12
51−75	14
76−100	16
101−125	17
126−150	25
151−175	25
176−200	20
201−225	12
226−250	3

2.

Grouped mileage data	Number of occurrences for *all* riders taken together	Cumulative frequency
1−25	6	6
26−50	12	18
51−75	14	32
76−100	16	48
101−125	17	65
126−150	25	90
151−175	25	115
176−200	20	135
201−225	12	147
226−250	3	150

Exercise 4.59

x	$y = x^2$
3	9
2	4
1	2
0	0
−1	2
−2	4
−3	9

345

Exercise 4.60

1. Whatever values for rise and run are taken, they should be approximately equal amounts (approximately, due to accuracy of your drawing) and the gradient should be 1. The line should be almost at 45° to the x axis at the point where $x = 1$.

2. $\tan 45° = 1$

3. At $x = 2$, $y = 2^2$ which is 4. The gradient should be approximately 4.

4. At $x = 3$, $y = 3^2$ which is 9. The gradient should be approximately 6. To obtain the gradient at $x = 6$ would require the scale for y extending as far as 12, or more.

Exercise 4.61

x value	Gradient at x value
0.5	1
1	2
2	4
4	8
7	14
10	20

You should have spotted, but just in case you did not, the gradient of $y = x^2$ for any value of x along the curve, is equal to $2x$. At $x = 2$, the gradient = $2 \times 2 = 4$; at $x = 3$, the gradient = $2 \times 3 = 6$, and so on.

This one is more difficult to spot, but the gradient of $y = x^3$ for any value of x along this curve, is equal to $3x^2$

Exercise 4.63

1. $y = x^{1/2}$ the gradient at any point = $1/2\, x^{-1/2}$ (do not forget the ½ − 1 = 1/2)

2. $y = x^4$ the gradient at any point = $4x^3$ (from $4 - 1 = 3$ for the index)

3. $y = x$ the gradient at any point = 1 so this is a straight line at 45°

    (remember x means x^1 so $1x^{1-1}$ − $1x^0$, which = 1)

4. $y = 3x^4$ the gradient at any point = $12x^3$ (being 4 multiplied by x, which is already multiplied by 3, so 4 times 3 = 12, and the index reduces by 1.

Exercise 4.64

All terms which include 'x' or some other variable are exactly that −variables, and a variable can be differentiated. 4 is a number which is constant and all constants are always equal to the same value, so their rate of change is obviously zero. If you drew the graph of $y = 4$, what would it look like?(Hint, the value of y would always = 4; hence it is a straight, horizontal line.) The gradient of a straight, horizontal line is zero, because it does not slope.

Exercise 4.65

2.

$y = x^2 + 3x + 4$	$dy/dx = 2x + 3$
At $x = -1.5$, there is a turning point, where the graph has no gradient	At $x = -1.5$, $dy/dx =$ zero
At $x = 1$, the gradient is approximately 22/5 (rise over run) = between 4 and 5.	At $x = 1$, $dy/dx =$ approximately 5
At any point where $x =$ less than -1.5, the gradient is negative	At any point where x is less than -1.5, the value of dy/dx is negative.

Exercise 4.66

The expression is multiplied by the index (n), then the value of n is reduced by 1. In other words, multiply by the power, then reduce the power by 1. (Note, when n and a are given numerical values, they can be multiplied out.)

Exercise 4.67

1. $dy/dx = 3x^2 + 6x$
2. $dy/dx = 9x^2 + 6x + 4$
3. $dy/dx = 3x^2 - 6x$
4. $dy/dx = 12x^3 + 1.5x^{-4}$ not forgetting that '−' times '−' = '+', and $-3 - 1 = -4$

Exercise 4.68

This should return the value of e (remember, e^x means 'antilog of the x value) so if you get a gradient of something between 2.2 and 3.0, you have done well. If your value is very close to 2.7, you have done extremely well.

Exercise 4.69

1. $y = 4e^x$

Find d/dx and evaluate this for the value $x = 2.5$.

$dy/dx = 4e^x$

At $x = 2.5$, $4e^x = 6.59$ to 2 d.p.

2. $y = 6.8e^x$

Differentiate this with respect to (w.r.t.) x. Evaluate the gradient when $x = 0.5$.

$dy/dx = 6.8e^x$

At $x = 0.5$, $6.8e^x = 11.21$.

Exercise 4.70

1. It is actually a cosine wave.

2. You should have plotted a few points and determined that the wave shape is that of a sine wave, but upside down (or a negative sine wave). It is also a sine wave which has been shifted by 180° along its horizontal axis.

Exercise 4.71

3. If $v = \sin\theta$, differentiate to obtain d/dθ (sinθ)

$d/d\theta (\sin\theta) = \cos\theta$

4. The instantaneous voltage of a waveform is

$$v = 24\sin\theta + 12\cos\theta$$

use differentiation to obtain the rate of change of v w.r.t , or dv/dθ

$$dv/d\theta = 24\cos\theta - 12\sin\theta$$

Exercise 4.72

You should have arrived at something like:

- add 1 on to the power
- divide by the new power.

Hence, $2x$ has power 1, add 1 onto this gives $2x^2$ and dividing by this power 2, gives

$$2x^2 /2 = x^2$$

Exercise 4.73

1. $3x^{(3+1)} / (3+1) = ¾ x^4 = 0.75x^4$

2. x^5

3. $⅓ x^3$

4. $3x$ (note $3 = 3$ multiplied by x^0 so adding 1 to the power creates $x^1 = x$

5. $\sin x$

6. $\sin x + \cos x$

7. $-\cos x - \sin x$

Exercise 4.74

Add 1 to $-1 = 0$. Divide by the new power (zero) and the answer would be infinity.

Exercise 4.75

2. You should find that

$$2 = \frac{3^3}{3} + c$$

returns a value of $c = -7$.

3.
$$y = \frac{x}{3} - 7$$

4. You should notice that the equations are all basically the same, having the same shape, but the intercept is different. This is because the information about the intercept disappears after differentiation, and the integral of any expression can lead to a family of curves, all with different intercepts. If you end up with a negative value, the intercept is located below the horizontal axis.

Exercise 4.78

$\frac{2}{3} x^3 - 5x^2 + C$

It is always good practice to differentiate the result to check your working.

Exercise 4.79

5. (a) area = 3.84 square units (note: -3.84 means area *under* the x axis)

(b) area = $26\frac{2}{3}$ square units

(c) area = 1.22 square units

Exercise 4.80

t (seconds)	v_c (volts)
0	0
0.1	7.9
0.2	13.2
0.3	16.77
0.4	19.15
0.5	20.75

2. The graph should provide an estimate of the value for (c) By calculation (rearranging and taking natural logs, $t = 0.24$s.

3. $dv/dt = 96e^{-4t}$

4. at t = zero, dv/dt = 96V

By estimating a rise over run from your curve, you should achieve a reasonable comparison, depending on the size and accuracy of plotting.

Unit 5

Activity 5.1

All angles are relative to the +ve horizontal (0°)

Figure 5.7:	Resultant = 4.43 N at 89°	Equilibrant = 4.43 N at 271°
Figure 5.8:	Resultant = 7.15 kN at 192°	Equilibrant = 7.15 kN at 12°
Figure 5.9:	Resultant = 2.16 kN at 157°	Equilibrant = 2.16 kN at 337°
Figure 5.10:	Resultant = 6.21 kN at 126°	Equilibrant = 6.21 kN at 306°

Activity 5.2

Figure 5.19: Resultant = 7.69 kN at 236°

Figure 5.20: Resultant = 2.09 kN at 105°

Figure 5.21: Resultant = 6.44 N at 163°

Figure 5.22: Resultant = 6.23 kN at 335°

Activity 5.3

Figure 5.26: Resultant = 1.66 kN at 264° Equilibrant = 1.66 kN at 84° the line of action is 1.17 m at 354° relative to point p.

Figure 5.27: Resultant = 3.69 kN at 325° Equilibrant = 3.69 kN at 145° the line of action is 1.50 m at 55° relative to point p.

Activity 5.4

Figure 5.33:	Left-hand reaction = 14.33 N	Right-hand reaction = 15.67 N
Figure 5.34:	Left-hand reaction = 21 kN	Right-hand reaction = 21 kN
Figure 5.35:	Left-hand reaction = 7.50 kN	Right-hand reaction = 10.50 kN
Figure 5.36:	Left-hand reaction = 4.29 kN	Right-hand reaction = 35.71 kN

Activity 5.5

1. 318.31 N/mm^2

2. 933.33 × 10^6

3. 14.65 mm

4. 14.8 × 10^{-3} mm

5. 28.27 kN

6. 5.4

Activity 5.6

1. v = 58 m/s   s = 588 m

2. a = −93 ×10^{-3} m/s^2 s = 666.67 m

3. t = 0.167 s   a = 72 m/s^2

Activity 5.7

1. h = 14.27 m
2. KE = 350.12 kJ d_1 = 291.8 m d_2 = 129.7 m

Activity 5.8

1. v = 24.38 m/s   Loss of KE = 19.69 kJ
2. Tractive effort = 96.45 kN   Work done = 2.60 MN   Average power = 57.78 kW
3. 31.86 kN

Activity 5.9

1. h = 1.53 m
2. Density = 441.73 kg/m³
3. p = 19.88 kPa

Activity 5.10

1. F = 1.47 MN
2. M = 2.83 MNm

Activity 5.11

1. v_1 = 0.26 m/s diameter_2 = 11.3 mm   mass flow rate = 0.41 kg/s
2. v_2 = 42.78 m/s   volumetric flow rate = 53.9×10^{-3} m³/s   mass flow rate = 53.9 kg/s

Activity 5.12

1. Total heat energy supplied = 7.52 MJ
2. Heat supplied /kg = 2.5 MJ   output power = 3.19 MW

Activity 5.13

Stress induced = 338.8 MN/m²

Activity 5.14

Final temperature = 296°C

Index